PENGUIN MODERN CLASSICS

A Fire on the Moon

Norman Mailer (1923–2007) was one of the great post-war American writers, both as a novelist and as one of the key inventors of the New Journalism. His books include the novels *The Naked and the Dead*, *The Deer Park*, *Why Are We in Vietnam?*, *The Executioner's Song* and *Harlot's Ghost* and the non-fiction works *The Armies of the Night*, *A Fire on the Moon* (published in the USA as *Of a Fire on the Moon*) and *The Fight*. He won the National Book Award and twice won the Pulitzer Prize.

Geoff Dyer is the author of many books including *But Beautiful*, *The Ongoing Moment* and, most recently, *Another Great Day at Sea*, about his time aboard an American aircraft carrier. His books have been translated into twenty-four languages.

NORMAN MAILER

A Fire on the Moon

With an Introduction by Geoff Dyer

PENGUIN BOOKS

PENGUIN CLASSICS

Published by the Penguin Group
Penguin Books Ltd, 80 Strand, London WC2R 0RL, England
Penguin Group (USA) Inc., 375 Hudson Street, New York, New York 10014, USA
Penguin Group (Canada), 90 Eglinton Avenue East, Suite 700, Toronto, Ontario, Canada M4P 2Y3
(a division of Pearson Penguin Canada Inc.)
Penguin Ireland, 25 St Stephen's Green, Dublin 2, Ireland (a division of Penguin Books Ltd)
Penguin Group (Australia), 707 Collins Street, Melbourne, Victoria 3008, Australia
(a division of Pearson Australia Group Pty Ltd)
Penguin Books India Pvt Ltd, 11 Community Centre, Panchsheel Park, New Delhi – 110 017, India
Penguin Group (NZ), 67 Apollo Drive, Rosedale, Auckland 0632, New Zealand
(a division of Pearson New Zealand Ltd)
Penguin Books (South Africa) (Pty) Ltd, Block D, Rosebank Office Park,
181 Jan Smuts Avenue, Parktown North, Gauteng 2193, South Africa

Penguin Books Ltd, Registered Offices: 80 Strand, London WC2R 0RL, England

www.penguin.com

First published in the United States of America by Little, Brown and Company 1970
Published in Penguin Classics 2014
001

Several sections of this book appeared originally in *Life* magazine.

The author wishes to acknowledge his indebtedness to *First on the Moon* by Neil
Armstrong, Michael Collins and Edwin E. Aldrin, Jr, written by Gene Farmer and Dora
Jane Hamblin, for the field of quotation it offered. Quotations are reprinted by permission
of the publisher, copyright © 1970 by Little, Brown and Company.

Set in 10/12.5pt Dante MT Std
Typeset by Jouve (UK), Milton Keynes
Printed in Great Britain by Clays Ltd, St Ives plc

A CIP catalogue record for this book is available from the British Library

ISBN: 978-0-141-39496-1

www.greenpenguin.co.uk

For Susan, for Dandy, for Betsey and Kate,
for Michael and Stephen Mailer

The author wishes to acknowledge his indebtedness to *First on the Moon* by Neil Armstrong, Michael Collins and Edwin E. Aldrin, Jr, for the field of quotation it offered.

Contents

Introduction

Mailer starts with the news of Hemingway's death; we'll start with Ezra Pound's claim, in *The ABC of Reading*, that literature 'is news that STAYS news'. The appeal of having one of America's best-known writers cover the biggest news story of the decade – probably of the century, conceivably of all time – was obvious, and Mailer was a natural fit. Back then, a lot of people were quoting the opinion that he was the best journalist in America. One of those people was Mailer himself, who took umbrage at praise that tacitly downgraded his achievements as a *novelist*. This gets aired very early on in a book in which, sooner or later, most things get aired. The irony is that Mailer 'knew he was not even a good journalist' (p. 4). Unless, that is, he could succeed in redefining and enlarging journalism to cover pretty much everything, including the writing of the book in which the attempt would be made. Imagine Laurence Sterne with a huge subject, a big advance and a looming deadline and you have some sense of the conflicting pressures at work on *Of A Fire on the Moon* (the original American title).

The deadline needs emphasizing. Other writers had plenty to say about the moon landing – everyone had *something* to say about it – but few would have had the chops to bang out 115,000 words for publication in three issues of *Life* magazine, the first tranche of which, Mailer groans, was due less than three weeks after the astronauts splashed down in the Pacific. That, to put it mildly, is a lot of words in a very short time: not quite as challenging a task as the one set out by John F. Kennedy in 1961 – to land a man on the moon and return him safely to earth by the end of the decade – but a serious job of work all the same. So the question today, when no one under the age of forty-five was alive and able to experience the event, let alone read about it as news, is the extent to which the result is compromised or enhanced by the circumstances of its occasion and composition. Now that the subject matter is the stuff of history – when the word astronaut might be used in the context of historical (as opposed to *science*) fiction – does Mailer's book pass Pound's testing definition? And where does it stand within two quite different contexts, that of other books about the moon landings and within the large scope and wildly mixed quality of Mailer's work as a whole?

Coming warm on the heels of *The Armies of the Night* and *Miami and the Siege of Chicago* (both 1968), *A Fire on the Moon* was Mailer's third book of extended reporting. He features as a participating observer in both of the earlier works ('Mailer' in *Armies*, 'the reporter' in *Miami*), as he would again (as 'Norman') in his masterly account of the Muhammad Ali–George Foreman bout in Zaire in 1975. *The Fight*'s curt titular nod – almost a butt! – to a famous 1822 essay by Hazlitt makes us wonder if the roots of the New Journalism extend more deeply into the past than is sometimes thought. *Life* magazine proudly announced that the first excerpt from Mailer's report on the 'the moon venture' was the longest non-fiction piece they'd ever published (forcing into second place an article on 'the Woodstock Rock Festival', and featuring a cover photo not of Neil Armstrong but 'Aquarius', as Mailer termed himself this time around). By Martin Amis's estimation some of the excursions of the New Journalism seem 'as long as *Middlemarch*', but freedom from restrictive word or page counts could also be extended to the old journalism, if the subject matter demanded it. The *New Yorker* had published John Hersey's *Hiroshima* in its entirety in 1946. What's new is the undisguised presence of a generic growth hormone whereby the previously self-effacing disposition of the journalist gives way to the swagger and confidence of the performance-enhancing novelist. As such it's difficult to think of a work that better exemplifies the strengths and shortcomings of what we now think of as New Journalism (which might be different to what the term designated at the time it was coined) than *A Fire on the Moon*. Hunter S. Thompson's *Hell's Angels* (1966) seems, by comparison, a model of calm and dispassionate economy. The unfettered deployment of a participating voice, abandonment of the idea of disinterested recording in favour of subjectivity so extreme that it threatens to displace or usurp the subject appear unprecedented – because the precedent is hiding in plain view. A flickering but frequent presence in Mailer's writing, D. H. Lawrence operated along similar lines in his non-fiction and journalism, albeit on a smaller scale. As Rebecca West realized after his death, Lawrence only ever wrote 'about the state of his own soul', using whatever subject was symbolically at hand at any given moment. The occasion for Lawrence might have been the death of a porcupine or his first encounter with native Americans; for Mailer it is a Saturn V rocket, the launch of which is evoked in stirringly Lawrentian rhythms and images. Other passages could almost have been written *by* Lorenzo, at his ranch in New Mexico: 'In that long-ago of prairie spaces when the wind was the message of America, Indians had lived in greater intimacy with the moon than any European' (p. 68).

More generally, Mailer shares Lawrence's hieratic belief that he might be able to offer some kind of solution to the sickness of his times (of which both men are, according to a different series of tests, symptoms). For Mailer the success or failure of the Apollo 11 mission is constantly elided with the project of writing his book about it: each, in its different way, exhibits signs of pathology and cure. The moon landing hails both a new age and the end of an era, while the book about it (which is often thought of as an embodiment of the new) is shot through with a dread of impending obsolescence – of a very recent kind. How can the printed word compete with television in an age when events are made for TV, when the author is reduced to *watching* the events he is covering on telly?* In one respect the two turn out to be complementary. Those pages evoking the launch of the Saturn V might have been helped by watching the filmed footage, just as the blow-by-blow account of the rumble in the jungle would have been impossible without reviewing playback of the fight. Given that the moon landing was a dramatic extension of the possibilities of communication (as President Nixon said when he was patched through to the men on the moon, this had to be 'the most historic phone call ever made' (p. 108)) the fact that it could be broadcast live on TV was itself part of the story.

---

* This was particularly acute in the case of the aptly-named space *programme* in which even the primary experience of the astronaut could feel like a re-enactment of something that had occurred hundreds of times previously, in simulators. As Tom Wolfe points out in *The Right Stuff*, America's first manned space flight in 1961 saw astronaut Alan Shepard 'introducing the era of precreated experience'.

If the landing could not be filmed there was no point going. At the risk of the tail wagging the dog, they went to the moon *in order* to film it. (To that extent people who claimed that the whole thing was a set-up were right; it's just that it was filmed on location, not in a studio.) In the age of *Gravity*, of simulated cinematic immersion in space, it is more striking than ever that footage of the greatest technological feat of all time looked no better 'than a print of the earliest silent movies . . . Ghost beckoned to ghosts and the surface of the moon looked like a ski slope at night' (p. 105). These blurry images herald a brave new world of satellite communications but the era that Mailer foresees, when reporters' work involves 'rewriting publicity handouts', is close to the situation sketched in *Flat Earth News* (2009), Nick Davies's analysis of the parlous state of contemporary journalism. Prophecies can come true in ways and circumstances very different to how they were originally envisaged. And since no one minds whether they turn up exactly as and when predicted they're immune to the kind of fact-checking that might cast retrospective doubt on Mailer's extravagancies of style and method. Did the moon adventure really 'help to disclose the nature of the Lord and the Lucifer who warred for us' (p. 415)? Is jaundice the 'infectious disease beyond all other which comes to strong people when they live too long in an environment alien to their will, work with all their power to solve the complexities of that environment, and fail' (p.286)? Does the fact that John Borman was so afflicted by diarrhoea and vomiting, that Apollo 8 (not 9 as Mailer writes) became, in Andrew Chaikin's words, 'a flying toilet' mean that he was not 'hard as hand-forged nails' (p.272)? Such questions miss the point, especially since one of Mailer's many big points is that the world of NASA, of computers and data, is so distinct from the realm of imaginative thought where the likes of Aquarius reside and reign. *A Fire on the Moon* is a plea on behalf of the unverifiable, of metaphysics (or a novelist's idea of metaphysics, that is to say, 'the fiction of unspoken evidence') over physics (p. 282). Since Mailer's book is partly a psychological portrait and *reaction* to the moon shot anything in it is accurate ('there is no psychological reality like a man's idea of himself') by virtue of the fact that it occurred to him. He was, after all, one of the 'sensors in the currents of the churn' (p. 102).

The suspicion that, in the process of churning out all these words,

Mailer barely had time to stop and think is not always assuaged by the way that Aquarius is constantly telling us that he's skulking off to brood, cogitate and mull over. But it is perhaps countered by his commitment to the idea that the quality of creative thought might be enhanced by the pace of composition being forced. *A Fire on the Moon* incarnates the conviction that the finished work can have something of the feel of a work in progress, one that shares the minute-by-minute uncertainty and drama of the moon shot (a quality that remains strong even now, when the outcome of the mission is no longer in doubt). The lingering traces of Kerouac's creed of 'first thought best thought' manifest themselves, at worst, as a stream of filibustering. Describing the multiple and overlapping safety measures at work in the spacecraft Mailer writes that redundancy 'was built into every aspect of every system' (p. 206). The book echoes with repeated rhythmic growls, self-reflecting asides and recurring observations: Armstrong's 'face within the space helmet as lashless in appearance as a newborn cat in a caul', for example, gets replayed a few pages later (p. 72). But just as Mailer acknowledges and thereby deflects the reader's potential complaint that he's taken on 'this grim tough job of writing for enough money to pay his debts' (p. 40) so he acknowledges the potential complaint that 'your style is redundant' by appeal to the subject matter: by following NASA's example and making 'redundancy a virtue'. A work which would seem to have benefitted from careful editing is conceptually and rhythmically at odds with the idea. The conundrum whereby the massive weight and bulk of the Saturn V is made up mainly of the fuel necessary to hurl it into space finds equivalent expression in the power-base of Mailer's prose. The money-driven, time-pressed mode of composition generates a torrent of insights, large and small feats of perception, analysis and phrasing. Of course Aquarius is self-indulgent but a more temperate author would have drawn more attention to – would practically have framed and exhibited – passages of brilliance, intimacy and delicacy that he allows to rush past without a backward glance. All the more remarkable, then, that in Part II, 'Apollo', we are treated to an abrupt increase in the proportion of diligent recording, reporting and transcribing in which the minutiae of the journey to the moon, the landing and the re-docking of the astronauts in space, are painstakingly itemized, with considerable scientific expertise, no loss of momentum and an escalation of tension and intensity. It's a reminder of how Mailer could do so much writing that is not of the kind we associate with him – even while doing much that is Maileresque to the point of parody. Indeed, the tonal range within the overall narrative trajectory is striking. Again there is perhaps a parallel with the moon journey itself, with the way that the Command Module is constantly rotating in flight so that it's kept nicely toasted all-round rather than scorched by solar rays on one side while the other freezes. Mailer, by turns, is analytical, hyperbolic, delirious, homey (that wonderful evocation of the moon's surface as 'a barnyard trod by countless hooves') and lyrical ('the moon was a voice which did not speak') (pp. 253, 240). He may not have been a wit but that did not stop him being uproariously funny. One of Armstrong's first jobs after the small step–giant leap is to scoop up a sample of moon dust so that 'if the unmentionable yak or the Abominable Snowman were to emerge from the crater', he could leap back aboard the Lunar Module and make his escape. Buzz Aldrin has to remind Armstrong to do this, causing the first man on the moon to snap back – understandably, since 'nagging was nagging even on the moon' (p. 105).

Having studied aeronautical engineering at Harvard, Mailer had the knack – it's something he shares with Len Deighton – of conveying complex technical specifications and design imperatives with a confidence and ease that enables ignorant readers feel capable of a level of understanding to which they had not hitherto been privy. That 'curious creature' the Lunar Module, or LEM, for example, 'had been designed from the inside, and so was about as ugly as a human body which had shaped itself around the excessive development of a few special organs. Conceive of a man whose only function in life was to win pie-eating contests – what a stomach would he develop, large as a steamer trunk' (p. 293).

These registers, of course, are not entirely distinct, are all the time merging and overlapping so that to read Mailer is like being rigged up to a piece of machinery that pre-articulates our own reactions to momentous events: 'A reductive society was witnessing the irreducible. But the irreducible was being presented with faulty technique. At that they [the Press] could laugh. And did again and again. There were moments when Armstrong and Aldrin might just as well have been Laurel and Hardy in space suits' (p. 106). Note how one tone facilitates rather than cancels out another. We've stressed the way that various writers from the past make themselves felt in the book but perhaps it's Whitman's American imperative, to 'contain multitudes' that is best exemplified by Mailer's prose. If the astronauts 'came in peace, for all mankind' then Mailer aims to speak for all Americans – and is obliged to recognize that, in a time of such fraught divisions, he can't. After meeting with an African-American professor of his acquaintance, Mailer is confronted by a version of the view comically and economically expressed by another enduring cultural response to Apollo 11: Gil Scott-Heron's song, 'Whitey on the Moon'. It's one of several encounters that leave Aquarius troubled, depressed, unsure of how best to take the measure and sound the depths of what's happening. While the astronauts are still on their way back to Earth, he too will leave Houston and return home, 'mental digestions churning', recording the churn and adding to it while hurtling towards both a deadline and a marital break-up.

Clearly, *A Fire on the Moon* was the first work of literature to be devoted to the moon landings. Tom Wolfe would have the luxury of almost a decade of hindsight and research before publishing a kind of prequel, *The Right Stuff*, in 1979. Reading the two books in tandem serves, among other things, to validate much of Mailer's analysis, including his understanding of the psychology of astronauts, of how the reticence of 'men with a sense of mission so deep it could not be communicated' co-existed with a profound passivity: 'They were virile, but they were done to' (pp. 268, 39). It is probably not the first book to which a reader interested in the history of the space programme will turn. That honour must surely go to Andrew Chaikin's gripping and comprehensive *A Man on the Moon* (1994). A sceptical, even grudging alternative is provided by Gerard Degroot's *Dark Side of the Moon* (2006). What Mailer's book manages, uniquely, to do is evoke events that will soon be half-a-century distant in such a way that they unfold again before our eyes. Within the mixed and lumpy bag of his own work, *A Fire on the Moon* is a lesser book than the story of the life and death of Gary Gilmore in *The Executioner's Song* (1979). Inevitably lacking the focus and tight cohesion of *The Fight* it fully deserves to join the latter among the ranks of Modern Classics. Its belated elevation, in this regard, is part of a continuing extension of the *kind* of book deemed worthy of admittance. If it is difficult to imagine when the resources might again be marshalled or the will found to push through a venture on the scale of the Apollo programme, it is equally hard to think of anyone taking on the task of describing such an undertaking with Mailer's gusto and urgency. As an extravagant and immediate response to 'the most expensive gesture ever made' (p. 292), it is not just a stunning achievement; it is also an appropriate one.

PART I

Aquarius

I

A Loss of Ego

Now sleeps he with that old whore death . . . Do thee take this old whore
death for thy lawful wedded wife?

Ernest Hemingway

Norman, born sign of Aquarius, had been in Mexico when the news
came about Hemingway. He had gone through the *New York Times* to
read the well-turned remarks of notables who for the most part had
never cared about Papa, not that much! and had one full heart-clot of
outraged vanity that the *Times* never thought to ask *his* opinion. In fact,
he was not certain he could have given it. He was sick in that miasmal
and not quite discoverable region between the liver and the soul. Hemingway's suicide left him wedded to horror. It is possible that in the
eight years since, he never had a day which was completely free of
thoughts of death.

Of course, he finally gave a statement. His fury that the world was
not run so well as he could run it encouraged him to speak. The world
could always learn from what he had to say – his confidence was built
on just so hard a diamond. Besides, a British lady columnist passing
through Mexico with him thought it would be appropriate to get his
remarks on the demise. This, after all, was special stuff – the reactions
of one of America's best-known young novelists would certainly be
appropriate to the tragic finale of America's greatest living writer. So
with thoughts of Hemingway's brain scattered now in every atmosphere –
what a curse to put upon his followers! – Norman coughed up what was
in effect a political statement. He had no taste in such matters, and a
pedagogic voice for public remarks; leave it that he inveighed gracelessly

on how the death would put secret cheer in every bureaucrat's heart for they would be stronger now. He had, of course, been thinking that Hemingway constituted the walls of the fort: Hemingway had given the power to believe you could still shout down the corridor of the hospital, live next to the breath of the beast, accept your portion of dread each day. Now the greatest living romantic was dead. Dread was loose. The giant had not paid his dues, and something awful was in the air. Technology would fill the pause. Into the silences static would enter. It was conceivable that man was no longer ready to share the dread of the Lord.

ii

Are we poised for a philosophical launch? There may be no way to do anything less. We will be trying after all to comprehend the astronauts. If we approach our subject via Aquarius, it is because he is a detective of sorts, and different in spirit from eight years ago. He has learned to live with questions. Of course, as always, he has little to do with the immediate spirit of the time. Which is why Norman on this occasion wonders if he may call himself Aquarius. Born January 31, he is entitled to the name, but he thinks it a fine irony that we now enter the Age of Aquarius since he has never had less sense of possessing the age. He feels in fact little more than a decent spirit, somewhat shunted to the side. It is the best possible position for detective work.

Forgive him, then, if he takes mild pleasure in conjunction of dates. John F. Kennedy had made his declaration concerning the moon not six weeks before Hemingway was dead. The nation, Kennedy decided, 'should commit itself to achieving the goal, before this decade is out, of landing a man on the moon and returning him safely to the earth . . . This is a new ocean, and I believe the United States must sail upon it.' Presumably, the moon was not listening, but if, in fact, she were the receiving and transmitting station of all lunacy, then she had not been ignoring the nation since. Four assassinations later; a war in Vietnam later; a burning of Black ghettos later; hippies, drugs and many student uprisings later; one Democratic Convention in Chicago seven years later; one New York school strike later; one sexual revolution later; yes,

eight years of a dramatic, near-catastrophic, outright spooky decade
later, we were ready to make the moon. It was a decade so unbalanced
in relation to previous American history that Aquarius, who had begun
it by stabbing his second wife in 1960, was to finish by running in a
Democratic Primary for Mayor of New York during the hottest May
and June he could ever recall. In sixty days he must have made three
hundred speeches, he appeared on more radio and television than he
could remember, walked streets, shook hands, sometimes two or three
thousand hands a day, worked fourteen hours a day, often sixteen, went
on four and five hours sleep, and awoke on many a morning with the
clear and present certainty that he was going to win. Norman was lazy,
and politics would make him work for sixteen hours a day the rest of his
life. He was so guilty a man that he thought he would be elected as a fit
and proper punishment for his sins. Still, he also wanted to win. He
would never write again if he were Mayor (the job would doubtless
strain his talent to extinction) but he would have his hand on the rump
of History, and Norman was not without such lust.

He came in fourth in a field of five, and politics was behind him. He
had run, when he considered it, no very remarkable race. He had obviously not had any apocalyptic ability to rustle up huge numbers of
votes. He had in fact been left with a huge boredom about himself. He
was weary of his own voice, own face, person, persona, will, ideas,
speeches, and general sense of importance. He felt not unhappy, mildly
depressed; somewhat used up, wise, tolerant, sad, void of vanity, even
had a hint of humility. Somewhat disembodied spirit. He burned something in his soul those eight weeks of campaigning, but he was not
certain just what he might have squandered. Nonetheless, he might be
in superb shape to study the flight of Apollo 11 to the moon. For he was
detached this season from the imperial demands of his ego; he could
think about astronauts, space, space programs, and the moon, quite
free of the fact that none of these heroes, presences, and forces were by
any necessity friendly to him. No, he felt like a spirit of some
just-consumed essence of the past, and so finally took the liberty to
christen himself Aquarius. It was the perfect name for a man who
would begin the study of rockets. The water-bearer traversed the earth
and breathed the air: three elements were his medium, solid, liquid, and
gas. That was kin to the rocket. Apollo 11 would leave the earth, travel

on the combustion of its liquids, and traverse a space. What indeed was space but the final decompression of a gas? On such unscientific thoughts did Norman, sign of Aquarius, travel.

iii

In the middle of his Mayoralty campaign, a story had appeared whose small headlines stated that he would receive a million dollars for doing a book about the astronauts. It was a peculiar story, because the sums listed in the journalistic details added up to $450,000, and this second figure, while certainly too generous, was not vastly inaccurate. Actually, Aquarius would be lucky if he were left with any real money at all, for he was in debt from having made three movies (for which he had put up the cash himself) and he calculated that with the restitution of consequent borrowings, and the payment of taxes, he would have enough to live and think for a year. Not so bad. He had only to write a book about the moon shot. Small matter. It would be as easy to go to the Amazon to study moon rocks as to write a book about these space matters, foreign to him, which everyone would agree is worth a million dollars. In fact everyone thought he was worth a million dollars already. Contributions for his campaign to the Mayoralty stopped on the instant the story appeared. He did not know whether to bless the gods, the *Times*, or somebody in the office of his agent.

Of course, he was not displeased that everyone thought a quick book by him – magazine, hard-cover, paperback, foreign rights, and syndication – was worth a million. While Aquarius had never been accorded the respect he thought he deserved as a novelist, he had been granted in compensation the highest praise as a journalist. People he had never met were forever declaring in print that he was the best journalist in America. He thought it was the superb irony of his professional life, for he knew he was not even a good journalist and possibly could not hold a top job if he had to turn in a story every day. He had known such journalists, and their work was demanding. They had first of all to have enormous curiosity, and therefore be unable to rest until they found out the secret behind even the smallest event. Since Aquarius had long built his philosophical world on the firm conviction that nothing was finally knowable (an exact and proper recompense to having spent his formative years and young manhood in searching for the true nature of women) he had almost no interest in the small secret behind a small event. (There was invariably another secret behind that.) He preferred to divine an event through his senses – since he was as nearsighted as he was vain, he tended to sniff out the center of a situation from a distance. So his mind often stayed out of contact with the workings of his brain for days at a time. When it was time, lo and behold, he seemed to have comprehended the event. That was one advantage of using the nose – technology had not yet succeeded in elaborating a science of smell.

But calculate for yourself the small ails and woes which came upon Aquarius when he went to visit the NASA Manned Spacecraft Center in Houston two weeks after the conclusion of his Mayoralty campaign. The first and most unhappy truth was that there were no smells coming out of NASA. It was hardly the terrain for Aquarius.

He had grown up in New York. He understood cities, particularly big cities, he had looked forward to getting to know a little of Houston – now, draw near to his vast pleasure in discovering that the Manned Spacecraft Center was not in Houston at all, but located about twenty-five miles south in the middle of that flat anonymous and near to tree-impoverished plain which runs in one undistinguished and not very green stretch from Houston to Galveston. Farther east as he would soon discover was Seabrook, Kemah, and Texas City south of that, then Galveston on the Gulf. Raunchy, sexy, hot and brooding, houses on stilts and old shacks – that was the Gulf of Mexico. He liked it. If he lived there, he too would write like Tennessee Williams. Tennessee, he discovered by this visit, was a natural and simple recorder of the elements.

All that, however, was miles away. MSC (the Manned Spacecraft Center) was located on a tract of many acres, flat and dry as a parking lot, and at the moment of entering the gate past the guard, there was no way to determine whether one was approaching an industrial complex in which computers and electronic equipment were fashioned, or traveling into a marvelously up-to-date minimum-security prison, not a clue to whether one was visiting the largest insurance and financing corporation which had ever decided to relocate itself in the flatlands behind a fence, or if this geometrically ordered arrangement of white modern buildings, severe, ascetic, without ornament, nearly all of two or three stories but for an Administration Building of eight stories, was indeed the newest and finest kind of hospital for radiological research. But, perhaps it was a college campus, one of those miserable brand-new college campuses with buildings white as toothpaste, windows set in aluminum casements, paths drawn by right angle or in carefully calculated zigzag to break the right angle, and a general air of studies in business administration, a college campus in short to replace the one which burned in the last revolution of the students.

In fact, it was the Manned Spacecraft Center, MSC, the home of the astronauts, the place where they were given the bulk of their training in Mission Simulators and Docking Simulators, the Center from which Mission Control would direct and collaborate on their flights, the astronauts' brain on earth, to nail it thus crudely, when they were up in space. And if this assembly of buildings looked as we have said like the worst of future college campuses, all-but-treeless, milk-of-magnesia white, and composed of many windowless buildings and laboratories which seemed to house computers, and did! why the error was in fact natural. For when Lyndon Johnson, then Vice President, succeeded in getting the unmistakable plum of the new Manned Spacecraft Center located in Texas on land he just happened to know about south of Houston owned by some nice fellows named Humble (Humble Oil & Refining) and ready for the Federal Government to purchase reasonable – reasonable a word capable of being reasoned and expanded with upon occasion – why this purchase might even have a clause inserted that the buildings to be constructed must be capable, in the event of the demise of NASA and the Space Program, of being converted without difficulty into an adjunct of Rice University in Houston. Could it be a crypto-campus after all! Let no one say that Lyndon Johnson was not a super local patriot always working for TALC (Texas Association for the Advancement of Local Culture).

Recognize then how much this Manned Spacecraft Center would honor Aquarius' sense of smell. Outside the Spacecraft Center, he could not say that his situation was improved. The immediate suburb, Nassau Bay, which housed many of the technicians, engineers, and executives in NASA, was situated on the other side of NASA Highway 1 from MSC, and was built around a body of water called Clear Lake. Nassau

Bay and adjoining suburbs like it were all new, their roads laid out in winding turns so absent of surprise that you could recognize they came off the French curve of the draftsman. If these homes were architecturally reasonable, built in sedate earth colors for the most part, charcoal browns, subdued clay-orange, stone-colored tans, houses which were modern but restrained adaptations for the most part of Swiss chalets, Tudor and Elizabethan, with hints of hacienda and ranch corral, they were nonetheless without flavor or odor. Aquarius was discovering that we cherish the sense of smell because it gives us our relation to time. We know how old something is by its odor; its youth, its becoming and its decay are subtly compounded to tell us at once – if we dare to contemplate mortality – how much time has been appropriated by such a life.

Nor were the people who worked for NASA bound to help him, since they were also by every evidence part of that vast convocation of Americans, probably a majority, whom one saw in New York only on television. They were, in short, Wasps, and it was part of the folklore of New York that Wasps were without odor. From the vantage point of New York, Wasps were already halfway to the moon, and devoted their efficiency to earning enough money to purchase large amounts of deodorant, depilatory, mouthwash, hair spray, and if they were ladies – Arrid. But these jokes are not very good. It would be tasteless to dwell on anybody's insulation from odor but for the fact that if this thesis is correct, if we honor or fear the presence of odors because they are a root to the past and an indication of the future, are indeed our very marriage to time and mortality, why then it is no accident that the Wasps were, in the view of Aquarius, the most Faustian, barbaric, draconian, progress-oriented, and root-destroying people on earth. They had divorced themselves from odor in order to dominate time, and thereby see if they were able to deliver themselves from death! No less! It is fiendish to get into such exaggeration so early, but think where Dr Christiaan Barnard would be today if on the threshold of his first heart transplant, he had declared, 'Nope, this organ ain't funky enough to make its new home happy!'

Obviously, then, if the great brain of NASA were attached to any particular sense, it was the eye. The eye was the collector of incontrovertible facts (which at MSC they called data-points). So the men who worked off NASA Highway 1 at the Manned Spacecraft Center were all clear-eyed and bullet-eyed and berry-eyed (pupils no larger than hard small acidic little berries) and they all seemed to wear dark pants, short-sleeve button-down white shirts and somber narrow ties. They all had identification badges pinned to their shirt pockets and they wore them with pride. Practically all had straight hair, and most of them cut it close. Whether they were tall or short, they were rarely overweight, and the only distinction between them which enabled Aquarius to differentiate these engineers, technicians and young executives from one another was that many wore horn-rimmed glasses with dark frames, and these fellows were usually smaller, more sallow, and with that absolute lack of surface provocation, or idiosyncrasy of personality, which characterizes physicists, engineering students, statisticians, computer technicians, and many a young man of science. By accent, appearance, and manner they could have come from any part of America, although most, Aquarius judged, were from the Midwest.

The other category belonged in general to men who were taller, more athletic, meaner-looking, sunburned upon occasion – despite their hours of work in air-conditioned rooms – and had the contained anger and cool crisp manner of men who have domiciled their unruly and bust-out impulses: so they emit a sense of discipline, order, and unmistakably virile, if controlled, determination. Aquarius who, for all his forty-six years and wretched inability to lose weight, liked to keep a sense of his own virility – what more valuable possession had an artist? – was obliged somewhat ruefully to recognize that this second category of men were tough. They reminded him of the officers and enlisted men of the Texas outfit, the 112th Cavalry, in which he had served overseas during the war. So he took it for granted that these executives, athlete-engineers, hondos on Mission Control, aides or instructors for various astronaut training courses, and general troubleshooters were in the main from the Southwest. They had a lot of morale. They were so proud of NASA, the astronauts, the Command Module, Lem, the United States of America that their voices went husky a hint when they talked about such topics.

Yet both categories of men were absolutely helpful in every way. But in such a way that they were no help at all to Aquarius. There was a style at NASA he had begun to divine. Every question you asked was answered and the truth so far as he knew was always told. It was as if NASA, unlike other Government bureaus, had recognized why honesty is the best policy – it is simply because no intriguer will ever believe the truth which is presented to him, but will rather interpret it as a lie which only he can transform into the buried fact. The assumption is that honest men will come to recognize your truth can make them strong. So everybody at NASA was courteous, helpful, generous of information, saintly at repeating the same information a hundred times, and subtly proud of their ability to serve interchangeably for one another, as if the real secret of their discipline and their strength and their sense of morale was that they had depersonalized themselves to the point where they were true Christians, gentle, helpful, replaceable, and serving on a messianic mission. The only flaw was that the conversation could only voyage through predetermined patterns. They would do their best to answer any technical question in the world, and voluminous mimeographings of NASA literature, often valuable enough to be classified, were available to all the Press. It was just that there was no way to suggest any philosophical meandering. Like real Americans, they always talked in code. It happened to be technological code. 'The whole philosophy of power descent monitoring is that when the Pings [PGNCS] have degraded . . .' or 'The bulk of Delta V is to kill his retrograde component.' These were notes Aquarius picked out for himself after a half hour of talking to the Chief of Flight Operations Division, who would help to bring the Lem down to the surface of the moon, a hard green-eyed crew-cut man in his thirties named Gene Kranz who looked and talked like a professional football quarterback. And in fact his problems were not dissimilar. They arrived at the same rate of speed and were as massive. 'During the first five minutes of descent,' Kranz said, 'the landing will be almost luxurious. But during the last three minutes, he'll be coming like Whistling Dixie.' Behind Kranz as he spoke were the twenty-odd consoles and the forty-plus screens, the dull gray-green walls, the thirty-five square lights inset in the ceiling – the gray controlled environment of the Mission Control room. Kranz lived with phrases like Primary Guidance and Navigation Section and Abort Guidance Section (Pings and Ags), Service Modulator Controllers, Power Descent Information, Program Descent Rates, Sequential Events Control System, Time of Ephemeris Update, Transponder, he spoke of

T Eff Em, and Reference Stable Member Matrix, of SMC, and PDI, SECST, the names and their related initials were used interchangeably – Kranz lived in a world of instruments and concepts which would take years for Aquarius to command well enough to make judgments on the other's character. Yes, real Americans always spoke in code. They encapsulated themselves into technological clans. Codes were like bloodlines. So they could be friendly and helpful and polite but they quietly separated themselves when their codes did not flourish. Aquarius was obliged to recognize that if the machine seemed a functional object to the artist, an instrument whose significance was that it was there to be used – as a typewriter was used for typing a manuscript – so to the engineer it was the communication itself which was functional. The machine was the art.

Perhaps for that reason, relations with these engineers reminded Aquarius of how he felt when he looked at the windowless walls of new buildings now sprouting all over the mean dry fields of the Space Center and the corporation developments outside the fence. These windowless buildings were as sinister to him as the arbitrary growth of ugly species of mushrooms in the middle of nowhere. These architectural fungoids were there to say: 'Lo, we work in the electronics computeroid complex, and need no windows, for we are the architectural skull case for a new kind of brain.'

Windowless, they also lack ears, so he cannot tell them, 'My eyes are my windows.'

'Recognize,' the windowless walls say, 'that something is taking over from you, kid.'

He stayed in a motel surprising in its luxury on this Texas plain. He had two rooms, and one room had a private indoor pool four feet deep, seven feet long, and five feet wide, with a green light overhead. The other room had a full king-sized circular bed with a red velvet cover. He discovered on inquiry that the motel had been decorated by a new owner who hoped to attract honeymoon couples to memories of the deluxe in the middle of the flatlands. But the clientele continued to consist of engineers visiting MSC from corporations which did business with NASA. Aquarius had a picture of some of the engineers he had met, the ones with the lunar pallor, sleeping in the round red velvet-covered king-sized bed. As if to emphasize this conjunction of the two centuries, the red velvet of the Nineteenth and the gray transistors of the Twentieth, there was a club in the motel with two go-go girls and one of them walked off abruptly one night and went to the bar. When the bartender whispered to her, she went back to the platform, turned on the jukebox again, giggled and said to the technology-ridden air of her audience, 'Shucks, I plumb forgot to take off my clothes.'

She was a round sullen country girl. Aquarius saw her dance another night when she was full of relish for her work, slinging her breasts, undulating her belly on a river of cogitating promise – the voracity of her hip-sock suggested she was one real alligator, but then six of her friends were in from Houston and sitting in the center seats, and they looked to have just gotten off their motorcycles. They were hardly from NASA.

There were exceptions to these uniform varieties of experience. He spent a night talking to Pete Conrad – Charles Conrad, Jr, the astronaut who would command Apollo 12 on the flight to the moon after Apollo 11 – and it was not a bad night. Conrad was wiry, he was feisty, he could rap without too much of a look over his shoulder for the proprieties, and his wife Jane was sensationally attractive in a quiet way. They had four young and handsome sons, one of whom, Tommy, aged twelve, became famous forever in Aquarius' mind because he obliged a photographer by riding his bicycle off the slope of the garage roof right into the swimming pool. Norman was invited back to a party the Conrads gave for their neighbors, and he had a good time – it was a party like a night in Westchester, except that it was Texas, so he finally got into a bathing suit in order not to wrestle up and down the edge of the pool when enthusiasts were ready to throw him in. Agreeably drunk, he stood under the hot Texas night in the hot Texas pool, laughing with two Texas ladies – it was at least an approach to the sensate experience of the East. And the next day he remembered Conrad saying to him over the outdoor steak grill – 'For six years I've been dreaming of going to the moon,' and the moon – as a real and tangible companion of the mind – was suddenly there before him.

He saw Armstrong, Aldrin, and Collins through much of a long day they spent in press conferences with the newspapers, magazines, and the television networks, and he learned much. (In the absence of a sense of smell, the hairs in his nostrils began to quiver at clues.) He thought about astronauts often. He would probably be able to produce an interesting thought or two on the psychology of astronauts. He felt as if he had begun the study of a new world so mysterious to his detective's heart (all imaginative novelists, by this logic, are detectives) that he could only repeat what he had said on the day the assignment was first offered to him: it was that he hardly knew whether the Space Program was the noblest expression of the Twentieth Century or the quintessential statement of our fundamental insanity. It was after all the mark of insanity that its mode of operation was distinguished by its logic – insanity was often more logical than sanity when it came to attacking a problem.

Something of this question was in his mind when he talked to Dr Gilruth, Robert R. Gilruth, Director of the Manned Spacecraft Center, but of course he did not pose the question directly and if he had, would not have gotten an answer. Speculation was on nobody's program at NASA. In any case, Gilruth was hardly one of the new technicians. A man in his late fifties, he had worked as a student under Piccard, the old balloonist, and had discovered the jet stream when a balloon built by his wife and himself was sent up in Minnesota and came down in Mississippi. This was the sort of story Gilruth had obviously told before to make an item in many a feature story, it was a way of keeping the interviewer away, and Aquarius recognized after a while that Dr Gilruth was a man who had probably developed his official style in the Eisenhower period, in fact he looked like a mild version of Eisenhower in the mid-Fifties, he was half bald in about the same way, and had deep gentle sympathetic eyes which gave him almost a saintly appearance; he talked in a quiet voice in his large office high up in the Administration Building and therefore facing down on the rectilinear play of the campus walks and buildings. Aquarius looked for something charitable to say about the view, but that proved too hard to produce, so he tried to win Gilruth's confidence in other ways. But the good doctor was not particularly responsive to questions, which is to emphasize that he would take an ordinary question and go on at such length in his reply, rambling through such hesitancies – as if the act of speech were painful to him – that the next question was hardly spurred to appear. He was remarkably gentle and determinedly undistinguished, as if his deepest private view suggested that good administration and public communication were best kept apart. In this sense, he was certainly no proper representative of the NASA style, much rather like a Chinese mandarin – completely pleasant, altogether remote – it occurred that Eisenhower had also been a mandarin.

Just once did Aquarius reach him. He asked: 'Are you ever worried, Dr Gilruth, that landing on the moon may result in all sorts of psychic disturbances for us here on earth?' At the look of pain in Gilruth's eyes at the thought of mustering NASA-type answers for this sort of question, Aquarius went on quickly, 'I mean, many people seem to react to the full moon, and there are tides of course.'

He was not mistaken. As he stammered into silence, there was the breath of dread in the room. Just a hint, but his nostril quivered. Gilruth was feeling the same silence; he could swear to that. And Gilruth, when he answered, spoke gratefully of the tides and yes, they had an effect on geography and men's industry by the sea – no answer could have been more Eisenhooverian – but then as if the question held him also in its grip, Gilruth came out of this long divagation to say that – *yes*, he had looked at some figures on the subject, and there seemed to be a higher incidence of hospital commission reports of admission to mental institutions during the full moon. Dread in the room again, and a silence between the two men which was exactly opposite to the silence of expectation when sex is near, no, now it was the opposite, how rather to move off this point, this continuing mounting silence. Who would be most implicated by breaking it? Now silence became the palpable appearance of the present, that breath of the present which holds all ultimates in its grip. Gilruth took responsibility by saying at last, 'I expect the moon is many things to many men. From Frank Borman's description on Apollo 8 we thought of it as rather a forbidding place' – he looked gloomy in recollection – 'whereas Stafford and Cernan and Young give us the idea from Apollo 10 that the moon is agreeable, so to speak, and not at all unpleasant but perhaps kind of a nice place to be,' and he smiled gently, hopefully, but perhaps a little regretfully for filling his share of the silence. They nodded at one another.

2

The Psychology of Astronauts

Well, let us make an approach to the astronauts. Aquarius sees them for the first time on the fifth of July, eleven days before the launch. They are in a modern movie theater with orange seats and a dark furrowed ceiling overhead, much like marcelled waves in a head of hair, a plastic ceiling built doubtless to the plans of one of the best sound engineers in the country. Sound is considerably ahead of smell as a fit province for scientific work, but since the excellence of acoustics in large and small concert chambers seems to bear more relation to old wood and the blessings of monarchs and bishops than to the latest development of the technical art, the sound system in this movie theater (seats 600) is dependably intolerable most of the time. The public address system squeals and squeaks (it is apparently easier to have communication with men one quarter of a million miles away) and one never gets a fair test of the aural accommodations. The walls and overhead are of plastic composition, and so far as one can tell, the tone is a hint sepulchral, then brightened electronically, finally harsh and punishing to that unnamed fine nerve which runs from the anus to the eardrum. As the sound engineers became more developed, the plastic materials provided for their practice by corporations grew acoustically more precise and spiritually more flattening – it was the law of the century. One was forever adjusting to public voices through the subtlest vale of pain.

Still this movie theater was the nearest approach to a diadem in the Manned Spacecraft Center. The theater was part of the visitors' center, where tourists could go through the space museum, a relatively modest affair of satellites, capsules, dioramas, posters and relics, now closed and given over to the installation of monitors and cables for the television networks, even as the gallery to the rear of the theater was now being converted into the Apollo News Center and would consist finally of endless aisles of desks, telephones and typewriters, plus one giant Buddha of a coffee urn. (Coffee is the closest the Press ever comes to *satori*.)

In the theater, perhaps eight rows back of the front seats, was a raised platform on which television cameras and crews were mounted. From the stage they must have looked not unrelated to artillery pieces on the battlement of a fort – in the front row were fifty photographers, which is to say fifty sets of torsos and limbs each squeezed around its own large round glass eye. Little flares of lightning flashed out of bulbs near their heads. The astronauts did not really have to travel to the moon – life from another planet was before them already. In the middle ranks, between the front row and the barricade of television cameras, were seated several hundred newspaper men and women come to Houston for the conference this morning. They were a curious mixture of high competence and near imbecility; some assigned to Space for years seemed to know as much as NASA engineers; others, innocents in for the big play on the moon shot, still were not just certain where laxatives ended and physics began. It was as if research students from the Institute of Advanced Studies at Princeton had been put in with a group of fine young fellows from an Army class in remedial reading. Out of such a bag would questions come to the astronauts. Wait! There will be samples.

The astronauts entered from the wings wearing gas masks, gray snout-nosed covers which projected out from their mouths and gave their profiles the intent tusk-ready slouch of razorback hogs. They were aware of this – it was apparent in the good humor with which they came in. In fact, a joke of some dimensions had been flickering for a few days – the Press had talked of greeting them with white hospital masks. In the attempt to protect the astronauts as much as possible from preflight infection they were being kept in a species of limited quarantine – their contacts with nonessential personnel were restricted. Since journalists fit this category, today's press conference had installed Armstrong, Aldrin and Collins up on the stage in a plastic box about twelve feet wide, ten feet deep and ten feet high. Blowers within this three-walled plastic room blew air from behind them out into the audience: thereby, the breath of the astronauts could enter the theater, but the airborne germs of journalists would not blow back. It made a kind of sense. Of course the cause of the common cold was still unknown, but gross studies of infection would surmise a partial quarantine might be effective partially. However, the instrumentation of this premise was not happy. The astronauts looked a bit absurd in their plastic box, and the few journalists who had actually fleshed their joke by putting on masks caused the astronauts to grin broadly as though to dissociate themselves from the pyramids of precaution they were in fact obeying.

Once they sat down, their manner changed. They were seated behind a walnut-brown desk on a pale blue base which displayed two painted medallions in circles – NASA and Apollo ii. Behind them at the rear of the plastic booth stood an American flag; the Press actually jeered when somebody brought it onstage in advance of the astronauts. Aquarius could not remember a press conference where Old Glory had ever been mocked before, but it had no great significance, suggesting rather a splash of derision at the thought that the show was already sufficiently American enough. In fact, between the steady reporters who worked out of Houston and the astronauts, there was that kind of easy needling humor which is the measure of professional respect to be found among teams and trainers.

So the entrance went well. The astronauts walked with the easy saunter of athletes. They were comfortable in motion. As men being scrutinized by other men they had little to worry about. Still, they did not strut. Like all good professional athletes, they had the modesty of knowing you could be good and still lose. Therefore they looked to enjoy the snouts they were wearing, they waved at reporter friends they recognized, they grinned. A reporter called back to Collins, 'Now, you look good.' It all had that characteristically American air which suggests that men who are successful in their profession do best to take their honors lightly.

Once they sat down, however, the mood shifted. Now they were there to answer questions about a phenomenon which even ten years ago would have been considered material unfit for serious discussion. Grown men, perfectly normal-looking, were now going to talk about their trip to the moon. It made everyone uncomfortable. For the relation of everyone to each other and to the event was not quite real. It was as if a man had died and been brought back from death. What if on questioning he turned out to be an ordinary fellow? 'Well, you see,' he might say, 'having visited death, I come back with the following conclusions . . .' What if he had a droning voice? There was something of this in the polite unreality of the questioning. The century was like a youth who made love to the loveliest courtesan in Cathay. Afterward he was asked what he thought and scratched his head and said, 'I don't know. Sex is kind of overrated.' So now people were going to ask questions of three heroes about their oncoming voyage, which on its face must be in contention for the greatest adventure of man. Yet it all felt as if three young junior executives were announcing their corporation's newest subdivision.

Perhaps for this reason, the quiet gaiety of their entrance had deserted them as they sat behind the desk in the plastic booth. Now it was as if they did not know if they were athletes, test pilots, engineers, corporation executives, some new kind of priest, or sheepish American boys caught in a position of outlandish prominence – my God, how did they ever get into this? It was as if after months in simulators with knowing technicians geared to the same code languages, they were now debouched into the open intellectual void of this theater, obliged to look into the uncomprehending spirits of several hundred media tools (human) all perplexed and worried at their journalistic ability to grasp more than the bare narrative of what was coming up. Yaws abounded. Vacuums in the magnetism of the mood. Something close to boredom. The astronauts were going to the moon, but everybody was a little frustrated – the Press because the Press did not know how to push into nitty-gritty for the questions, the astronauts because they were not certain how to begin to explain the complexity of their technique. Worse, as if they did not really wish to explain, but were obliged out of duty to the program, even if their privacy was invaded.

So the conference dragged on. While the focus of attention was naturally on Armstrong for commanding the flight, he seemed in the beginning to be the least at ease. He spoke with long pauses, he searched for words. When the words came out, their ordinary content made the wait seem excessive. He minted no phrases. 'We are here' . . . a pause . . . 'to be able to talk about this attempt' . . . a real pause, as if the next experience were ineffable but with patience would yet be captured . . . 'because of the success of four previous Apollo command flights' . . . pause, as if to pick up something he had left out . . . 'and a number of unmanned flights.' A shy smile. 'Each of those flights' – he was more wooden than young Robert Taylor, young Don Ameche, young Randolph Scott – 'contributed in a great way' . . . deprecatory smile . . . 'to this flight.' As a speaker he was all but limp – still it did not leave him unremarkable. Certainly the knowledge he was an astronaut restored his stature, yet even if he had been a junior executive accepting an award, Armstrong would have presented a quality which was arresting, for he was extraordinarily remote. He was simply not like other men. He would have been more extraordinary in fact if he had been just a salesman making a modest inept dull little speech, for then one would have been forced to wonder how he had ever gotten his job, how he could sell even one item, how in fact he got out of bed in the morning. Something particularly innocent or subtly sinister was in the gentle remote air. If he had been a young boy selling subscriptions at the door, one grandmother might have warned her granddaughter never to let him in the house; another would have commented, 'That boy will go very far.' He was apparently in communion with some string in the universe others did not think to play.

Collins and Aldrin followed with their opening remarks, and they had personalities which were more comfortable to grasp. Aldrin, all meat and stone, was a man of solid presentation, dependable as a tractor, but suggesting the strength of a tank, dull, almost ponderous, yet with the hint of unpredictability, as if, eighteen drinks in him, his eyes would turn red, he would arm-wrestle a gorilla or invite you to join him in jumping out a third-story window in order to see who could do the better somersault on the follow-through out of the landing. This streak was radium and encased within fifty psychical and institutional caskings of lead, but it was there, Aquarius thought, perhaps a clue in the way he dressed – very dressy for an astronaut – a green luminous silk suit, a white shirt, a green luminous tie. It clashed with the stolid presentation of his language. Aldrin spoke in a deep slow comfortingly nasal tone – a mighty voice box – his face was strong and grim. The movie director in Aquarius would have cast him on the spot for Major in Tank Cavalry. He had big features and light brown hair, almost gold. His eyes took a turn down like samurai eyes, the corners of his lips took a right-angle turn down – it gave him the expression of a serious man at home on a field of carnage, as if he were forever saying, 'This is serious stuff, fellows, there's lots of blood around.' So Aldrin also looked like the kind of jock who could be headmaster of a prep school. He had all the locker-room heartiness and solemnity of a team man. Although he had been a pole-vaulter at West Point, it would have been easy to mistake him for a shot-putter, a lacrosse player, or a baseball catcher. In football he would have probably been a linebacker. For this last, he was actually not big enough (since the astronauts were required to be no more than five feet eleven inches tall and could hardly be overweight), but he was one of those men who looked larger than his size for his condition was excellent – every discipline of his moves spoke of grim devoted unrelenting support given to all his body-world of muscle. From the back of the neck to the joints of the toes, from the pectorals to the hamstrings, the deltoids to the abdominals, he was a life given over to good physical condition, a form of grace, since the agony of the lungs when straining is not alien to the agony of the soul. Leave it that Aldrin was so strong he had a physical presence which was bigger than his bulk.

He talked like a hardworking drill. He had the reputation of being the best physicist and engineer among the astronauts – he had written a valuable thesis on Orbital Rendezvous Techniques at MIT, but he put no humor into his presentation, he was selling no soap. If you did not read technologese, you might as well forget every last remark for his words did not translate, not unless you were ready to jog along with him on technology road. Here is the way he gave himself to the Press: 'We do have a few items on the Lem side of the house on this particular mission. We'll be picking up where Apollo 10 left off when they did their phasing maneuver. And at this point after departing the Command Module, coming down in the descent orbit, we'll be igniting the descent engine for the first time under a long burn condition when it is not docked with the Command Module. And executing this burn under control of a computer, being directed towards the various targets that are fed into the computer will be new on this flight. Also we'll be making use of the landing radar and its inputs into the computer. Inputs in terms of altitude and velocity updates which will bring us down in the prescribed conditions as we approach the surface of the moon. Of course, the actual control of the touchdown itself will be a rather new item in that it will be testing this man-machine interface to a very sophisticated degree. The touchdown itself will be the ultimate test on the landing gear and the various systems that are in the spacecraft. The environment of one-sixth G will be seen for the first time by crews and spacecraft. We'll also be exposed to thermal conditions that have not been experienced before. The two-man EVA is something that is a first in our program. Sleeping in the Lem on the lunar surface, which we hope to be able to do, will be another new item in that flight.'

He went on to talk of star sightings and the powered ascent from the moon – that moment when, having landed successfully and recon- noitered the moon ground, they would be back in the Lem and ready to ascend – would the motor ignite or did the moon have a curse? Aldrin spoke of this as a 'new item,' then of rendezvous with the Command Module, which would return them to earth, of 'various contingencies that can develop,' of 'a wider variety of trajectory conditions' – he was talking about not being able to join up, wandering through space, lost forever to life in that short eternity before they expired of hunger and thirst. Small hint of that in these verbal formulations. Even as the Nazis and the Communists had used to speak of mass murder as liquidation, so the astronauts spoke of possible personal disasters as 'contingency.' The heart of astronaut talk, like the heart of all bureaucratic talk, was a jargon which could be easily converted to computer programming, a language like Fortran or Cobol or Algol. Anti-dread formulations were the center of it, as if words like pills were there to suppress emotional symptoms. Yet Aldrin, powerful as a small bull, deep as his grasp of Celestial Mechanics, gave off in his air of unassailable solemnity some incommunicable speech about the depth of men's souls and that razor's edge between the hero's endeavor and vainglory. Vainglory looked real to him, one might assume, real as true peril – he had the deep gloomy clumsy dignity of a man who had been face to face in some stricken hour with the depths of his own nature, more complex than he had hitherto known.

Collins, in contrast, moved easily; Collins was cool. Collins was the man nearly everybody was glad to see at a party, for he was the living spirit of good and graceful manners. Where Armstrong referred to Wapakoneta, Ohio, as his hometown, and showed a faint but ineradic- able suspicion of anyone from a burg larger than his own, where Aldrin protected himself from conversation with the insulations of a suburban boyhood and encapsulement among his incommunicable fields of competency, Collins had been born in a well-set-up apartment off the Borghese Gardens in Rome. His father, General James L. Collins, was military attaché (and could conceivably have been having a drink around the corner in the bar at the Hassler to celebrate the birth of his son). Since the year was 1930, Dick Diver could have been getting his going-over from the Fascisti police in the basement of *Tender Is the Night*. No surprise then if Collins had a manner. It was in part the manner of Irish elegance – a man must be caught dead before he takes himself seriously. It was as if Collins were playing a fine woodwind which had the merriment and the sadness (now that the madness was gone) of those American expatriates for whom culture began in the Year One of *The Sun Also Rises*. Indeed, if Collins was later to grow a mustache on the trip back, an act which increased his slight but definite resemblance to the young Hemingway, he had a personal style which owed more to Fitzgerald. It was Fitzgerald, after all, who first suggested that you could become the nicest man in the world. So Collins had that friendliness which promises it would be sacrilege to give offense in a social situation. It was apparently as unnatural for him not to make a small joke as it would have been offensive to Aldrin not to take on a matter in its full seriousness. Yet Collins had little opportunity to show his humor. It existed mainly in the fine light smiling presence he bestowed on the interview while the others were asked all the questions. Collins was the only one of the three not landing on the moon. So he would obviously be the one whose remarks would go into the last paragraph, where the layout man would probably lop them off. Therefore nobody had bothered to direct a question to him through all the interview.

Toward the end of the press conference, somebody asked of the astronauts at large, 'Two questions. Firstly, what precautions have been taken at your own homes to prevent you from catching germs from your own family? And secondly, is this the last period that you will spend at home with your families?' The Public Affairs Officer, Brian Duff, was quick to say, 'Take a crack at that, Mike.'

It could not have been easy to have waited so long for so little. But Collins came up smiling, and said, 'My wife and children have signed a statement that they have no germs and – and yes this will be the last weekend that we will be home with our families.' It was not much of a joke but the press conference had not been much of a joke either, and the Press brightened, they laughed, Collins, quick not to offend the man who had asked the question, now added, 'Seriously, there are no special precautions being taken.'

His conversational manner was easy. It was apparent that of the three, he was the only one you could drink with comfortably. Since the ability to drink with your material is as important to a journalist as the heft of his hammer to a carpenter, a sense of dismay passed through the press corps – why hadn't NASA had the simple sense of press relations to put Collins in command? What a joy it could have been to cover this moon landing with a man who gave neat quotes, instead of having to contend with Armstrong, who surrendered words about as happily as a hound allowed meat to be pulled out of his teeth. Collins would have been perfect. In combination with his manner, so obviously at ease with a martini, he had the trim build, the bald forehead, and economical features of a college boxer, or a shortstop, or a quarterback. (In fact he was the best handball player among the astronauts and had been captain of his wrestling team at St Albans.) He looked like copy, he talked like copy, and Armstrong had the sad lonely mien of a cross-country runner. Of course, since he also had the sly privacy of a man whose thoughts may never be read – what a vast boon was this to the Press! – one could, if picturing Armstrong as an athlete, see him playing end. He might, thus sly and private, be difficult to keep up with on pass patterns.

The story resided, however, with the two men who would land on the moon – it could reside nowhere else – but since Collins with a few smiles and a remark or two had become the favorite, a question and then another came his way at the end of the interview. Finally, the real question came.

'Colonel Collins, to people who are not astronauts, you would appear to have the most frustrating job on the mission, not going all the way. How do you feel about that?' The contradiction implicit in being an astronaut was here on this point – it was skewered right here. If they were astronauts, they were men who worked for the team, but no man became an astronaut who was not sufficiently exceptional to suspect at times that he might be the best of all. Nobody wins at handball who is not determined to win.

He answered quickly. 'I don't feel in the slightest bit frustrated. I'm going 99.9 percent of the way there, and that suits me just fine.' Growing up in Rome, Puerto Rico, Baltimore and Washington, Texas and Oklahoma, son of one of the more cultivated purlieus of the military grace, the code would be to keep your cool. The only real guide to aristocracy in American life was to see who could keep his cool under the most searching conditions of unrest, envy, ambition, jealousy and heat. So not a quiver showed. 'I couldn't be happier right where I am,' he concluded and the voice was not hollow, it did not offer a cousin to a squeak. Still nobody believed him. Somewhere in the room was the leached-out air of a passion submitted to a discipline. For a moment Collins was damnably like an actor who plays a good guy.

Armstrong came in quickly. 'I'd like to say in that regard that the man in the Command Module' . . . pause . . . 'of course by himself' . . . another pause . . . 'has a giant-sized job.' When Armstrong paused and looked for the next phrase he sometimes made a sound like the open crackling of static on a pilot's voice band with the control tower. One did not have the impression that the static came from him so much as that he had listened to so much static in his life, suffered so much of it, that his flesh, his cells, like it or not, were impregnated with the very cracklings of static. 'He has to run Buzz's job and my job' . . . static . . . 'along with his own job simultaneously' . . . static . . . 'in addition act as relay to the ground' . . . pause and static . . . 'It's at least a three-man job and – he murmured a few words – 'Michael is certainly not lacking for something to do while he's circling around.' Then Armstrong flashed a smile. One of his own jokes came. His humor was pleasant and small-town, not without a taste of the tart. 'And if he can't think of anything else, he can always look out the window and admire the view.'

Now came a question from a reporter who was new on the job: 'From your previous experience in the two and a half hours or so that you're atop the rocket before actual blast-off, is this a period of maximum tension, rather like being in a dentist's waiting room?'

A temporary inability to understand the question was finally replaced by this speech. 'It's one of the phases that we have a very high confidence in,' Armstrong answered with his characteristic mixture of modesty and technical arrogance, of apology and tight-lipped superiority. 'It's nothing new. It's the thing that's been done before,' now static while he searched for the appropriate addition, 'and done very well on a number of occasions, and we're quite sure this girl will go,' he said solemnly, pleasantly, lightly, carefully, sadly, sweetly. He was a presence in the room, as much a spirit as a man. One hardly knew if he were the spirit of the high thermal currents, or that spirit of neutrality which rises to the top in bureaucratic situations, or both, both of course – why should Armstrong have a soul less divided than the unruly world of some billions of men? Indeed contradictions lay subtly upon him – it was not unlike looking at a bewildering nest of leaves: some are autumn fallings, some the green of early spring. So Armstrong seemed of all the astronauts the man nearest to being saintly, yet there was something as hard, small-town and used in his face as the look of a cashier over pennies. When he stopped to think, six tired parallel lines stood out on his forehead, and his hair was very straight, small-town hair-colored humorless straight, his pupils were very small, hardly larger than buckshot, you could believe he flew seventy-eight combat missions off the *Essex* near Korea. He was very thin-mouthed, almost as thin and wide a mouth as Joe E. Brown, yet with no comic spirit, or better, or worse, the spirit of comedy gave orders to the mouth most of the time. Much like President Nixon or Wernher von Braun (whom we are yet to meet) he would smile on command. Then a very useful smile appeared – the smile of an enterprising small-town boy. He could be an angel, he could be the town's devil. Who knew? You could not penetrate the flash of the smile – all of America's bounty was in it. Readiness to serve, innocence, competence, modesty, sly humor, and then a lopsided yawing slide of a dumb smile at the gulfs of one's own ignorance, like oops am I small-town dumb! – that was also in it. Aquarius decided it was not easy to trust him then – the smile was a vehicle to remove Armstrong from the scene. But when he spoke, all ambition was muzzled. He spoke with the unendurably slow and triple caution of a responsibility-laden politician who was being desperately careful to make no error of fact, give no needless offense to enemies, and cross no conflicting zones of loyalty among friends. Add the static, and he was no happy public speaker. At communicating he was as tight as a cramped muscle.

Perversely, it became his most impressive quality, as if what was best in the man was most removed from the surface, so valuable that it must be protected by a hundred reservations, a thousand cautions, as if finally he had such huge respect for words that they were like tangible omens and portents, zephyrs and beasts of psychic presence, as if finally something deep, delicate and primitive would restrain him from uttering a single word of fear for fear of materializing his dread. So, once, men had been afraid to utter the name of the Lord, or even to write it in such a way as to suggest the sound, for that might be enough to summon some genie of God's displeasure at so disrupting the heavens. Armstrong of course did not brandish an ego one could perceive on meeting; where Aldrin gave off the stolid confidence of the man who knows that problems can be solved if properly formulated and appropriately attacked (which is to say attacked in good condition!) and where Collins offered the wiry graceful tension of a man who will quietly die to maintain his style, Armstrong could seem more like a modest animal than a man – tracer hints of every forest apprehension from the puma to the deer to the miseries of the hyena seemed to stalk at the edge of that small-town clearing he had cut into his psyche so that he might offer the world a person. But his thoughts seemed to be looking for a way to drift clear of any room like this where he was trapped with psyche-eaters, psyche-gorgers, and the duty of responding to questions heard some hundreds of times.

On the other hand, he was a professional and had learned how to contend in a practical way with the necessary language. Indeed, how his choice of language protected him!

'Mr Armstrong, at the time you are down on the moon, what will be your overriding consideration and what will be your main concern?'

'Well,' said Armstrong, 'immediately upon touchdown our concern is the integrity of the Lunar Module itself' . . . nnnnnnnhr went the sound of the static . . . 'For the first two hours after touchdown we have a very busy time verifying the integrity of the Lunar Module and all of its systems' . . . nnnnhr . . . 'A great deal of technical discussion . . . between spacecraft and ground during a time period when most people will be wondering, well what does it look like out there? . . . We will be eager to comment' . . . nnnnhr . . . 'but reluctant to do so in the face of these more important considerations on which . . . the entire rest of the lunar mission depends.'

Aldrin, the formalist, had said just previously, 'I think the most critical portion of the EVA will be our ability to anticipate and to interpret things that appear not to be as we expected them to be, because if we don't interpret them correctly then they will become difficult.' It was the credo of the rationalist. Phenomena are only possessed of menace when they do not accommodate themselves to language-controls. Or, better, to initial-controls. EVA stood for Extravehicular Activity, that is for action taken outside their vehicle, the Lem. EVA therefore referred to their walk on the moon; but the sound of the letters E, V, A might inspire less perturbation than the frank admission that men would now dare to walk on an ancient and alien terrain where no life breathed and beneath the ground no bodies were dead.

It was, of course, a style of language all the astronauts had learned. There were speeches where you could not tell who was putting the words together – the phrases were impersonal, interlocking. One man could have finished a sentence for another. 'Our order of priorities was carefully integrated into the flight plan . . . there is no requirement on the specific objectives that we're meeting on the surface to go great distances from the spacecraft, and to do so would only utilize time that we now have programmed doing things in the specific mission object- ives.' Sell newspapers with that kind of stuff! The quote could belong to any one of a dozen astronauts. In this case it happened to be not Aldrin but Armstrong.

Only on occasion did the language reveal its inability to blanket all situations. Mainly on personal matters. There came a question from one of the remedial readers. 'Tell us very briefly how your families have reacted to the fact that you're taking this historic mission.'

'Well,' Aldrin deliberated, 'I think in my particular case, my family has had five years now to become accustomed to this eventuality, and over six months to face it very closely. I think they look on this as a tre- mendous challenge for me. They look upon it also as an invasion somewhat of their privacy and removing of my presence away from the family for a considerable period of time.' He spoke glumly, probably thinking at this moment neither of his family nor himself – rather whether his ability to anticipate and interpret had been correctly employed in the cathexis-loaded dynamic shift vector area of changed field domestic situations (which translates as: attractive wife and kids playing second fiddle to boss astronaut number two sometimes blow group stack). Aldrin was a man of such powerful potentialities and iron disciplines that the dull weight of appropriately massed jargon was no mean gift to him. He obviously liked it to work. It kept explosives in their package. When his laboriously acquired speech failed to mop up the discharge of a question, he got as glum as a fastidious housewife who cannot keep the shine on her floor.

They could not, of course, restrain the questions which looked for ultimate blood. 'James Gunn, BBC. You had mentioned that your flight, like all others, contains very many risks. What, in view of that, will your plans be' – a British courtesy in passing – 'in the extremely unlikely event that the Lunar Module does not come up off the lunar surface?'

Armstrong smiled. His detestation of answering questions in public had been given its justification. Journalists would even ask a man to comment on the emotions of his oncoming death. 'Well,' said Armstrong, 'that's an unpleasant thing to think about.' If, as was quite possible, he had been closer to death than anyone in the room, and more than once, more than once, that did not mean the chalice of such findings was there to be fingered by fifty. 'We've chosen not to think about that up to the present time. We don't think that's at all a likely situation. It's simply a possible one.' He had, however, not answered the question. If he put in twelve and more hours a day in simulators, if there were weeks when they worked seventy and eighty hours a week at the abrasive grind of laying in still more hierarchies of numbers and banks of ratio in their heads, well, they were accustomed to hard work. So the grind today of being interviewed in full press conference, then by the wire services, then by magazine writers and finally for the television networks, a fourteen-hour day before it would all be done, and of the worst sort of work for them – objects on display to be chipped at by some of the worst word-sculptors ever assembled in southeastern Texas – well, that would still be work they must perform to the best of their duty. Being an astronaut was a mission. Since the political and power transactions of the age on which NASA's future was – put no nice word on it – hung, were not in spirit religious, the astronauts did not emphasize their sense of vocation. But being an astronaut was a mission and therefore you were obliged to perform every aspect of your work as well as you could. At a press conference you answered questions. So Armstrong now finally said in answer to what they would do if the Lunar Module did not come up off the lunar surface, 'At the present time we're left without recourse should that occur.'

When the conference was done, there was only a small pattering of applause from the Press. The atmosphere had been equal to any other dull press conference in which a company had unveiled a new and not very special product. Resentment in the Press was subtle but deep. An event of such dimensions and nothing to show for it. The American cool was becoming a narcotic. The horror of the Twentieth Century was the size of each new event, and the paucity of its reverberation.

But what if you're unable to get off the moon?

'Unpleasant thing to think about.'

ii

It was the answer Aquarius thought about after the conference was done, for that was the nearest anyone had come to saying that a man could get killed in the pits of this venture. And yes, they did think about it. A man who was in training for six months to go to the moon would be obliged to think about his death. Yet, if to contemplate the failure of the ascent stage of the Lunar Module to rise off the moon was unpleasant for Armstrong to think about, did that derive automatically and simply because it would mean death, or was it, bottomless taint of the unpleasant, a derivation deep out of the incommensurable fact that the moon ground would be the place where his body must rest in death? People who had nearly died from wounds spoke of the near death as offering a sensation that one was rising out of one's body. So had spoken Hemingway long ago, writing in Paris, writing in Spain, probably writing in apartments off the Borghese Gardens near where Collins had been born. Now was there to be a future science of death, or did death (like smell and sound and time – like the theory of the dream) resist all scientists, navigators, nomenclature and charts and reside in the realm of such unanswerables as whether the cause of cancer was a malfunction of the dream? *Did* the souls of the dead choose to rise? Was the thought of expiring on the moon an abyss of unpleasantness because the soul must rest in the tombless vacuums of a torso dead on the moon and therefore not able to voyage toward its star? A vertigo of impressions, but Aquarius had been living at the edge of such thoughts for years. It was possible there was nothing more important in a man's life than the hour and the route and the power of his death, yes, certainly if his death were to launch him into another kind of life. And the astronauts – of this he was convinced – would think this way, or at least would have that vein of imagination in some inviolate and noncommunicatory circuit of their brain; somewhere, far below the language of their communication, they must suspect that the gamble of a trip to the moon and back again, if carried off in all success, might give thrust for some transpostmortal insertion to the stars. Varoom! Last of all over the years had Aquarius learned how to control the rapid acceleration of his brain. Perhaps as a result, he was almost – in these first few days of covering the astronauts in Houston – fond of the banality of their speech and the anodyne of technologese.

But that press conference reserved exclusively for the magazine writers was about to begin – the writers would be working at least half as hard as the astronauts this day – and Aquarius on his way over to the Lunar Receiving Laboratory, where the interview was to be staged (for reasons soon explained) was wondering if the glints and notes of these cosmic, if barely sketched, hypotheses about earth, moon, life, death, the dream and the psychology of astronauts would be offered the ghost of a correlative. Aquarius was contemplating again the little fact that man had not done so very much with Freud's theory of the dream – had the theory of wish fulfillment shown a poor ability 'to anticipate and interpret things that appear to be not as we expected them to be'? Did that old Freudian theory of the dream bear the same relation to the veritable dimensions of the dream that a Fourth of July rocket could present to Saturn V?

iii

Since the astronauts were being guarded against infection, they were seen next behind the protection of a glass wall in the visitors' room at the Lunar Receiving Laboratory. An entire building had been constructed to quarantine them on their return, a species of hospital dormitory, galley and laboratory for the moon rocks. Since for twenty-one days after their return they would not be able to be in the same room with their families, or with the NASA technicians and officials who would debrief them, a chamber like the visitors' room in a prison had been built with a plate-glass partition hermetically sealed from floor to ceiling running down the middle. Dialogue through the glass wall proceeded through microphones.

Now, for the rest of the day, the astronauts would receive the other media layers here: TV, radio, wire service, magazines, etc. Now the magazine writers could sit within a few feet of their subjects, and yet – as if suggesting some undiscovered metaphysical properties of glass – they were obliged at the same time to feel a considerable distance away. Perhaps the full lighting on the astronauts and the relative gloom on the writers' side of the enclosure may have suggested the separation of stage and audience, but probably the effect was due most to the fact that laying-on of hands through that glass, so certainly shatterproof, could never occur, and so there was a dislocation of the sense of space. The astronauts were near enough to sit for a portrait, but – through the glass – they were as far away as history.

There was a new intimacy to the questions however. The setting was of aid, and besides, the magazine writers were in need of more. One of them took up immediately on the question which had bothered Aquarius, but the approach was practical now. How indeed would the astronauts spend their time if they found they could not get off the moon? Would they pray, would they leave messages for their family, or would they send back information on the moon? Such were the alternatives seen by the questioner.

Aldrin had the happy look of a linebacker who is standing right in the center of a hole in the line as the runner tries to come through. 'I'd probably spend it working on the availability of the ascent engine.'

That brought a laugh, and there would be others to follow, but the twenty or so magazine writers had the leisure to ask their questions out of a small group, and so there was not the itch of the newspaperman to look for a quick lead and therefore ask brutal or leading or tendentious questions. Indeed there was no need to ask any question whatever just so that the journalist and his newspaper could be identified as present at the conference. (Such identifications give smaller newspapers and their reporters a cumulative status over the years with public relations men.) No, here the magazine writers could take their time, they could pursue a question, even keep after the astronaut. Covertly, the mood of a hunt was on. Since they would have more time to write their pieces, by severer standards would they be judged. So they had to make the astronauts come to life whether the astronauts wished to exhibit themselves or not.

Will you take personal mementos? Armstrong was asked.

'If I had a choice, I guess I'd take more fuel,' he said with a smile for the frustration this might cause the questioner.

The magazine writers kept pushing for personal admission, disclosure of emotion, admission of unruly fear – the astronauts looked to give replies as proper and well-insulated as the plate glass which separated them. So Armstrong replied to a question about his intuition by making a short disclaimer, which concluded, 'Interpret the problem properly, then attack it.' Logical positivism all the way was what he would purvey. Don't make predictions without properly weighted and adequate inventories of knowledge. Surely he trusted his intuitions, the questioner persisted. 'It has never been a strong suit,' said Armstrong in a mild and honest voice. Obviously, the natural aim of technology was to make intuition obsolescent, and Armstrong was a shining knight of technology. But, in fact, he had to be lying. A man who had never had strong intuitions would never have known enough about the sensation to disclaim its presence in himself.

Would he at least recognize that his endeavor was equal in magnitude to Columbus' adventure?

He disclaimed large reactions, large ideas. 'Our concern has been directed mainly to doing the job.' He virtually said, 'If not me, another.' If they would insist on making him a hero, he would be a hero on terms he alone would make clear. There had been only one Columbus – there were ten astronauts at least who could do the job, and hundreds of men to back them up. He was the representative of a collective will.

Sitting in his drab gray-green suit, a suit as close to no color as possible, his shirt pale blue, his tie nondescript dark gray-blue, a blue-green wall behind him (perhaps to hint at empyreans of sky) his neck seemed subtly separated from his collar, as if – no matter how neatly he was dressed – his clothes felt like a tent to him, like a canvas drop out of which his head protruded through the hole of his collar. They were popping baseballs at him, he was dodging.

'Will you keep a piece of the moon for yourself?' asked a questioner.

It was a beautiful question. If he admitted desire, one could ask if the Armstrong house would sleep on nights of full moon when the piece of rock bayed silently to its distant mistress, and emanations wandered down the stairs. But Armstrong said stiffly, 'At this time, no plans have been made' . . . (Would he ever have the desire to steal a rock, Aquarius asked silently.) 'No,' Armstrong went on, 'that's not a prerogative we have available to us.' He could of course have said, 'We can't do it,' but in trouble he always talked computerese. The use of 'we' was discouraged. 'A joint exercise has demonstrated' became the substitution. 'Other choices' became 'peripheral secondary objectives.' 'Doing our best' was 'obtaining maximum advantage possible.' 'Confidence' became 'very high confidence level.' 'Ability to move' was a 'mobility study.' 'Turn off' was 'disable'; 'turn on' became 'enable.' It was as if the more natural forms of English had not been built for the computer: Latin maybe, but not simple Anglo-Saxon. That was too primitive a language – only the general sense could be conveyed by the words: the precise intent was obliged to be defined by the tone of the voice. Computerese preferred to phase out such options. The message had to be locked into a form which could be transmitted by pulse or by lack of pulse, one binary digit at a time, one bit, one bug to be installed in each box. You could not break through computerese.

Through it all, Collins would smile, turn his sensitive presence as eyes to the questioners, ears to the answer. His smile would flicker at the plastic obsidian impenetrability of computerese. 'Darn it all,' his smile would seem to say to the magazine writers, 'if I had to learn how to translate this stuff, I'm sure you fellows can do as well!' Once again, Collins was being asked few questions.

They turned after a while to Aldrin and began to draw some flecks of a true-blooded response. He was, of course, equally impenetrable in the beginning, but after a time he may have made the mistake of essaying a joke. Asked of his reactions to visiting the moon, he proceeded to build a wall of verbal brick, then abruptly with that clumsy odd sobriety, almost engaging, with which he was forever showing his willingness to serve, Aldrin made a remark about having been a boy scout. 'I attained the rank of tenderfoot,' he said. He gave a discomfited smile. 'I hope I don't have a tender foot after walking around the moon.' It was so bad a joke that one had to assume it was full of interior reference for him, perhaps some natural male anxiety at the thought of evil moon rays passing into one's private parts. A glum expression sat next to gloom – the damnedest things can happen to a good man.

Then they queried Aldrin on personal mementos. Would he be taking any along?

Well, yes, he admitted reluctantly, he would be taking a little family jewelry along. He stopped, he looked mulish. It was obvious he didn't want to go on. The primitive value of the objects, their power, their retention of charms, their position in the possible hierarchy of the amulets would be vitiated by describing them. On the other hand, a high quotient of availability-for-miscellaneous-unprogrammed-situations (known in the old days as charity, spontaneity, or generosity of spirit) also ranked high in good astronaut qualifications. So Aldrin gave answers even if he didn't want to.

Well, he admitted, the family jewelry were . . . *rings*. He had two heavy gold rings on two fingers. Yes, he nodded distrustfully, looking for a moment like a chow forced to obey a command he cannot enjoy, yes, on the flight, he would probably still be wearing them.

What else in the way of family jewelry?

But now Aldrin had had enough. 'Personal category,' he grunted.

A Viennese or German correspondent asked in a heavy accent of Armstrong, 'Have you had any der-reams?'

Dreams. Armstrong smiled. He couldn't say he did. The smile was as quick to protect him as the quick tail flick of a long-suffering cow standing among horseflies in a summer meadow's heat, yes, smile-and-flick went Armstrong, 'I guess after twenty hours in a simulator, I guess I sometimes have dreams of computers.'

Yet as the questions went on, the game was turning. The German might have asked his question about dreams with the happy anticipation that any material provided would offer a feast – the symbols of the dream were pot roast after all and gravied potatoes to the intellectual maw of a nice German head, but the answer, frustrating as nearly all the answers had been, now succeeded in working up a counterpressure. Slowly, unmistakably, the intellectuals and writers on the dark side of the glass were becoming a little weary of the astronauts. Collins' implacable cheerful cool, Aldrin's doughty monk's cloth of squaredom, Armstrong's near-to-facetious smile began to pique their respect. The questions began to have a new tone, an edge, the subtlest quivering suggestion that intellectual contempt was finally a weapon not to be ignored. Were these astronauts not much more than brain-programmed dolts? The contempt was a true pressure. For give an athlete brains, give an aviator brains, give an engineer a small concealed existence as presumptive poet, and whatever is not finished in the work of their ego, whatever is soft in their vanity, will then be exercised by the contempt of an intellectual. The writers were pushing Armstrong now.

Why, why ultimately, they were asking, is it so important to go to the moon? Man to man, they were asking, brain to brain, their leverage derived from the additional position of asking as writer to small-town boy: why is it important?

Armstrong tried to be general. He made a speech in fair computer-ese about the nation's resources, and the fact that NASA's efforts were now tapped into this root. Well, then, asked a dry voice, are we going to the moon only for economic reasons, only to get out of an expensive hole? No, said Armstrong.

Do you see any philosophical reason why we might be going? the voice went on, as if to imply: are you aware there is philosophy to existence as well?

Armstrong had now been maneuvered to the point where there was no alternative to offer but a credo, or claim that he was spiritually neuter. That would have violated too much in him. Yes, he blurted now, as if, damn them and damn their skills, they had wanted everything else of him this day, they had had everything else of him, including his full cooperation, now damn them good, they could have his philosophy too if they could comprehend it. 'I think we're going,' he said, and paused, static burning in the yaws of his pause, 'I think we're going to the moon because it's in the nature of the human being to face challenges.' He looked a little defiant, as if probably they might not know, some critical number of them might never know what he was talking about, 'It's by the nature of his *deep inner soul*.' The last three words came out as if they had seared his throat by their extortion. How his privacy had been invaded this day. 'Yes,' he nodded, as if noting what he had had to give up to writers, 'we're required to do these things just as salmon swim upstream.'

iv

That was a fair haul for a working day – Aquarius now had a catch to fry. Yet the day was hardly over for our astronauts. They still had to have their conversations with the television networks. Since each man would have his own half hour before the camera, that meant there would be three interviews for each man, or nine altogether. With breaks and dinner, their day would continue for another six hours.

Aquarius was invited to audit a filming and chose Armstrong's session with NBC. He had an idea Armstrong would be more comfortable in a TV interview and he was not wrong. But then Armstrong had indicated his concern for good television earlier at the full press conference when he had apologized for the program they would send from the moon. 'I don't mean to sound discouraging but I don't have high hopes that the picture that we will be able to send back from the surface will be nearly so good as those you have been looking at from the recent flights from the Command Module. The camera is somewhat different and is somewhat more restricted in the kinds of lenses that we can use, and the kinds of lighting we have available to us . . . And I suspect that you will be somewhat disappointed at those pictures. I hope that you'll recognize that it's just one of the problems that you face in an environment like the lunar surface and it'll be some time before we really get high quality in our lunar surface pictures back on TV link.'

It was the one time he had spoken without many pauses, almost as if he were talking already to the TV audience rather than to reporters, almost as if he just simply believed that Americans were entitled to good television – one of their inalienable rights. And now, up before the TV cameras, Armstrong looked not at all uncomfortable at the thought of being presented to some forty or fifty million viewers.

Indeed, Aquarius was to see cool pieces and parts of the half hour in thirty-second segments, minute segments and two-minute segments over the next few weeks, particularly during the days of the flight. During many a pause on the trip to the moon, the TV screen would cut to the face of Armstrong, Aldrin or Collins standing or sitting with the blue-green wall of the Lunar Receiving Laboratory behind him. Whether the filmed insertion was to elucidate some remark of the commentators, or merely to fill some frayed space in the ongoing hours of exposition and recapitulation, the effect after having seen ten concrete bits from this interview was to recognize that a new species of commercial was being evolved. NASA was vending space. Armstrong was working directly for his corporate mill. Despite the fact that this future audience of forty million would be listening and studying him, he spoke without long pauses, and seemed oddly enough to be at ease, a salesman with a clear modest mild soft sell. But, then, Aquarius decided, it was not really so very odd. If Armstrong's most recognizable passion was to safeguard his privacy, a desire which approached the force of sanctuary to him, then there was nothing on television he would be likely to reveal or betray. He came, after all, from that heartland of American life which had first induced the particular public personality now bequeathed to all TV viewers as the most viable decorum – that intolerable mixture of bland agreeability and dissolved salt which characterized all performers who appeared in public each day for years and prospered. That view of the world, if designing a face, would have snubbed the nose, faded out the color of the eyes, snugged the lips, slicked the hair and dispensed with the ears for they were protuberances with obscene interior curves – first cousins to the navel.

Armstrong was being interviewed by Frank McGee who turned in a good workmanlike job. McGee, a friendly fellow with a bony face and eyeglasses (whose frames whether tortoiseshell, plastic or pale gold would be remembered afterward as silver wire), had a personality all reminiscent of a country parson, a coach of a rifle team or the friendly investigator from a long-established high-minded insurance company. He was obviously the very ring-tailed hawk of Waspitude.

Their collaboration on the questions and responses had the familiar comfort of piety. Armstrong came near to chatting with him. It was implicit to Network Nugatory that a chatty tone went hand in hand with the pious. So the dullest but most functional, which is to say the most impermeable side of Armstrong, was naturally presented. He responded soberly, even chastely to questions about whether he had been elated when chosen – 'I have to say that I was' – but quickly added that there could have been many pitfalls during the waiting period (such as intervening flights, which might not succeed) and so he had not indulged any large excitement at any particular period.

He was determinedly modest, going clear out of his way to specify that he was certain the Apollo 12 crew was as competent as his own to make this first trip to the moon, and went on once again to give credit for success to all the Americans who had been working to back them up. 'It's their success more than ours,' said Armstrong, as if the trip had been completed already, or perhaps this was intended to be the commercial to be employed after touchdown, or lunar ascent, or splashdown. Queried about his private life and the fact that he would lose it after the achievement, he said diffidently, in a voice which would win him twenty million small-town cheers, 'I think a private life is possible within the context of such an achievement.' Aquarius' mind began to wander – he failed to make notes. Recovering attention at some shift in the mood he realized that Armstrong had finished this interview for he was saying '. . . to take man to another heavenly body . . . we thank all of you for your help and prayers.'

There was a hand from the TV crew when the cameras stopped. The trade unions once again were backing patriotic and muscular American effort. 'Godspeed and good luck, Neil' one of them actually cried out into the wall of the glass, and Armstrong smiled and waved, and there was more good feeling here than ever at the other conference with Press and magazine. It was apparent the television interview had added little to the store of Aquarius.

But by one detail it had. McGee, referring to a story in *Life* by Dora Jane Hamblin about Armstrong, spoke of a recurring dream the astronaut had had when a boy. In this dream, he was able to hover over the ground if he held his breath.

Aquarius always felt a sense of woe when he found himself subscribing to a new legend. Glut and the incapacity to absorb waste were the evils of the century – the pearls of one's legends were not often founded on real grains of sand. The moment he read the story in *Life*, Aquarius had become infatuated with Armstrong's recurrent dream. It was a beautiful dream – to hold one's breath and to levitate; not to fly and not to fall, but to hover. It was beautiful because it might soon prove to be prophetic, beautiful because it was profound and it was mysterious, beautiful because it was appropriate to a man who would land on the moon. It was therefore a dream on which one might found a new theory of the dream, for any theory incapable of explaining this visitor of the night would have to be inadequate, unless it were ready to declare that levitation, breath, and the moon were not proper provinces of the dream.

Because it was, however, awesome, prophetic, profound, mysterious and appropriate, Aquarius hated to loose the vigors of his imagination onto the meaning of this dream unless he could believe it had actually happened. It was too perfect to his needs to accept it when he read it. But after studying Armstrong this day, listening to his near-humorous admission that yes, he had had that dream when he was a boy, there was a quietness at the center of his reply which gave balm to the sore of Aquarius' doubt. He knew he had now chosen to believe the dream had occurred.

And this conviction was not without the most direct kind of intellectual intoxication, for it dramatized how much at odds might be the extremes of Armstrong's personality or for that matter the personality of astronauts. From their conscious mind to their unconscious depth, what a spectrum could be covered! Yes, Aquarius thought, astronauts have learned not only to live with opposites, but it was conceivable that the contradictions in their nature were so located in the very impetus of the age that their personality might begin to speak, for better or worse, of some new psychological constitution to man. For it was true – astronauts had come to live with adventures in space so vast one thought of the infinities of a dream, yet their time on the ground was conventional, practical, technical, hardworking, and in the center of the suburban middle class. If they engaged the deepest primitive taboos, they all but parodied the conventional in public manner; they embarked on odysseys whose success or failure was so far from being entirely in their own control that they must be therefore fatalistic, yet the effort was enterprising beyond the limits of the imagination. They were patriots, but they were moonmen. They lived with absolute lack of privacy, their obvious pleasure was to be alone in the sky. They were sufficiently selfless to be prepared to die for their mission, their team, their corporate NASA, their nation; yet they were willy-nilly narcissistic as movie stars. 'Sugar, I tried and couldn't make doo-doo,' says Lulu Meyers in *The Deer Park*. The heart pressure, the brain waves, the bowel movements of astronauts were of national interest. They were virile men, but they were prodded, probed, tapped into, poked, flexed, tested, subjected to a pharmacology of stimulants, depressants, diuretics, laxatives, retentives, tranquilizers, motion sickness pills, antibiotics, vitamins and food which was designed to control the character of their feces. They were virile, but they were done to, they were done to like no healthy man alive. So again their activity was hazardous, far-flung, bold, demanding of considerable physical strength, yet the work and physical condition called for the ability to live in cramped conditions with passive bodies, the patience to remain mentally alert and physically inactive for days. They lived, it was evident, with no ordinary opposites in their mind and brain. On the one hand to dwell in the very center of technological reality (which is to say that world where every question must have answers and procedures, or technique cannot itself progress) yet to inhabit – if only in one's dreams – that other world where death, metaphysics and the unanswerable questions of eternity must reside, was to suggest natures so divided that they could have been the most miserable and unbalanced of men if they did not contain in their huge contradictions some of the profound and accelerating opposites of the century itself. The century would seek to dominate nature as it had never been dominated, would attack the idea of war, poverty and natural catastrophe as never before. The century would create death, devastation and pollution as never before. Yet the century was now attached to the idea that man must take his conception of life out to the stars. It was the most soul-destroying and apocalyptic of centuries. So in their turn the astronauts had personalities of unequaled banality and apocalyptic dignity. So they suggested in their contradictions the power of the century to live with its own incredible contradictions and yet release some of the untold energies of the earth. A century devoted to the rationality of technique was also a century so irrational as to open in every mind the real possibility of global destruction. It was the first century in history which presented to sane and sober minds the fair chance that the century might not reach the end of its span. It was a world half convinced of the future death of our species yet half aroused by the apocalyptic notion that an exceptional future still lay before us. So it was a century which moved with the most magnificent display of power into directions it could not comprehend. The itch was to accelerate – the metaphysical direction unknown.

Aquarius, aware of the profundity of his natural bent for error, aware of the ineradicably romantic inclination of his mind to believe all those tales and legends he desired to believe, nonetheless came to a conclusion on this hot Saturday evening, July 5, on the southeastern rim of Houston, that Armstrong when a boy had indeed had a recurring dream in which he would hold his breath and rise from the ground and hover, and on this dream Aquarius, who had been reconnoitering for months through many a new thought (new at the very least to him) on the architecture and function and presence of the dream, would build his theory, on Armstrong's dream would Aquarius commit himself. Any notes toward a new psychology could take their departure from here, from this *fact*. And as this evening went on, and he continued to the party at Pete Conrad's house and talked to the future commander of Apollo 12 over the steaks at charcoal grill, and Conrad made his confession of dreaming for years of going to the moon, and now concluded somberly, manfully – one had to be manful when contemplating the cost of desire – 'now the moon is nothing but facts to me,' Aquarius felt confirmation building in his mood, his happiness and his senses, that this grim tough job of writing for enough money to pay his debts and buy his little plot of time, was going to be possibly, all passions directed, all disciplines flexed, a work whose size might relieve the chore. And as he thought of the little details he had picked up in the biographies of Collins, of Aldrin, of Armstrong, he thought that yes, the invasion of the moon was signal direct to commence his new psychology – he would call it, yes, beneath this Texas moon, full near the Fourth of July, he would call it The Psychology of Astronauts, for they were either the end of the old or the first of the new men, and one would have nothing to measure them by until the lines of the new psychology had begun to be drawn.

3

Some Origins of the Fire

We move on to Florida and the launch. If Aquarius had spent a week in Houston, he was to put in ten days on Cape Canaveral. He was loose in some real tropics at last with swamp and coconut palms. It was encouraging. Technology and the tropics were not built to hide everything from each other.

Let us take the tour. On Merritt Island and old Cape Canaveral, now Cape Kennedy, the Space Center has been installed, a twenty-mile stretch between the Intracoastal Waterway and the Atlantic, a terrain of marshland and scrub where raccoon, bobcat and alligator are still reported, and moors and truncated dunes lie low before the sea. It is country beaten by the wind and water, not dissimilar to Hatteras, Chincoteague and the National Seashore on Cape Cod, unspectacular country, uninhabited by men in normal times and normal occupations, for there are few trees and only occasional palms as ravaged and scabby as the matted backside of a monkey, a flat land of heat and water and birds, indeed birds no less impressive to Aquarius than ibis, curlew, plover and tern, hawks and vultures gliding fine as squadrons in formation, even bald eagles, ospreys and owls. In the brackish water are saltwater trout, redfish, largemouth bass, and bream. It is country for hunting, for fishing, and for men who seek mosquitoes; it was next to uninhabited before the war. Now, first spaceport – think on it! first *space*port – of an industry which pays salaries to perhaps so much as half a million men and some women before it is through, and has spent more than four billion dollars a year for average the last few years, a spaceport which is focus to the aerospace industries, a congeries of the richest corporations supplying NASA. Yet this port to the moon, Mars, Venus, solar system and the beyond is a first clue to space, for it is surprisingly empty, mournful beyond belief for the tropics, and its roads through the Air Force Base and the Space Center pass by empty marshes, deserted dune grass, and lonely signs. Every quarter-mile or so along that low grassy ridge toward the side of the sea is a road sign pointing to an old launch complex which on exploration turns out to consist of an unoccupied road and a launching tower for rockets no longer fired, and so left to commune by itself on a modest field of concrete, a tall, rust-red vertical structure of iron girders surrounded by abandoned blockhouses and utility sheds. To Aquarius the early history of the Space Program is contained in these empty launch towers, now as isolated and private as grain elevators by the side of railroad tracks in the flat prairies of Nebraska, Kansas, and the Dakotas, the town low before them, the quiet whine of the wind like the sound of surf off a sea of wheat. It was the grain elevator which communed on prairie nights with the stars. Here in the cricket-dinning tympani of Florida's dunes and marshes, the launching towers of rockets now obsolete give that same sense of the sentinel in a field of space, stand already as monoliths and artifacts of a prehistoric period when rockets usually exploded in the first few hundred feet of their flight.

Yes, the Cape has given a turn to Aquarius. If at Houston he still remained attached to a somewhat disembodied ego (which felt like a balloon on a tether) – if for all his extorted admiration at the self-sufficiency of NASA and its world, he could still not quite like it, quite rid himself of the idea that finally space travel proposed a future world of brains attached to wires, his ego was therefore of use. He would pull in the string from time to time to criticize what he saw. If he were heard to utter 'This is not unimpressive,' when encountering some perfection of cooperation or technique, he was also ready to whisper – in his heart at least – that the Manned Spacecraft Center was not the coziest home for the human heart. Indeed, it was so cold that one could finally walk away from it like from a chill corridor in a dream. The beauties of MSC went on in the minds of technicians, but the soul of a visitor felt locked in the vault with an air conditioner. So it was attractive to think that one could end the dream, unlock the door, and walk away.

That was hardly possible on the Cape. If the abandoned launch towers and the hot lonely ocean breeze opened vistas of the West and thoughts of how many of the most important events in America seemed to take place in all the lonely spaces – as if the Twentieth

Century had become the domain of all the great and empty territories (the Saharas, the Siberias, and the Minutemen in the buried silos of the West) – that was forced to give way to a sense of huge activity and gargantuan dimensions. If MSC near Houston was a brain, then Cape Kennedy was the body, and at Launch Complex 39, up twenty miles to the north of Cocoa Beach and Canaveral, were found the bones and muscles of a Colossus. Here the big components of Saturn V came in by cargo plane, came by ship through the Panama Canal and by barge through the Gulf, came from Los Angeles and Sacramento, from Huntsville in Alabama to Michoud in Louisiana, and from Michoud to the Cape; here at Complex 39 the parts were assembled in a mammoth cube of an edifice with a smaller box attached, the Vehicle Assembly Building, 526 feet high, a building just about as large as the combined volume of the Merchandise Mart in Chicago and the Pentagon. Covering eight acres, enclosing 129 million cubic feet, the Vehicle Assembly Building was nonetheless windowless, and decorated from the outside in huge concentric rectangles of green-gray, and charcoal-gray, ivory-gray and light blue-gray; it looked like a block of wood colored by an Op Art painter, but since it was over fifty stories high, it also looked like the walls of a gargantuan suburban department store. If by volume it was when built the largest building in the world, the Vehicle Assembly Building, as one saw it standing on the flat filled-in marshes of the Cape, had to be also a fair candidate for the ugliest building in the world. Viewed from any external approach it was the architectural fungoid of them all.

Once inside, however, it was conceivably one of the more beautiful buildings in the world. Large enough to assemble as many as four moongoing Apollo-Saturn vehicles at once, it was therefore open enough to offer interior space for four tall bays, each of these niches tall enough to house the full rocket, which was thirty-six stories high. Since the rocket in turn sat on a transporter, called a crawler, of some dimension itself, the doors to the four bays were each over forty stories and therefore high enough and wide enough to take in through their portals the UN Building or the Statue of Liberty. Yet for all its size, the VAB was without decoration inside, rather a veritable shipyard and rigging of steel girders which supported whole floors capable of being elevated and lowered, then rolled in and out like steel file drawers in order to

A Fire on the Moon

encircle each rocket with adjustable working platforms from either
side. Since some of these platforms had three complete stories contained within them, the interior of the VAB was a complexity of
buildings within buildings which had been first maneuvered then suspended ten and twenty and thirty stories above the ground. Because the
sides were usually open, one could look out from the platforms to other
constellations of girders and buildings and could look down from
whichever great height to the floor of the VAB, sometimes as much as
forty stories below. Note however: one was still inside a closed space,
and the light which filtered through translucent panels rising from floor
to ceiling was dim, hardly brighter than the light in a church or an old
railroad terminal. One lost in consequence any familiar sense of
recognition – you could have been up in the rigging of a bridge built
beneath the dome of some partially constructed and enormous subterranean city, or you could have been standing on the scaffolding of an
unfinished but monumental cathedral, beautiful in this dim light, this
smoky concatenation of structure upon structure, of breadths and vertigos and volumes of open space beneath the ceiling, tantalizing views
of immense rockets hidden by their clusters of work platforms. One did
not always know whether one was on a floor, a platform, a bridge, a
fixed or impermanent part of this huge shifting ironwork of girders and
suspended walkways. It was like being in the back of the stage at an
opera house, the view as complex, yet the ceiling was visible from the
floor and the ceiling was more than fifty stories up, since above the rockets were yet some massive traveling overhead cranes. To look down
from the upper stages of the rocket, or from the highest level where the
crew would sit, was to open oneself to a study of the dimensions of
one's fear of heights. Down, down, a long throw of the soul down,
down again, still falling was the floor of the building, forty floors below.
The breath came back into the chest from an abyss. And in one corner
of the floor like a stamp on the edge of a large envelope was a roped-in
square of several hundred tourists gawking up at the yellow cranes and
the battleship-gray girders.

Taken originally on a tour by a guide, Aquarius had spent the good
part of a day in this building, and was back again twice to be given a
more intimate trip and a peek into the three stages and the Command
and Service Module of Apollo 12, which was then being prepared for its

flight in November. Looking into any portion of the interior of a rocket was like looking into the abdominal cavity of a submarine or a whale. Green metal walls, green and blue tanks, pipes and proliferations of pipes, black blocks of electrical boxes and gray blocks of such boxes gave an offering of those zones of silence which reside at the center of machines, a hint of that ancient dark beneath the hatch in the hold of the bow – such zones of silence came over him. He could not even be amused at the curtained walls of white and the in-sucking wind of the dust collectors and the electrical shoe polishers, the white smocks and the interns' caps they were obliged to put on before they could peer through the hatch of the Command Module, and see the habitation of the astronauts. A gray conical innerland of hundreds of buttons and switches looked back at him, and three reclining seats vaguely reminiscent of instruments of torture. Three dentists' chairs side by side! Yes, he could have found the white outfits they were wearing a touch comic – if dust they were to protect the machine against, then garments they could wear, but why white, why the white hospital walls? And thought that of course they would keep it like the sterile room in a delivery ward, for indeed there was something about space which spoke of men preparing to deliver the babies they would themselves bear. The aim of technique was to parallel nature, and the interior of the VAB was the antechamber of a new Creation.

So, it was probably the Vehicle Assembly Building which encouraged Aquarius to release the string of the balloon and let his ego float off to whatever would receive it. It was not that he suddenly decided to adopt the Space Program, or even approve it in part, it was just that he came to recognize that whatever was in store, a Leviathan was most certainly ready to ascend the heavens – whether for good or ill he might never know – but he was standing at least in the first cathedral of the age of technology, and he might as well recognize that the world would change, that the world *had* changed, even as he had thought to be pushing and shoving on it with *his* mighty ego. And it had changed in ways he did not recognize, had never anticipated, and could possibly not comprehend now. The change was mightier than he had counted on. The full brawn of the rocket came over him in this cavernous womb of an immensity, this giant cathedral of a machine designed to put together another machine which would voyage through space. Yes, this emergence of a ship to travel the ether was no event he could measure by any philosophy he had been able to put together in his brain.

Yet all the signs leading to the Vehicle Assembly Building said VAB. VAB – it could be the name of a drink or a deodorant, or it could be suds for the washer. But it was not a name for this warehouse of the gods. The great churches of a religious age had names: the Alhambra, Santa Sophia, Mont-Saint-Michel, Chartres, Westminster Abbey, Notre Dame. Now: VAB. Nothing fit anything any longer. The art of communication had become the mechanical function, and the machine was the work of art. What a fall for the ego of the artist. What a climb to capture the language again! It occurred to him on the instant that one's fear of height must be at least a partial function of the importance of one's ego. Or was it a direct function of that part of one's ego which was useless? A man was presumably ready to take any drop when the ego was finally congruent to the soul and all the signs said go. Yes, one would have to create a psychology to comprehend the astronaut. For a beginning, however, it would be good to recognize how simple he must become. Do not dominate this experience with your mind was the lesson – look instead to receive its most secret voice. He would be, perforce, an acolyte to technology. What a gruel. By whatever measure, he was now forced to recognize the ruddy good cheer and sense of extraordinary morale of the workers in the VAB. As they passed him in the elevators, or as he went by them in the halls and the aisles, a sense of cooperative effort, of absorption in the work at hand, and anticipation of the launch was in the pleasure of their faces. He had never seen an army of factory workers who looked so happy. It was like the week before Christmas. As at the Manned Spacecraft Center they seemed to be ranked by the number of admission badges they wore. The smiles of the ones who wore the most seemed to thrive the most, as if they were not identification tags which reduced them to parts of a machine, but rather were combat ribbons, theater-of-war ribbons. Trade-union geezers, age of fifty, with round faces and silver-rimmed spectacles strutted like first sergeants at the gate for a three-day pass.

So Aquarius began to live without his ego, a modest quiet observer who went on trips through the Space Center and took in interviews, and read pieces of literature connected to the subject, and spent lonely nights not drinking in his air-conditioned motel room, and thought – not of himself but of the size of the feat and the project before him, and by the night before the launch, he was already in orbit himself, a simple fellow with a mind which idled agreeably, his mind indeed out in some weightless trip through the vacuum of a psychic space, for a mind without ego he was discovering is kin to a body without gravity. He was there now merely to observe, to witness. And the days went quietly by. We would pick him up on the night before the launch, but we may not be able to. He is beginning to observe as if he were invisible. A danger sign. Only the very best and worst novelists can write as if they are invisible.

ii

Saturn V would take off from a plain of gray-green moor and marsh, no factory or habitation within three and a half miles. Saturn V had almost six million pounds of fuel. So it would take the equivalent of thirty thousand strong men to raise it an inch. It would take liquid oxygen, liquid hydrogen and a very high grade of kerosene called RP-1. It had hydrazine, unsymmetrical dimethyl-hydrazine, and nitrogen tetroxide in the Service Module. It was in effect a bomb, thirty-three feet wide – the length of a long living room. Corporation executives earning $50,000 a year just begin to think of a thirty-three-foot living room for themselves. And it was the height of a football field set on end. Sometimes they described it as a thirty-six-story building (ten feet to a floor) but a football field was clear measure of size, and this bomb, 363 feet high, 33 feet wide at the base, would blow if it blew with a force kin to one million pounds of TNT. That was like an old-fashioned bombing raid in World War II – one thousand planes each carrying one thousand pounds of bombs. So Saturn V would devastate an area if ever it went. Flight Control, the Press Site and the VIP stands were located therefore three and a half empty miles away across barren moors which, having been built by dredging fill into marshland, looked as if a bomb had gone off on them already.

On the night before the launch of Apollo 11, in the heart of Brevard County, in that stretch which runs from Melbourne through Eau Gallie, and Cocoa, to Titusville, on the coastal strip from Patrick Air Force Base through Cocoa Beach and Cape Canaveral to the Cape Kennedy Air

Force Station and above to the Space Center and Launch Pad 39, through all that several hundred square miles of town and water and flat swampy waste of wilderness, through cultivated tropical gardens, and back roads by rivers lined with palms, through all the evening din of crickets, cicadas, beetles, bees, mosquitoes, grasshoppers and wasps some portion of a million people began to foregather on all the beaches and available islands and causeways and bridges and promontories which would give a clear view of the flight from six miles and ten miles and fifteen miles away. Tomorrow most of them would need field glasses to follow the flight up from the pad and out of sight over the sea down a chain of Caribbean isles, but they would have a view – they knew tonight that if the skies were clear they would have their view because they were encamped only where the line of sight was unimpeded to Launch Pad 39 on the horizon. There one could certainly see Apollo 11 on her Saturn V, see her for seven, nine, eleven miles away; she was lit up. A play of giant arc lights, as voluminous in candlepower as the lights for an old-fashioned Hollywood premiere, was directed on the spaceship from every side. On U.S. 1 in Titusville, eleven miles from Cape Kennedy across Merritt Island and the Banana and Indian rivers, all that clear shot across the evening waters, at an artillery range of twenty thousand yards, two hundred football fields away, by an encampment of tourists up from southern Florida, Everglades, Miami, and the Keys; in from Tampa, and Orlando; down from Daytona, St Augustine, Gainesville, and Jacksonville; come from Fort Myers and Fort Lauderdale, from Sarasota, St Petersburg, Lakeland, Ocala and Tallahassee, come from all the towns of Georgia and points farther north and west as well as every itinerant camper in the area from all of the ambulatory camping-out families of the fifty states, and tourists down on economy flights for a week in cheap hot summer Florida and now slung out in the back seats of rented cars, on U.S. 1 in Titusville, in an encampment of every variety of camper, was a view of the spaceship across flat land and waters, and she looked like a shrine with the lights upon her. In the distance she glowed for all the world like some white stone Madonna in the mountains, welcoming footsore travelers at dusk. Perhaps it was an unforeseen game of the lighting, but America had not had its movie premieres for nothing, nor its Rockettes in Radio City and fifty million squares tooling the tourist miles over the years to Big Town to buy a ticket to spectacle and back home again. If you were going to have a Hollywood premiere and arc lights, a million out to watch and a space-ship which looked across the evening flutter like the light on the Shrine of Our Lady outside any church in South Brooklyn or Bay Ridge, then by God you might just as well have this spectacle on the premiere trip to the moon. That deserved a searchlight or two! And the campers stared across the waters in their bivouac off Route 1 in Titusville, campers sat on the banks of the Indian River at twilight and waited for the tropical night to pass its hold on the hours.

There were new industries in America these years. After five decades of suspense movies, and movies of the Wild West, after the adventures of several generations of men in two world wars and Korea and Viet-nam, after sixteen years of *Playboy* and American iconization of the gravity-defying breast and the sun-ripened buttock, after ten years of the greatest professional football, after a hundred years and more of a tradition that the frontier was open and would never close, and after twenty more perplexing technological years when prosperity came to nearly every White pocket, and technology put out its plastic, its super-highways, its supermarkets, its appliances, its suburbs, its smog, and its intimation that the frontier was damn shut, shut like a boulder on a rabbit burrow, America had erupted from this pressure between its love of adventure and its fear that adventure was not completely shut down; America had spewed out on the road. The country had become a nation of campers, of cars toting trailers, of cars pulling tent-trailers, of truck-campers, top-of-car tent packs, Volkswagen buses converted to ambulatory bedrooms, jeeps with Chic Sale houses built on the back, Land-Rovers with bunks, Broncos with more bunks – any way a man could get out of the house with his buddies or his family or his grand-mother, and take to the road and find some ten by twenty feet of parking grass not posted, not tenanted, and not too muddy, he would camp. All over America in the summer the night fields were now filled with Americans sleeping on air mattresses which reposed on plastic cloth floors of plastic cloth tents – what a sweet smell of Corporate Chemical, what a vat and void to mix with all the balmy fermy chloro-phylls and pollens of nature! America the Sanitary, and America the

Wild, went out to sleep in the woods, Sanitary-Lobe and Wild-Lobe nesting together neatly, schizophrenic twins in the skull case of the good family American.

So they were out tonight, some portion of a million, all drawn on the lines of sight in Brevard County, and on every highway and causeway in the area the ground was covered with cars and campers, the shelter-roof extension of one family's tent near to topping the picnic blanket spread out behind the tailgate of the next station wagon, and the open trunk lid of a twelve- or fifteen-year-old Dodge convertible (rusty, top all rent, peeling friction tape and dirty white adhesive tape chasing a flap of a patch) stood next to both, part of the family sleeping in the trunk, the others with their good dirty feet out the windows. It was hardly just middle-class America here tonight, rather every echo of hard trade-union beer-binge paunch-gut-and-muscle, and lean whippy redneck honkytonk clans out to bird-watch in the morning with redeye in the shot glass. There were tourists and not inelegant campers which spoke of peanut butter and jelly, watercress, and cucumber – suburban campers – but there was also the raw gasoline of expectation in the air, and families of poor Okies. One felt the whole South stirring on this night. Quiet pious Baptists, out somewhere on their porches (kin to some of the redneck and Okie – Okie for Okee-chobee! – and working class here) seemed to be waiting over an arc of a thousand miles, certainly all the way from Cape Kennedy across Florida along the Gulf of Mexico to Houston and the Manned Spacecraft Center and back again, all across that belt of Fundamentalist piety, hot dry tempers burning like closed-up balefire against the humidity of the swamps, religion and lust to work their combat in the tropical nights, yes all over the South they had to be praying, yes even more than everywhere for the safety of this shot, the astronauts part of that family of concern which White Southerners could share with each other out of the sweet deep wells of their Christian hearts, what was left of them. It was not hard to have a vision of mothers and grandmothers looking like spinsters, silver-rimmed glasses to shield your skin from their eyes of burning faith, predictable turkey wattles on the neck: they would be praying for America tonight – thoughts of America served to replace the tender sense of the Virgin in Protestant hearts. And out here on the camp-grounds of Brevard County, out on all the scorched shoulders and oil-coated grass of the available highway, were men getting ready to drink with their wives, middle-aged, green-eyed Southern mill workers with sunburned freckled skin, reddish hair, hard mechanic's muscles in their forearms, wife a trinity of worrying mother, fattening slattern, and give-me-a-drink-and-I'll-holler-happy sort of bitch. Dutiful work, devotion to family and property, their sloven property! mingy propriety, real raucous bust-outs – that sort of South, married out of high school, oats half-sown like three quarters of all America over thirty, and their boys on the hunt through the encampment looking for opposite numbers, other boys or – Gods of fornication with them! – girls without bank locks on their bloomers. You can expect nothing less on a night so filled with heat, human meat, bubbles of fear, prayer soft as love, and tropical sex in every sauce. And that mill worker with red hair and gray-green eyes, red sunburn, red peeling skin on his knotty forearms – he could be an astronaut in another life. He looks like an older version of Neil Armstrong, maybe, he looks like some of them, like Gordon Cooper for sure, or Deke Slayton, or Walt Cunningham of Apollo 7, yeah, the mill worker is tonight an American all drenched in pride and fear and sorrow – his wild rebel yell guaranteed to diminish each year, is riding the range with awe tonight. He has worked with machines all his life, he has tooled cars to the point where he has felt them respond to his care, he has known them and slept beside them as trustingly as if they were hunting dogs, he knows a thousand things about the collaboration between a man and a machine, and he knows what can go wrong. Machines – all the old machines he has known – are as unreasonable as people. And here, tomorrow, going up three men of whom he could have been one if he had a) not been a drunken fool half his life b) not married young c) had an education d) had twice as many guts and e) been full of real luck rather than cursed family luck, could have been going up with them in a machine with millions of parts, eighty percent electronics which he does not put his hand to, no grease, nut, wrench and arm in electronics. Yes, could have been going up in a machine no man could ever sleep next to, or trust, not a machine with millions of parts, and ten million fingers worked on them – how much evil, error and deception in millions of fingers? – he is thinking of his wife, why she alone when drunk and in full lunatic cohabitation with the all-out rays of the full moon was a hundred thousand fingers of evil herself!

and his bowels come near to dropping out of him with awe at the daring of the act in the morning. He has spent his life with machines, they are all he has ever trusted with affectionate trust for he has had a nose for their treacheries (more than he can say for women), and now, twelve hours from now, in the full light of nine-thirty in the morning, that Apollo-Saturn is going to go up. He will see a world begin where machines are king and he does not know whether to cry from pride or the all-out ache that he does not really comprehend the new machinery.

And the wife sipping the booze, hot and much too funky is her flesh on this hot night, listening to the transistor radio Red has bought her, is moony and full of tears at the heroism of male craziness tomorrow. She wishes – floodgates of middle-age sorrow – some crack of that holy lightning in her womb: too late now! and trying to love up a warmth for Red, married all these nineteen years, she rears up on the very pinpoint of spite. She has powers, her family has powers, there's Indian blood in both her grandmas, and bruises, sorrows, slights, and nights of the loveliest now lost in the disappearing wahoo of studs she will never see again, our redneck Molly abloom in this encampment, she thinks on that pinpoint of spite of a curse she could put and will not on the launch tomorrow. For if Saturn were ever to burst and explode – she sees the flames across the sky: all witch and bitch on a holiday such pictures encourage the lapping of gentle waters in her. Whereas putting wax on the tip of the needle, and capping the curse, leaves dull lead in her chest. Not to mention the future woman troubles of her gut. She stoppers the bottle and looks on a slant at Red – there'll be an angle soon by which to pick a fight.

And men and women, tired from work and travel, sat in their cars and sat outside their cars on aluminum pipe and plastic-webbing folding chairs, and fanned themselves, and looked across the miles at the shrine. Out a car window projected the sole of a dirty foot. The big toe pointed straight up to Heaven in parallel to Saturn V.

iii

Aquarius passes these sights like a stranger. He feels in such surroundings a foreigner equally as much as he feels American. It is his country, but he merely traverses it. His feet do not take root.

Studying the encampments on the roadside this plump and burgeoning night, he is thinking of the party he has attended at the Royal Oak Country Club in Titusville. A full occasion. Wernher von Braun has spoken there earlier. Wernher von Braun has in fact arrived and left by helicopter. Von Braun, the *deus ex machina* of the big boosters! He is of course a legend. If you ask the man on the street: Who is the head of NASA, he will probably not quite know what NASA is. If you ask: Well, who heads the Space Program, he is not likely to tell you James E. Webb was Administrator of NASA from 1961 to 1968, nor does he figure to know that in the summer of 1969 it is Dr Thomas O. Paine who is ultimately responsible for all the NASA installations: Headquarters in Washington; Ames Research Center for space flight research at Moffet Field, California; Electronic Research Center at Cambridge; Flight Research Center at Edwards, California; Goddard Space Flight Center for unmanned satellites at Greenbelt, Maryland; Jet Propulsion Laboratory at Pasadena; Langley Research Center at Hampton, Virginia; Lewis Research Center at Cleveland; Nuclear Rocket Development Station at Jackass Flats, Nevada; KSC Western Test Range Operations Division at Lompoc, California; Wallops Station, Wallops Island, Virginia; NASA Pasadena Office; and of course the Manned Spacecraft Center at Houston; the George C. Marshall Space Flight Center at Huntsville, Alabama; and Kennedy Space Center itself on Cape Kennedy and Merritt Island. No, he will not have heard of Dr Paine, nor of Dr George E. Mueller who heads the Manned Spacecraft Program, which is to say is in charge of everything to do with men in rockets as opposed to unmanned rockets, and so has authority over the directors of every space center and laboratory concerned with manned flight. Nor is anyone too likely to have heard of Dr Gilruth, nor necessarily of Dr Kurt H. Debus at Kennedy Space Center where he is Director and so in charge of every mammoth launching as well as planning, designing, developing and utilizing the launching facilities. No, they have heard only of Von Braun. Since his formal title is Center Director of George C. Marshall Space Center at Huntsville, Alabama, and he is therefore on an organizational level equal only to Debus and Gilruth, whereas Dr Mueller and Dr Paine are his most definite superiors in this hierarchy of NASA stations, divisions, laboratories, operations, facilities, centers and hegemonies, he can hardly by any organizational measure be the Boss, but to the public sense of these affairs, to the Press, and to a corps of space workers, he is the real engineer, the spiritual leader, the inventor, the force, the philosopher, the genius! of America's Space Program. Such is his legend in the street. That is the positive side of his reputation; it is enormous; say, rather it is immense. Yet he has that variety of glamor usually described as fascinating, which is to say, the evocation of his name is attractive and repellent at once, because no one forgets for an instant that he worked on the V-2 rockets at the German Rocket Research Center at Peenemünde, second only to General Dornberger, and so was implicated on one occasion by giving an orientation lecture to the Leader himself, who stood and stared and did not say a word when rockets were later fired for him on test stands. It was expected that Hitler with his love of the cosmic, the primitive, the apocalyptic, and the more audible wars of Hell and Heaven would be enthusiastic about the extraordinary sound of rocket motors. The future of the rocket program at Pennemünde was indeed even dependent in 1939 upon just such hopes as Hitler's ecstatic reaction. But the Fuehrer did not say a word until lunch, when he stated, '*Es war doch gevaltig*,' which may be translated as, 'That was sensational.' (Göring, who visited a week later, was openly enchanted. Rocket propulsion for railroad and passenger cars, airplanes, airships! and ocean liners! was what he saw next.)

Then in 1943, after an audience with Hitler, Von Braun was granted the very high honor of a titular professorship. That much was ineradicable from Von Braun's record, but he had also had the opposite honor of being arrested and jailed for two weeks in an SS prison by Heinrich Himmler himself. One of the accusations: Von Braun was not really interested in rockets for war so much as for space exploration. It took General Dornberger's intercession with Hitler to spring Von Braun from Stettin Prison. Without Von Braun, said Dornberger to Hitler, there would be no V-2. Then in 1945 Von Braun had managed with considerable skill to move about five thousand employees and their families, and some of their papers, documents and drawings, to the Harz Mountains in the south of Germany where they could be captured by Americans rather than Russians.

Von Braun had not been out of higher headquarters since. While the U.S. Army test-fired V-2's at the White Sands Proving Grounds in New Mexico, he served as adviser. Five years later, still working for the

Army, he directed the development of the Redstone and Jupiter missiles. For NASA he had created the launch vehicle for the Apollo program, the famous, the monumental, the incomparable three-stage Saturn V, that launch vehicle we have already glimpsed at the VAB, a booster the size and weight of a full Navy destroyer, a rocket to deliver seven and a half million pounds of thrust at blast-off, Saturn V, 281 feet long, 33 feet wide, designed to put – we may be Germanic in metaphor here – designed to put its *little brother*, the 82-foot Apollo spacecraft implanted on top of it, into Trans-Lunar Injection, which is to say: on its way to the moon. In terms of size, the Apollo spacecraft was no more than a witch's hat perched on Saturn V's Instrument Unit of a head!

Yet since this launch vehicle in all its three stages did not have fuel to burn for even eighteen minutes, all six million pounds of fuel consumed in bursts of two and a half minutes and six and a half minutes, then two minutes and six minutes, near to five million pounds of fuel being burned in the first 150 seconds, whereas, in contrast, the Command Module would be in flight for eight days; since Saturn V in relation to the complexity of the electronic vitals and conceptions on the Command Module was relatively simple in design, Saturn V hardly more by the severest measure than a mighty mortar of a firework to blast an electric brain into space, why then was Von Braun so worshiped, why, if the true technology, the vertiginous complexity of the engineering feat of putting a man on the moon and back belonged rather in sum of work and intimate invention to echelons of electronic engineers out at MSC, North American and Grumman and too many other places to name? Well, the brute but inescapable answer if one studies the morphology of rockets is that man worships his phallus in preference to a drop of his seed. Yeah and yea. Saturn V was guts and grease, plumbing and superpipes, Lucifer or the Archangel grinding the valves. Saturn V was a furnace, a chariot of fire. One could witness some incandescent entrance to the heavens. But Apollo 11 was Command Module and therefore not to be seen. It spoke out of a crackling of static, or rolled like a soup can, a commercial in a sea of television, a cootie in a zoo of oscillating dots.

We may, then, absorb the lesson: Electricity is an avatar of hate which gives pain to the senses, emits static, electronic hum, neon flicker, light glare, shock, heat radiation. Whereas thoughts of the sun and royal spectacle are in the mystery of a flame. So Von Braun was the heat in rocketry, the animal in the program. By public estimate he had been a Nazi – that was glamor enough. Who could begin to measure the secret appeal of the Nazis by now? It was a fit subject for Aquarius to begin to brood upon: America was this day mighty but headless, America was torn by the specter of civil war, and many a patriot and many a big industrialist – they were so often the same! – saw the cities and the universities as a collective pit of Black heathen, Jewish revolutionaries, a minority polyglot hirsute scum of nihilists, hippies, sex maniacs, drug addicts, liberal apologists and freaks. Crime pushed the American public to give birth to dreams of order. Fantasies of order had to give way to lusts for new order. Order was restraint, but new order would call for a mighty vault, an exceptional effort, a unifying dream. Was the conquest of space then a potential chariot of Satan, the unique and grand avenue for the new totalitarian? Aquarius was not certain. It was possible that neo-Nazism and technology were finally inimical to each other, but it was all to be considered again and again. It was complex. At this instant, he would not have minded the return of his ego.

Meanwhile, here was Von Braun for study. Yes, he had come in by helicopter to the Royal Oak Country Club in Titusville. The roads were crowded and it was incontestable that on this night, this night above all, hours before the mightiest launching of his life, Von Braun's hours were of value to him and to others. Still, the impression had to arise that he would have arrived by helicopter in any case. The helicopter had become the vehicle of status of that Praetorian Guard now forming of generals, state troopers, admirals, Republican congressmen with wives-on-junket, governors from he-man states, he-man senators, law-and-order mayors, traffic-crisis monitors, and VIPs on state visits to troublesome cities. The helicopter was there to signify: a man engaged in *flag* activity was dropping in on the *spot*. So the helicopter was a status symbol as special as a Junior League Ball. Not everybody who was moderately rich and powerful in American life would necessarily want to go to the ball or ride in the bubble, but for that matter not everybody who was thus rich and powerful was welcome at either.

Under whose auspices then had Von Braun descended? We can pretend to investigate. A large publishing corporation long associated with the Space Program had invited corporation presidents of important firms to voyage out for a few days on a trip to Houston to meet astronauts, then on to Kennedy to see the launching. A private speech by Von Braun was one of the features of the junket, and they waited for him now in a hexagonal banquet room finished in varieties of walnut-colored wood, a fit meeting place for American gods and cousins of the gods, since the shape of the chamber gave an echo of clans meeting in a wooded glen. Talismans in the form of intricate hex signs were inlaid in the wood of the walls around the room below the ceiling. Yet the walls, as though aware the gods were American, their powers corporate, were finished pale in stain, and therefore not excitative to the bottled emotions of business leaders. In any case, the golf course abutted the premises, and some of the guests left the bar and waited for the helicopter outside, standing in the steamy air of evening on that stiff rubbery thick-bladed Florida grass so much an overnight product of hyperfertilizer, turf-planting, and the tropics that it felt like plastic underfoot.

It was a not untypical American gathering. Doubtless, equivalent Soviet meetings were similar. It did not matter how high or prominent these people had become, how far some of them had traveled from their beginnings. There was still the same awkward, embarrassed, well-scrubbed air of a church social. Americans might yet run the world, they were certainly first on the way to the stars, and yet they had never filled the spaces between. Americans were still as raw as an unboiled potato. It hardly mattered if Americans were rich or poor. When they got together, they did not know what to say to each other. It is part of the double life of Americans, the unequal development of the lobes in the national schizophrenia. Men whose minds worked with an admirable depth of reference and experience in their business or occupation were less interesting in a social gathering, at least in this social gathering where they were plucked up from a more familiar core of small talk and deposited on the rubber-mat turf of the Royal Oak. It was almost a reflection of the national belief that a man who worked thoroughly at his job was given dispensation from the obligation to have a good time. So conversation took overloaded steps over successive hills, and that was all right, the point of the evening was that they would hear Von Braun and be able to refer to it afterward. The American family travels to strange states and places in order to take their photographs and bring them back, as if the photographs will serve in future years as data-points, crystals of memory to give emotional resonance to experience which was originally without any. The data-point will give warmth in old age. So Von Braun would be a data-point tonight. It would not matter if a good time was not otherwise had. Aquarius' mind, brooding through these familiar thoughts, was brought up short with the radically new idea that perhaps some instinct in American life had been working all these decades to keep the country innocent, keep it raw, keep it crude as a lout, have it indeed ready to govern the universe without an agreeable culture to call its own – for then, virgin ore, steadfastly undeveloped in all the hinterworld of the national psyche, a single idea could still electrify the land. Culture was insulation against a single idea, and America was like a rawboned lover gangling into middle age, still looking for his mission.

Since Aquarius on evenings like this would look for the nutrient in liquor the way a hound needles out marrow from a bone, he was nose-deep into his second drink, and hardly saw the helicopter come in. A sense of presence overhead, fore and aft lights whirruping like crickets in the dusk, a beating of rotors in a wheat-flattened gust, and it was down, a creature. Nothing inspired so fine a patriotic cocktail of mild awe, mild respect, and uncorrupted envy as the sight of Praetorians emerging from an insect the size of an elephant which *they* commanded.

The guests immediately made their way inside. Von Braun, dressed in a silver-gray suit, white shirt, and black tie looked more impressive tonight than the day before at a press conference. That had taken place in front of several hundred correspondents with movie cameras, television, and ushers in the audience holding portable microphones to amplify and record all questions the Press might ask for posterity. Von Braun had been on a panel with Dr Mueller, Dr Debus, Dr Gilruth, and a director from Langley, but half the questions had gone to Von Braun. He seemed sensitive to the fact that the Press made jokes about his past. There was one tale every reporter had heard – 'Tell me, Dr von Braun,' a correspondent said, 'what is there to keep Saturn V from landing on London?' Von Braun walked out of the room. But the story was doubtless apocryphal; it smacked of reporters' bile. Journalists were often vicious in their prior comments about VIPs they were going to interview, as if to compensate for the uxorious tone of the restrained questions they would finally ask. Aquarius had been with a small pack who had gone to talk to Dr Debus, director of all launching operations at Kennedy and a former colleague of Von Braun's. 'Just give the Nazi salute and he'll holler "Heil Hitler!" ' they all promised each other, but Debus to their consternation proved out a pleasant Junker gentleman with dueling scars on his mouth and bags under his eyes – the sort of aristocratic face and gracious if saturnine manner which belongs to an unhappy German prince from a small principality. The questions of the Press were predictably unctuous, and trading notes afterward, they quoted Debus respectfully. He had given them the best of lines; when asked if he were planning a celebration while the astronauts were on the moon surface he had smiled and cleared his throat with a cultivated sound. 'No,' he had said, 'no champagne in the refrigerator.' Debus was not afraid of the Press.

But Von Braun was too prominent, and had – although his official position was nominally no more elevated than his countryman's – much too much to lose. A press conference, no matter how many he had had, was a putative den of menace. So his eyes flew left and right as he answered a question, flicking back and forth in their attention with the speed of eyes watching a Ping-Pong game, and his mouth moved from a straight line to a smile, but the smile was no more than a significator, a tooth-filled rectangle. Words were being mouthed like signal flags.

Since he had, in contrast to his delivery, a big burly squared-off bulk of a body which gave hint of the methodical ruthlessness of more than one Russian bureaucrat, Von Braun's relatively small voice, darting eyes, and semaphoric presentations of lip made it obvious he was a man of opposites. He revealed a confusing aura of strength and vulnerability, of calm and agitation, cruelty and concern, phlegm and sensitivity, which would have given fine play to the talents of so virtuoso an actor as Mr Rod Steiger. Von Braun had in fact something of Steiger's soft voice, that play of force and weakness which speaks of consecration and vanity, dedication and indulgence, steel and fat.

Still he did not do badly at his press conference. If he had started nervously, there was an exchange where he encountered his opposition. A correspondent from East Berlin asked him in German to answer a question. There had been a silence. For an instant Von Braun had not known exactly what to do, had in fact stolen a look at Mueller. NASA was sensitive about origins. Two of the three directors in the center of the Manned Spacecraft Program were, after all, German. And there was no joy in emphasizing this, since those few liberal congressmen who were sympathetic to the needs of the space budget would only find their way harder if Von Braun and Debus were too prominent.

Von Braun fielded the difficulty as follows: He translated the question into English. Then he gave a long detailed answer in English (which succeeded in boring the Press). Then, taking an equally long time, he translated his answer back to German. Finally, he took a nimble step away from this now somnolent situation by remarking, 'I must warn the hundred and thirty-four Japanese correspondents here at Cape Kennedy that I cannot do the same in Japanese.' The remark drew the largest laugh of the afternoon, and thereby enabled him to prosper. The contest in press conferences is to utter the remark which will be used as the lead quotation in wire-service stories, and Dr George Mueller, anxious to establish his centrality on this panel, and his eminence over his directors, answered every question helpfully, giving facts, figures, prognostications of future activity. He was a one-man mine of pieces of one-line information with follow-up suitable for heads, leads, paragraph leads, and bottom-of-the-page slugs, but Von Braun picked up the marbles. In fact he had the subtle look of a fat boy who has gathered the shooters in many a game.

When asked how he evaluated the importance of the act of putting a man on the moon, Von Braun answered, 'I think it is equal in importance to that moment in evolution when aquatic life came crawling up on the land.' It drew a hand of applause. It would get the headline. Some of the Press literally stood up.

Thus, he was sound, sensible, and quick as mercury. Yet his appearance had been not as impressive then as now tonight at the Royal Oak. Then he had been somehow not forceful enough for the public image, small-voiced, almost squeaky for a man with so massive a frame. Whereas, here at the Country Club, shaking hands, he had obvious funds of charisma. 'You must help us give a *shove* to the program,' he said to Aquarius on greeting. (This was virtually what Debus had said on parting.)

Yes, Von Braun most definitely was not like other men. Curiously shifty, as if to show his eyes in full would give away much too much, he offered the impression of a man who wheeled whole complexes of caution into every gesture – he was after all an engineer who put massive explosives into adjoining tanks and then was obliged to worry about leaks. Indeed, what is plumbing but the prevention of treachery in closed systems? So he would never give anything away he did not have to, but the secrets he held, the tensions he held, the very philosophical explosives he contained under such supercompression gave him an air of magic. He was a rocketeer. He had lived his life with the obsession of reaching other planets. It is no small impulse. Immediate reflection must tell you that a man who wishes to reach heavenly bodies is an agent of the Lord or Mephisto. In fact, Von Braun, with his handsome spoiled face, massive chin, and long and highly articulated nose, had a fair resemblance to Goethe. (Albeit none of the fine weatherings of the Old Master's head.) But brood on it: the impulse to explore the universe seems all but to suppose a divine will or a divine displeasure, or – our impurities matched only by our corruptions – some mixture of the two. What went on in Von Braun's mind during a dream? 'Yes,' he said with a smile, 'we are in trouble. You must help us.'

'Who are you kidding?' said Aquarius, the good American. 'You're going to get everything you want.'

Whether the intimacy was too abrupt, or Von Braun's reaction disclosed too much – his eyes gleamed with sudden funds of pleasure at the remark – he quickly looked discomposed, and as quickly left the conversation by failing to forward a remark in return. Then he waved some ambiguous good-by and moved quickly across the room. If his sense of friend and foe was good – a reasonable assumption to make about a man like Von Braun – it was obvious he did not think Aquarius would make such a good friend.

The banquet was roast beef. Ice creams and sauces for dessert. Coffee. The spoon on the glass. The publisher of the publishing corporation was talking. 'We are signally honored tonight to have with us,' he began, 'one of the true fathers of space, Dr Hermann Oberth, who, with Dr Goddard and a Russian named Tsiolkovsky, is really and truly one of the originators of the whole concept of the exploration of space.' An old man, seventy-five at least, with white hair and a white bushy mustache stood up. He had the birdlike self-sufficiency of the old, a vain sly white old bird, as if he were not only cousin to an old condor but had bought the nest as well. There was a smattering of applause.

His name meant little outside of rocketry circles, but the speaker had used his presence for a joke. 'I heard earlier tonight that at least two men in this room, one of whom you'll hear from later, are students of his. They said in response to his comment that he was a very good teacher, that at least they weren't dropouts; one of them is Dr von Braun.' The laugh came on that. Of all the nations in the world, America had possessed the firmest patriotic firmament; the common culture had never been rich enough to corrode it. Not till recently. Now, dropouts were pits in the shining surface. This then offered the suggestion that Von Braun was a regenerator of the shining surface. Therefore, the audience was not to be at ease during his introduction, for the new speaker, who described himself as a 'backup publisher,' went into a little too much historical detail. 'During the Thirties he was employed by the Ordnance Department of the German government developing liquid fuel rockets. During World War II he made very significant developments in rocketry for his government.'

A tension spread in this audience of corporation presidents and high executives, of astronauts, a few at any rate, and their families. There was an uneasy silence, an embarrassed pall at the unmentioned word of Nazi – it was the shoe which did not drop to the floor. So no more than a pitter-patter of clapping was aroused when the speaker went quickly on to say: 'In 1955 he became an American citizen himself.' It was only when Von Braun stood up at the end that the mood felt secure enough to shift. A particularly hearty and enthusiastic hand of applause swelled into a standing ovation. Nearly everybody stood up. Aquarius, who finally cast his vote by remaining seated, felt pressure not unrelated to refusing to stand for *The Star-Spangled Banner*. It was as if the crowd with true American enthusiasm had finally declared, 'Ah don' care if he is some kind of ex-Nazi, he's a good loyal patriotic American.'

Von Braun was. If patriotism is the ability to improve a nation's morale, then Von Braun was a patriot. It was plain that some of these corporation executives loved him. In fact, they revered him. He was the high priest of their precise art – manufacture. If many too many an American product was accelerating into shoddy these years since the war, if planned obsolescence had often become a euphemism for sloppy workmanship, cynical cost-cutting, swollen advertising budgets, inefficiency and general indifference, then in one place at least, and for certain, America could be proud of a product. It was high as a castle and tooled more finely than an exquisite watch.

Now the real and true tasty beef of capitalism got up to speak, the grease and guts of it, the veritable brawn, and spoke with fulsome language in his small and well-considered voice. He was with friends this occasion, and so a savory, a gravy of redolence came into his tone, his voice was not unmusical, it had overtones which hinted of angelic superpossibilities one could not otherwise lay on the line. He was when all was said like the head waiter of the largest hofbrau house in Heaven. 'Honored guests, ladies and gentlemen,' Von Braun began, 'it is with a great deal of respect tonight that I meet you, the leaders and the captains of the mainstream of American industry and life. Without your success in building and maintaining the economic foundations of this nation, the resources for mounting tomorrow's expedition to the moon would never have been committed . . . Tomorrow's historic launch belongs to you and to the men and women who sit behind the desks and administer your companies' activities, to the men who sweep the floors in your office buildings and to every American who walks the street of this productive land. It is an American triumph. Many times I have thanked God for allowing me to be a part of the history that will be made here today and tomorrow and in the next few days. Tonight I want to offer my gratitude to you and all Americans who have created the most fantastically progressive nation yet conceived and developed.' He went on to talk of space as 'the key to our future on earth,' and echoes of his vision drifted through the stale tropical air of a banquet room after coffee – perhaps he was hinting at the discords and nihilism traveling in bands of brigands across the earth. 'The key to our future on earth. I think we should see clearly from this statement that the Apollo 11 moon trip even from its inception was not intended as a one-time trip that would rest alone on the merits of a single journey. If our intention had been merely to bring back a handful of soil and rocks from the lunar gravel pit and then forget the whole thing' – he spoke almost with contempt of the meager resources of the moon – 'we would certainly be history's biggest fools. But that is not our intention now – it never will be. What we are seeking in tomorrow's trip is indeed that key to our future on earth. We are expanding the mind of man. We are extending this God-given brain and these God-given hands to their outermost limits and in so doing all mankind will benefit. All mankind will reap the harvest . . . What we will have attained when Neil Armstrong steps down upon the moon is a completely new step in the evolution of man.' (Which would lead Aquarius days later to wonder at the origin of Armstrong's first speech on the moon.) 'It will cause a new element to sweep across the face of this good earth and to invade the thoughts of all men.'

He was almost done with his formal remarks. Out of his big bulk and his small voice he would offer miracles. That was his knowledge of America, no mean knowledge. Prosperity satisfies those who are rich in culture. But in lands where the geography like the people is filled with empty space, then faith in miracles is the staple of the future.

'Every man achieves his own greatness by reaching out beyond himself, and so it is with nations. When a nation believes in itself as Athenians did in their Golden Age, as Italians did in the Renaissance, that nation can perform miracles. Only when a nation means something to itself can it mean something to others. We are truly faced with the brightest prospects of any age of man. Knowing this, we can watch the launch tomorrow with a new dimension of hope. We can cheer the beginning of a new age of discovery and the new attainment that spans the space distances and brings us nearer to the heavens.'

His speech while quietly apocalyptic was not without gloom for the audience. They were heavy with food, and a band or a jukebox was blasting in some big room next to the hexagonal room. The country club couples of Titusville and Brevard County were having a good time next door. But it seemed to Aquarius that Von Braun's remarks had plunged the collective intelligence of these corporation men back into some of their problems.

Yes, Aquarius was thinking, ideas were what Americans cared about, and the biggest ideas were doubtless the best, but what a price had been paid. For now manufacturers and consumers chased frantically after fashion. It did not matter how cheap and shameful the execution. The bargain stereo could not last a month, the washing machine with the plastic console would break in a week – all that had been purchased was the idea. Something was happening to Americans. They were a guilty crew, guilty of new ideas, new license, sacrilege, cynicism, bad faith. As a result, they were always in a rush to purchase a new idea. When people were not willing to die for an old idea, they would rush to a new one. Guilty to the nose, guilty to the ears, they were even apathetic about blaming the fabricators, for they were guilty themselves. Everybody had been cheated so many times; everybody had cheated others so often. It was hard to remain angry that one had been defrauded. It was even hard to get angry. So food and ruminating drink lowered the audience from the excellence of Von Braun's achievement to the shoddy dimensions of their own.

The question period cheered this same audience however. Then Von Braun could speak of lunar jeeps, and space costs reduced by flying stages which could be used over and over in travel from earth to manned laboratories in earth orbit. He would be eloquent a little later about nuclear rockets the size of battleships which might be assembled in earth orbit and then voyage to Mars. While he talked of other planets, the audience grew warm again. It was the moon which was cold.

When applause subsided, the publisher cried out in his cheerful voice, 'I have a question. Will you be fired if you don't get on that helicopter and greet the senators and Cabinet members who are waiting?' Von Braun made a point of staying for two more full questions, then took his departure. The sound of helicopters rose over the room.

Aquarius would have thought the evening concluded, but as he was learning again, he would never understand Americans. Another speaker, a representative of American business, rose and gave a humorous introduction to a man as massive and slow-speaking as Lyndon Johnson who proceeded to get up and tell jokes in an absolutely assured drawl. The audience seemed happy with them. 'I was in an airport not long ago and sitting next to a woman smoking a cigar. I asked her how long she'd been smoking cigars. And she said ever since her husband had come home and found one in her ashtray in the bedroom.'

The couple in front of Aquarius, young, stingy, ambitious, and very respectable, were laughing. The husband scowled at the wife and said with existential humor, 'I wouldn't laugh at that joke if I was you.'

'Why not?' responded the wife with the serenity of total practicality. 'I've never done it.'

Yes, they were all good Americans and they would listen to jokes and be a little relieved Von Braun was gone (although they would treasure the experience), and as new jokes came along, Aquarius began to look again into his drink and brood on Von Braun's remarks. He had declared that reaching the moon would be the greatest event in history since aquatic life had moved up onto land, and that was a remark! for it passed without pause over the birth and death of Christ. Indeed Von Braun had said even more in a newspaper interview. 'Through a closer look at Creation, we ought to gain a better knowledge of the Creator.' Man was voyaging to the planets in order to look for God. Or was it to destroy Him?

Of course, in the interview, Von Braun had been careful to add, 'It could very well be that the Lord would . . . send His Son to the other worlds to bring the gospel to them – I believe the good Lord is full of such tremendous compassion that He will take whatever steps are necessary to bring the truth to His Creation.' While Aquarius had always assumed that compassion did not move by steps but seemed rather to bathe the wounded with its grace, this was after all no ordinary piety. It was possible Von Braun was sincere. Still what a grip he had on the jugular of the closet missionary in every Wasp. If he had dumped his private finds on American religious opinion into a computer and cranked it up for response, the words could not have come back better. On the other hand, Aquarius had believed for years in ideas not altogether dissimilar. Once, tentatively, he too had undertaken the doubtful liberty to state in an interview what he thought of God. God, he had presumed to suggest, was an embattled vision: God had created man in order that man might fulfill God's vision, but His vision of the future was at war with other visions of existence in the universe. Some of those other visions were not only out in the stars, and in the galaxies, but were right here, intimate, on earth. God was, for instance, at war with the Devil. Certainly the Devil had a most detailed vision of existence very much opposed to His own. In any case the war had gone on for so long that nearly everything human was inextricably tangled. Heroism cohabited with technology. Was the Space Program admirable or abominable? Did God voyage out for NASA, or was the Devil our line of sight to the stars?

NASA. The word had derived from NACA – National Advisory Committee for Aeronautics, which became the National Aeronautics and Space Administration, or NASA. It was an unhappy sound. Just think of NASA-ism. NASA would have no deliberate relation whatsoever to Nazi. But we are not a schizophrenic land for nothing. Deep in the unconscious where each sound leaves first its murmur and then its roar at a combustion of hitherto unconnected meanings, NASA had to stand for something. You bet.

Listening to the jokes, Aquarius was still brooding about Nazism. For the philosophy of the Folk, detesting civilization, claiming to be in love with the primitive, had nonetheless killed millions of men in the most orderly technological fashion yet devised. Nazism had been not one philosophy, but two – and each philosophy was utterly opposed to the other. It was primitive, it was vertiginously advanced. It gave brave men a sense of nobility in their hearts – it had been utterly heartless. It spoke of clean futures and buried Germany (for a time!) in vomit and slime and swill. Now its ghosts were pacing on every battlement of every surviving palace, now its ghosts were bubbling in the tubes of every laboratory, burning in the wires. Nazism had been an assault upon the cosmos – why think of it as less? That is why it moved as the specter behind every civilized transaction. For it had said: civilization will stifle man unless man is delivered onto a new plane. Was space its amputated limb, its philosophy in orbit?

Now the speaker was telling a joke about a Texan in Alaska who had mixed up his respective missions with a woman and a bear. Big was the laugh from the audience. And out on the beaches and the causeways and riverbanks, another audience was waiting for the launch. America like a lazy beast in the hot dark was waiting for a hint in the ringing of the night. Questions drowned Aquarius. In bed by two in the morning, he would be up by four. An early start was necessary, for traffic on the road to the Press Site would be heavy.

iv

The two hours of sleep were more worthy of Atlas than Aquarius, for in his dreams he held up a portion of the State of Florida. Aquarius had been covering the moon shot from July first to this morning of the sixteenth. The mildest form of purgatory was to spend two weeks in a motel.

Everybody complained that the tension accompanying the preparations for Apollo 11 had been less than launchings in the past. Cocoa Beach itself had altered. The old days of honky-tonks on the strip of highway and rockets threatening to carve a furrow down the beach were now gone. Money had come in and industry, space technicians and their families, supermarkets, motels, churches, and real estate developments had been put up. In the restaurants, public address systems broke into the mood of a meal to call patrons to the phone. The paper doilies under the plate carried legends: AMERICA'S SPACE PROGRAM BENEFITS ALL MANKIND – *Your Souvenir of Apollo 11 Lunar Landing.* Better Color Television. Water Purification at Less Cost. New Paints and Plastics. Lunar Walker for the Handicapped. Laser Surgery. Solar Power. So forth. Cocoa Beach had been one of the five places on the Atlantic Seaboard which deserved the title 'Wild West of the East,' but Cocoa Beach deserved it no longer. As the doily revealed, it was part of the Brevard Economic Development Council. Until the last few days when the Press arrived in hundreds then thousands, there had been monotonous hours when it was necessary to remind oneself that three men were leaving for the moon in less than a week.

But in the dark morning before dawn on the sixteenth, in the black hour of 4 A.M., the night air a wet and lightless forest in the nose, one was finally scared. It was not unlike awakening in a convoy with invasion of a foreign beach scheduled for the hours ahead, an awakening in the dark of the sort one will always remember, for such nights live only on a few mornings of one's life. Somewhere not so far from here, the astronauts were getting up as well. And the ghosts of old Indians.

In that long-ago of prairie spaces when the wind was the message of America, Indians had lived beneath the moon, stared at the moon, lived in greater intimacy with the moon than any European. Who could say the ride of the Indian with whisky in his veins was not some conflagration of messages derived from the silences of the moon? Now tonight were the ghosts of old Indians awakening in the prairies and the swamps? Did the echo of the wind through the abandoned launch towers of the Cape strike a resonance across two thousand miles to the grain elevators by the side of railroad tracks in the mournful empty windings of the West? The country had been virgin once, an all but empty continent with lavender and orange in the rocks, pink in the sky, an aura of blue in the deep green of the forest – now, not four centuries even spent, the buffalo were gone, and the Indians; the swamps were filled; the air stank with every exhaust from man and machine. All the while we had been composing our songs to the moon and driving the Indian onto the reservation, had we also been getting ready to go to the moon out of some deep recognition that we had already killed the nerve which gave life to the earth? Yet the moon by every appearance knew more about disease and the emanations of disease than the oldest leper on earth. 'Of what can you dream?' said the moon. 'I am battered beyond belief and you think to violate me now?'

Driving through the night, passing again the families and tourists who were waiting for morning on the banks of the causeway, showing a Press Pass to the guard at the gate, and being waved on in silence, yes near to conspiratorial silence, there was the tangible sense of time running in parallel, the million-headed witness now traveling to a point where the place would cross the time and the conscious eye of the nation would be there to witness this event. By television would they witness it. That would be an experience like getting conceived in a test tube.

Out at the Press Site, Saturn V was visible across the near distance of three and a half miles. It was the nearest Aquarius had approached on this long night, and it was indeed the nearest anyone would come but the launching crew and the three astronauts. As Apollo-Saturn stood on its concrete pad six thousand yards away across a lagoon from the small grandstand built for the Press, its details now visible, it looked less like a shrine, but all the more a presence. A squad of floodlights played upon it, and their beams reflecting from the thin night haze displayed a fan of separate lights across the sky on down into the surface of the lagoon, a bending and glancing of rays worthy of a diamond upon a mirror. In the black wet night, back of the floodlamps, lightning flickered, so regularly that one might have been looking at a lighthouse rotating its flare – somewhere down the horizon, a potency of storm was speaking up in the response of the Caribbean. Long far-off rolls of thunder.

Staring across the water, Aquarius took a long study. His binoculars appeared to lock him into collaboration with Saturn V, as if the rocket had the power to keep those binoculars pressed to his nose, as if finally Aquarius and Saturn V were now linked into some concupiscence of mission like a one-night stand which might leave its unexpected consequence upon him. What a vehicle was the spaceship! A planet-traveler massive as a destroyer, delicate as a silver arrow. At the moment it lifted off from the earth it would be burning as much oxygen as is consumed by half a billion people taking their breath – that was twice, no, more than twice the population of America. What a deep breath must then have been concentrated into the liquid oxygen they were passing into its tanks right now, a liquid oxygen cooled to 297 degrees below zero and thereby turning air to cloud at every hint of contact with the pipes two feet thick to insulate the fuel. Model of an ogress, umbilical cords of every thickness and sinuosity, snakes and cables and ropes and constrictors thick as tree trunks passed in cluster from swing-arms and the walkways of the launching tower into the thin-skinned walls of the rocket; a Medusa's head of umbilicals loading fuel, charging batteries, testing circuits; a complexity of interrelation between the launching tower and the rocket so simplified by the swing-arms that Apollo-Saturn resting on her pad did not look distraught but calm, like a silver-white ship standing erect by an iron tree with nine horizontal branches. There were clouds about both, strings of small firm well-puffed little clouds drifting off at right angles from the rocket at each place where an umbilical from an oxygen pipe had been disconnected; the clouds gave Saturn V the brow of a philosopher in contemplation above his thought, yes, the cloud belonged to Saturn V, it nuzzled at it, a new cloud not many hours old. And the light from the floodlamps reflected from the white icy skin of the wall. Sainted Leviathan, ship of space, she was a planetary traveler.

But now the Press was loading onto buses to have one last look at the astronauts, who had been invisible for the last eleven days. It was idiotic to leave the Press Site now, leave the unique pleasure of communing in the dark across the water with that technological gem on the horizon; it was idiotic to pack into buses and sweat in the dark night heat and crawl up photographers' backs for a glimpse of an astronaut's face, but he got into the bus along with the others and ground five miles along Space Center highways through the dark, a full load of observers transported on an irritating underpowered whine, and then debouched at the front entrance of the Manned Spacecraft Operations Building, MSOB, a cold white structure, impersonal as the offices of a factory complex hundreds of feet long and several stories high, as impersonal and architecturally undistinguished as the spirit of its design (which derived from the implicit architecture contained in the cost-apportionments of a multimillion-dollar bill).

The journalistic cargo of the bus was now ushered through long narrow corridors painted in institutional green to take turnings by water coolers and vending machines, passed down other corridors, filed past empty rooms and galleries and were loosed finally into a courtyard. Just over their heads, a story above, was a covered bridge which ran from the building they had just been in to the building from which the astronauts would exit.

Beneath this bridge a white van was parked, ready to receive the astronauts. They would pass for twenty steps down an aisle protected by policemen and by a fence from the hydraulic pressure of these excited journalists, and would then enter the van to be transported the nine miles to Apollo 11, mounting by elevator in the mobile launch structure to the ninth swing-arm where they would cross an iron gangplank to their seats. There they would wait while the countdown proceeded. There they would be when the lift-off occurred.

This gathering of the Press, this congestion of the Press – for taking the sum of the TV personnel, still cameras and movie men, there must have been several hundred people standing on tiptoe at every bad angle and hopeless vantage point for a good look – would remain more than an hour on this chance of the briefest glimpse, hardly more than a wave of the arm as it finally turned out. The magic of the long night would leave them as they waited. The gray of a new morning would diminish the high theater of their mood before finally a single astronaut was seen. Yet no one left. Men and women stood on every ledge and railing and stone and fence they could climb in the courtyard, and dozens periodically clambered onto the flat roof of a special TV truck only to be chased down by special police. Photographers wrestled themselves up wire fences, kept themselves in position by wedging their bodies into angles, waited out cramped positions like mountain climbers in a chimney for twenty minutes, thirty minutes, finally an hour. It was imperative for everyone there to have a look at the men who would come through the door as if that were equal in value to a piece of the moon. There seemed some mysterious booty residing in the literal aura of the astronauts, yes, the Press crowded forward, were ready to claw one another for a crack of the concrete, for one data-point, one photograph when the time came, as if the value of these astronauts revealed the illimitable value of condemned men. They were men who would walk past their eyes to a conceivable execution in space, and so like all condemned were close to some state of more valuable existence, as if men who were about to enter death or at the least to chance its dimensions were men of inestimable refinement. Clues to the richest profit lived somehow next to death – so the moments of their life as they passed by were like the passing of a new current. The moon. Their persons would touch the moon! That was why the atmosphere in the courtyard was like a prison on the night of an execution.

And from time to time, with the washing out of the night, and the oncoming of dawn to this dark courtyard with its waiting mob, a crack of lightning would be seen again in the distance, a crack which to Aquarius' nerve-heightened eyes looked like a literal crack in the sky, as if electricity were now revealed as the absence of some material of the firmament.

The mob cheered when the astronauts came through the door. Since guards and directors and technicians were periodically looking down the hall, or signaling, or making an abrupt move, the Press had been on the alert for the astronauts a dozen times. Now, they really came, and people throughout the crowd had experiences. '*Fenomenal, fenomenal, fenomenal!*' an Italian girl with a camera kept repeating, almost as if helpless to stop, and a worker from MSOB yelled, 'Go get' em' – the target was at last revealed – the moon like every other foreign body was an enemy, an intimate competitor. Armstrong, plastic helmet on, carrying his life-support system connected by a hose to his white space suit, white and luminous as Saturn V out on the pad, stopped just long enough to give a wave, his face within the space helmet as lashless in appearance as a newborn cat in a caul. He had never looked better. He stepped into the van, and the others, waving, stepped into the van, the doors were shut, the police keyed as always into a state of ultraresponsibility – what rumors of plots had they exchanged this morning over breakfast? – pushed everyone back as if the very departure of the van were in danger, and the vehicle of the astronauts drove down the last nine miles.

That was all the Press saw, and to add to this frustration, they had a bad ride back. It was after six-thirty now, the launch was not three hours off, and the heavy traffic they had escaped by getting up early now engorged the road for the last mile to the Press Site. If the night had been hot, the morning heat was fierce. It took the caravans an hour to travel the five and a half miles.

They were riding in school buses rented by the Space Center for the occasion. Like all old machines, their vehicle had developed a personality. It was hardworking, all but crippled, yet spoiled. Accustomed to carrying children, its load this morning was excessive.

Tall thin Swedish journalists were packed in with short stubby London and South American reporters, the seats were occupied, the aisles were loaded with writers and cameramen who would stagger every time the bus stopped or started, which was every twenty feet, and the differential clanged and slapped in its case as if it would break out at any moment. The bus was simply not used to being made to work so hard. Everybody was sweating. There had never been a hint of air conditioning in this vehicle.

It was the first and only discomfort the Press had been obliged to suffer on the job. They had been treated with courtesy at the Press Center, conducted on tours, taken in small groups for interviews with nearly every prominent executive and director at Kennedy, they had even been furnished with transcripts of each large press conference so there was no need to take notes on what was said unless they were filing a story that day, and in such event they usually had their own tape recorder. Yet with every vocational aid handed to them, they had still been in a depressed state, for the relation of their news stories to the complex truth of this moon shot might be analogous to a comparison between the school bus and Saturn V.

One of the cess-filled horrors of the Twentieth Century slowly seeping in on the journalists was that they were becoming obsolete. Events were developing a style and structure which made them almost impossible to write about. If a reporter did his homework for space, which is to say went figuratively back to school and got himself up again on forgotten physics and learned near-unpronounceable engineering terms, he could still hardly use this language in stories for popular consumption. Yet if he tried to do features on the people in the Space Project, he encountered the familiar difficulty that engineers who worked for NASA seemed to pride themselves on presenting personalities which were subtly faceless and interchangeable. A process was taking place that was too complex to be reported for daily news stories by passing observers, and so the process itself began to produce the news for the reporters. Their work had come down to rewriting

73

publicity handouts. When they interviewed a personality in the Space Program, the eminent figure gave them quotes which sounded exactly like the handouts, except the handouts, being freed of the vicissitudes of communication between brain and tongue, were more detailed and more quotable. Computers could have written their pieces.

This was of course happening everywhere. It was the signature of the century. Soon newspapers would be qualified to write only about fashion, theater, murder, movies, marriage and divorce. What a humiliation for these tall Swedes and stocky Britons, these hardworking Japanese, to travel so far and be able to tell so little. No wonder they had taken those unhealthy reporters' bodies, long sacked by bad food, bad hours, much whisky and not much sex, and had jolted and jammed into one another to get a peek at the astronauts. To write about men without ever seeing them was like covering an assignment by watching television – you could do it, but your liver would have to pipe up the warmth your lack of confrontation had lost. Yet Aquarius had come to the conclusion that in this case it hardly mattered. In dealing with men who were enormously complex or with men whose passions were buried in the depths of their work, an interview could be misleading. He told himself that when the time came, he would have material enough – what kind of detective was he, if he could not divine the depths of their character by the depths of his own experience and the few clues the astronauts had already provided in their shielded public interviews? Nonetheless, he had been as eager as the rest to get a glimpse of their faces, and had been more than pleased with the sight of Armstrong looking lashless and like a newborn cat in its caul. That was compensation for the endless creeping annoyance of taking an hour to go a mile in a bus. Yes, now when the time came to write about Armstrong, he would be better able from these few clues to reconstruct him, like the dinosaur from the fossil bone.

They got back to the Press Site at seven-thirty. The sun was clearly up and burning through a haze. In the distance, through binoculars, Saturn V now looked gray, a near-white gray, a gray palpability among other grays. Everything was gray, the Launch Pad, the tower, the vehicle, the sky. The rocket looked as if it were already on the moon. Only the sun gave a platinum edge to the outline of Saturn and the cryogenic cloud.

Aquarius, still brooding over the astronauts, came to the gloomy conclusion that even if you comprehended them (came to some whole notion that they were finally good and noble men, or men who were brave but not without malignity), you would still be inhibited from saying in confidence that the Space Program was for good or ill since History often used the best of men for the worst of purposes and discarded them when the machines of new intent were ready. As often History had used the worst of men to convert an unhealthy era to a new clime. But like a dive into a dream, explorations of these questions would only open into deeper questions. Standing in the early morning heat he had an instinct that before all was done the questions would travel through the unmapped continent of America's undetermined heart.

V

On this morning, with two hours to spend before the launch, he chooses not to get into another bus – not another bus this day! – to travel to the other side of the Vehicle Assembly Building where the VIPs will be sitting in bleachers. He knows there is a good haul there – all of two hundred congressmen, Sargent Shriver, Mr and Mrs James E. Webb, William W. Scranton, Jack Benny, Cardinal Cooke, Daniel Patrick Moynihan, Johnny Carson, Gianni Agnelli, Senator Javits, Leon Schacter of the Amalgamated Meat Cutters and Butcher Workers, Prince Napoleon of Paris, four hundred foreign ministers, attachés and military aviation officials, two hundred and seventy-five leaders of commerce and industry, Vice President Agnew, Lady Bird and former President Lyndon B. Johnson, plus Barry Goldwater in slacks and a red golfing shirt. The last two will shake hands for the cameras before this is all over – space will prove bigger than both of them. The voice of duty has suggested to Aquarius that he should be there to study them, record their expressions, comment on the part of history they command and their relation to the part of history now being born, but his liver will simply not permit it. He is here to see the rocket go up, not to stand and look at Very Important People and take notes in a notebook while he sweats in the heat. No, some sense of his own desire to dwell near the rocket, to contemplate its existence as it ascends, and certainly some sense of his own privacy, some demand of his vanity – aware of how grubby he looks and feels – now bids him to stay with his own sweaty grubs, the Press and photographers gathered in the grandstand bleachers and out on the small field before the lagoon which separates them from the blast-off of Apollo 11. Besides he dislikes the VIPs, dislikes most of them taken one by one, and certainly dislikes them as a gang, a Mafia of celebrity, a hierarchical hive. He is still sufficiently a Manichean to believe that if Saturn V goes up in perfect launch, it will not be the fault of the guests. No, some of the world's clowns, handmaidens, and sycophants and some of the most ambitious and some of the very worst people in the world had gotten together at the dignitaries' stand. If this display of greed, guilt, wickedness, and hoarded psychic gold could not keep Saturn V off its course, then wickedness was weak today. Or did wickedness crowd to the witness stand to cheer evil on its flight? It did not matter how the reporter's mind turned, it was filled with nothing but the most fruitless questions today. He discovered he was thirsty.

So in preference to taking the bus and taking notes on the guests, he stood in line for another large part of an hour waiting for a cold drink. In back of the Press Site, more than a hundred radio and TV trailers were now arrayed behind one another in ranks and rows of huge white ruminants, the very sacred cows of American technology. Yet there was only one trailer reserved for food. It was next to heartwarming to discover another piece of poor planning in the icy efficiencies of the Space Program – small surprise that it had to do with a creature comfort like food. The trailer was inadequate to the needs of the Press – over a hundred waited in line, more than a hundred walked away in disgust. The line drifted forward about as fast as a tide works up a beach. The trailer interior consisted of a set of vending machines for chiliburgers, hamburgers, pastries – all people wanted were cold drinks. So the line crawled, while everyone waited for the same machine. Nobody was about to have machine-vended chiliburgers at half-past eight in the morning. But so many demands on the iced-drink machine caused malfunctions. Soon, two vending machine workers were helping to service the machine. Still it took forever. Coins had to go into their slot, change be made, cups filled, tot of cracked ice dropped, syrup poured, then soda. Just one machine. It was pure American lunacy. Shoddy technology, the worst kind of American shoddy, was replacing men with machines which did not do the work as well as the men. This crowd of a hundred thirsty reporters could have been handled in three minutes by a couple of countermen at a refreshment stand in a ball park. But there was an insidious desire to replace men everywhere with absurd machines poorly designed and abominably put together; yes, this abominable food vending trailer was the proper opposite number to those smug and complacent VIPs in their stands a half mile away; this was the world they had created, not the spaceship. They knew nothing about the spaceship but its value in the eyes of the world – that was all they had to know. The food vending trailer was their true product. When they mouthed their portions of rhetoric, when they spoke, lo! their mouths poured forth cement – when they talked about poverty and how poverty could be solved by the same methods and discipline and effort devoted to space, he would have liked to say to them: Solve your food vendors first! Solve your shoddy appliances first! you planned obsolescences! – then you may begin to think of how to attack the poverty of others. He was in a fury at the complacency of their assumption that they could solve the problems of the poor. His favorite man, Lyndon Johnson, was telling Walter Cronkite on television, 'There's so much that we have yet to do – the hunger in the world, the sickness in the world. We must apply some of the great talent that we've applied to space to these problems.' Yes, his mouth poured forth cement.

Once Aquarius had gone – in payment for a professional debt and so against his will – to a show at the Jewish Museum in New York. Over the years, a photographer named Clayton had taken photographs of the poor, of the very poor, of Southern Black faces so poor they were tortured by hunger. Remarkable faces looked back at him, the faces of saints and ogres, of emaciated angels and black demons, martyrs, philosophers, mummies and misers, children with the eyes of old vaudeville stars, children with faces like midgets and witches, children with eyes which held the suffering of the lamb. But they were all faces which had gone through some rite of passage, some purification of their good, some definition of their remaining evil – how loyal could evil be to people so poor? Aquarius retreated from these photographs with no unhappy sense of shock – these poor to whom he like every other middle-class mother of a professional gave money from time to time (or to their charities at least) were a people on a plane of subsistence which had tortured their flesh, but delivered some essence of their nature, some delineation which never arrived to the face of men and women who were comfortable. They had survived, they looked indeed as if they had passed under one of the four corners of the winding-sheet of the dead, and so knew more than he would ever know. Yes, remarkable faces looked back at him, more beautiful by far than his face would ever be, or Lyndon Johnson's face, or the long uncorrugated snoot of Barry Goldwater. 'Well,' said Aquarius softly of these faces of the poor, 'they look better than most of my friends.'

Yes, the poor were not a gaggle to be collected by a bulldozer, shoved, mashed, annealed, and poured into plastic suits they could wear in New NASA Subsidiary Poverty Suburbs. No, they were rather part of the remaining resource of a spiritually anemic land, and so their economic deliverance was a mystery which would yet defy the first and the last of the social engineers. The truth was that their faces were better than the three perfectly pleasant and even honorable faces of the three astronauts.

But the voice of the Public Affairs Officer came out of the loud-speaker mounted on a speaker's platform on the grass in front of the grandstand.

This is Apollo-Saturn Launch Control. T minus 61 minutes and counting – T minus 61 minutes on the Apollo 11 countdown, and all elements are GO at this time. Astronaut Neil Armstrong has just completed a series of checks on that big Service Propulsion System engine that sits below him in the stack. We want to assure ourselves before lift-off that that engine can respond to commands from inside the spacecraft. As Neil Armstrong moved his rotational hand controller we assured ourselves that the engine did respond by swiveling or gimbaling.

Aquarius felt chopped into fragments. The combination of waiting in line at the mechanical vendor, of undergoing Lyndon B. Johnson homiletics, batting his eyes against the heat, subduing his rage, passing through the disembodied experience of recall! – those Black faces in the photographs had been a sure planet apart from the faces of his fellow grubs in the Press Site at the white-painted grandstand – and now the voice on the public address system, bringing to the journalists of the world some of the intimate details of the countdown. Somewhere – not very far away in fact – some giant conception was being delivered, and the doctors were careful to provide many a detail.

We'll also check out the tracking beacons in the Instrument Unit that travels as a guidance system for the Saturn V during the powered phase of flight. Now 59 minutes 48 seconds and counting.

Yes, his brain was so chopped in fragments that he felt as if he had awakened with a hangover (which he had not quite) and had done nothing since but smoke and drink coffee, neither of which he had gone near, for he had not taken a cigarette in years and coffee was never his drink until night.

The Test Supervisor, Bill Schick, has advised all hands here in the Control Center and spacecraft checkout people that in about thirty seconds that big swing-arm that has been attached to the spacecraft up to now will be moved back to a parked position some five feet away from the spacecraft. It's coming up now in five seconds, the swing-arm will come back. Mark. The swing-arm now coming back from the spacecraft. Countdown proceeding satisfactorily . . .

The voice was clear only if one forced oneself to listen to it. He tried to picture the scene in the Launch Control Center with hundreds of men scanning hundreds of consoles and computers, but there was not a real interest. He found himself going for a walk along the grass. Between the grandstand and the lagoon was a field about the size of a Little League baseball park and the photographers had all set themselves up at the edge of the water, their cameras with telephoto lenses set on tripods so that they looked from behind like a whole command of Army surveyors taking a lesson in their instrument. And the object on which they were focused, Apollo-Saturn, looked gray and indistinct across the air waves of heat shimmering off the lagoon.

Astronaut Buzz Aldrin in the middle seat. He's been working with the Spacecraft Test Conductor on setting up proper switch settings in preparation for pressurizing the reaction control system. These are these big thrusters on the side of the Service Module.

There are actually sixteen of them in four quadrants around the Service Module. They are used for maneuvers in space.

To the right of the photographers was a small grove of pure jungle. Recollections of his platoon on a jungle trail, hacking with machetes entered his head. A hash of recollections. He had thought he would be concentrating on the activities at the Launch Pad, the Control Center, in the Command Module, he had expected to be picturing the vitals of the rocket, and the entrance of the fuels into it, but he was merely out of sorts and with a headache, and waiting for the time to pass. He felt somehow deprived that he could feel so little. People had told him over and over that the sight of a large rocket going up was unforgettable and the sound would be remarkable, the ground would shake. It had begun to sound like an experience as mysterious and thoroughgoing as those first reverberations which had come to him in his early teens that there was a world of sexual intercourse out there, and it was unlike anything you and I had ever done. So, ridiculous as this, just as an adolescent would want his first real sexual hour to be at least approximately equal to his imagined pageant, so now Aquarius was cranky that nothing in his mood was remotely ready for the experience he had been promised. The damn astronauts weren't even real to him. He had no sense at all of three psyches full of awareness on the edge of the horizon. Just that gray stick out there.

Coming up shortly will be a key test here in the firing room as far as the launch vehicle people are concerned. It's some final checks of the destruct system aboard the three stages of the Saturn V launch vehicle. In the event during powered flight that the vehicle strayed rather violently off course, the range safety officer could take action to destroy the vehicle which obviously would occur after the astronauts were separated by their escape tower from the faulty vehicle.

He remembered the questions he had asked in the VAB. How would they destroy the rocket if something went wrong? Why a longitudinal set of explosives lined up, as if in the seam of a tin can, would burst the stages. The belly of the rocket would open like a Caesarian. A flood of burning fuel would burst into the sky.

This is Apollo-Saturn Launch Control. We've passed the eleven-minute mark. Now T minus 10 minutes 54 seconds on our countdown for Apollo 11.

He began to look for a place from which to watch. The grandstand had a roof which would obstruct the view once the rocket was high in the air. But, standing on the field, he felt a hint too low. Finally he took up a careful position a few steps from the ground in a wing of additional bleachers. He was still not properly ready for the spectacle. Yes, the future spoke of a human species which would live on diets and occasional feasts, and would travel to spectacles to feel extraordinary sensations. They might even look at photographs of starving Black faces in order to generate some of their deepest thoughts.

But he knew now why he was so irritated with everything and why he could not feel a thing. It was simple masculine envy. He too wanted to go up in the bird.

This is Apollo-Saturn Launch Control. We've passed the six-minute mark in our countdown for Apollo 11. Now 5 minutes 52 seconds and counting. Spacecraft Test Conductor Skip Chauvin now has completed the status check of his personnel in the control room. All report they are GO for the mission. Launch Operations Manager Paul Donnelly reports GO for launch. Launch Director Rocco Petrone gives a GO. We're 5 minutes 20 seconds and counting. We took a good look at Eagle, and it looks good. The Spacecraft Test Conductor for the Lunar Module reported that Eagle was GO. The swing-arm now coming back to its fully retracted position as our countdown continues. T minus 4 minutes 50 seconds and counting. Skip Chauvin informing the astronauts that the swing-arm now coming back.

Nobody was talking a great deal. If something went wrong, they would all be implicated. Who would know which evil had entered the ripe oven of space technology? Aquarius took note of himself. Yes, his throat was dry.

We're now passing the four-minute 30-second mark in the countdown – still GO at this time. Four minutes 15 seconds – the Test Supervisor now has informed Launch-Vehicle Test Conductor Norm Carlson you are GO for launch. We're now hitting the four-minute mark. Four minutes and counting. We are GO for Apollo 11. We'll go on an automatic sequence at 3 minutes and 7 seconds. Three minutes 45 seconds and counting. In the final abort checks between several key members of the crew here in the Control Center and the astronauts, Launch Operations Manager Paul Donnelly wished the crew on behalf of the launch team 'Good luck and Godspeed.' Three minutes 25 seconds and counting. We're still GO at this time. We'll be coming up on the automatic sequence in about 10 or 15 seconds from this time. All still GO at this time. Neil Armstrong reported back when he received the good wishes, 'Thank you very much. We know it will be a good flight.' Firing command coming in now. We are on the automatic sequence. We're approaching the three-minute mark in the count. T minus 3 minutes and counting. T minus 3 – we are GO with all elements of the mission at this time.

He had his binoculars to his eyes. A tiny part of him was like a penitent who had prayed in the wilderness for sixteen days and was now expecting a sign. Would the sign reveal much or little?

. . . all is still GO as we monitor our status for it. Two minutes 10 seconds and counting. The target for the Apollo 11 astronauts, the moon. At lift-off we'll be at a distance of 218,096 miles away. Just passed the two-minute mark in the countdown. T minus 1 minute 54 seconds and counting. Our status board indicates that the oxidizer tanks in the second and third stages now have pressurized. We continue to build up pressure in all three stages here at the last minute to prepare for lift-off. T minus 1 minute 35 seconds on the Apollo mission, the flight that will land the first man on the moon. All indications coming in to the Control Center at this time indicate we are GO. One minute 25 seconds and counting. Our status board indicates the third stage completely pressurized. Eighty-second mark has now been passed. We'll go on full internal power at the fifty-second mark in the countdown. Guidance system goes on internal at 17 seconds leading up to the ignition sequence at 8.9 seconds. We're approaching the sixty-second mark on the Apollo 11 mission. T minus 60 seconds and counting. We have passed T minus 60. Fifty-five seconds and counting. Neil Armstrong just reported back, 'It's been a real smooth countdown.' We have passed the fifty-second mark. Forty seconds away from the Apollo 11 lift-off. All the second-stage tanks now pressurized. Thirty-five seconds and counting. We are still GO with Apollo 11. Thirty-five seconds and counting. Astronauts reported, 'Feels good.' T minus 25 seconds. Twenty seconds and counting. T minus 15 seconds, guidance is internal, 12, 11, 10, 9, ignition sequence start, 6, 5, 4, 3, 2, 1, zero, all engines running, LIFT-OFF. We have a lift-off, 32 minutes past the hour. Lift-off on Apollo 11.

But nobody watching the launch from the Press Site ever listened to the last few words. For at 8.9 seconds before lift-off, the motors of

Apollo-Saturn leaped into ignition, and two horns of orange fire burst like genies from the base of the rocket. Aquarius never had to worry again about whether the experience would be appropriate to his measure. Because of the distance, no one at the Press Site was to hear the sound of the motors until fifteen seconds after they had started. Although the rocket was restrained on its pad for nine seconds in order for the motors to multiply up to full thrust, the result was still that the rocket began to rise a full six seconds before its motors could be heard. Therefore the lift-off itself seemed to partake more of a miracle than a mechanical phenomenon, as if all of huge Saturn itself had begun silently to levitate, and was then pursued by flames.

No, it was more dramatic than that. For the flames were enormous. No one could be prepared for that. Flames flew in cataract against the cusp of the flame shield, and then sluiced along the paved ground down two opposite channels in the concrete, two underground rivers of flame which poured into the air on either side a hundred feet away, then flew a hundred feet further. Two mighty torches of flame like the wings of a yellow bird of fire flew over a field, covered a field with brilliant yellow bloomings of flame, and in the midst of it, white as a ghost, white as the white of Melville's Moby Dick, white as the shrine of the Madonna in half the churches of the world, this slim angelic mysterious ship of stages rose without sound out of its incarnation of flame and began to ascend slowly into the sky, slow as Melville's Leviathan might swim, slowly as we might swim upward in a dream looking for the air. And still no sound.

Then it came, like a crackling of wood twigs over the ridge, came with the sharp and furious bark of a million drops of oil crackling suddenly into combustion, a cacophony of barks louder and louder as Apollo-Saturn fifteen seconds ahead of its own sound cleared the lift tower to a cheer which could have been a cry of anguish from that near-audience watching; then came the earsplitting bark of a thousand machine guns firing at once, and Aquarius shook through his feet at the fury of this combat assault, and heard the thunderous murmur of Niagaras of flame roaring conceivably louder than the loudest thunders he had ever heard and the earth began to shake and would not stop, it quivered through his feet standing on the wood of the bleachers, an apocalyptic fury of sound equal to some conception of the sound of your death in the roar of a drowning hour, a nightmare of sound, and he heard himself saying, 'Oh, my God! oh, my God! oh, my God! oh, my God! oh, my God! oh, my God!' but not his voice, almost like the Italian girl saying *'fenomenal,'* and the sound of the rocket beat with the true blood of fear in his ears, hot in all the intimacy of a forming of heat, as if one's ear were in the caldron of a vast burning of air, heavens of oxygen being born and consumed in this ascension of the rocket, and a poor moment of vertigo at the thought that man now had something with which to speak to God – the fire was white as a torch and long as the rocket itself, a tail of fire, a face, yes now the rocket looked like a thin and pointed witch's hat, and the flames from its base were the blazing eyes of the witch. Forked like saw teeth was the base of the flame which quivered through the lens of the binoculars. Upwards. As the rocket keened over and went up and out to sea, one could no longer watch its stage, only the flame from its base. Now it seemed to rise like a ball of fire, like a new sun mounting the sky, a flame elevating itself.

Many thousands of feet up it went through haze and the fire feathered the haze in a long trailing caress, intimate as the wake which follows the path of a fingerling in inches of water. Trailings of cloud parted like lips. Then a heavier cloud was punched through with sudden cruelty. Then two long spumes of wake, like two large fish following our first fish – one's heart took little falls at the changes. 'Ahhh,' the crowd went, 'Ahhh,' as at the most beautiful of fireworks, for the sky was alive, one instant a pond and at the next a womb of new turns: 'Ahhh,' went the crowd, 'Ahhh!'

Now, through the public address system, came the sound of Armstrong talking to Launch Control. He was quieter than anyone else. 'Inboard cutoff' he said with calm in his voice.

Far in the distance, almost out of sight, like an all-but-transparent fish suddenly breaking into head and tail, the first stage at the rear of the rocket fell off from the rest, fell off and was now like a man, like a sky diver suddenly small. A new burst of motors started up, some far-off glimpse of newborn fires which looked pale as streams of water, pale were the flames in the far distance. Then the abandoned empty stage of the booster began to fall away, a relay runner, baton just passed, who slips back, slips back. Then it began to tumble, but with the slow tender dignity of a thin slice of soap slicing and wavering, dipping and gliding on its way to the floor of the tub. Then mighty Saturn of the first stage, empty, fuel-voided, burned out, gave a puff, a whiff and was lost to sight behind a cloud. And the rocket with Apollo 11 and the last two stages of Saturn V was finally out of sight and on its way to an orbit about the earth. Like the others he stayed and listened to the voices of the astronauts and the Capcom through the PA system.

PUBLIC AFFAIRS OFFICER: *At 3 minutes, downrange 70 miles, 43 miles high, velocity feet per second 9,300 feet per second.*

ARMSTRONG: *We've got skirts up.*

CAPSULE COMMUNICATOR: *Roger, we confirm. Skirts up.*

ARMSTRONG: *Tower is gone.*

CAPCOM: *Roger, tower.*

PAO: *Neil Armstrong confirming separation and the launch escape tower separation.*

ARMSTRONG: *Houston be advised the visual is GO today.*

On the way back to Cocoa Beach there was a monumental traffic jam, and Aquarius had time to look at objects by the road. There was a parked trailer with a twelve-foot inflated rubber rocket – it looked like a condom with a painted tip. Down its length ran a legend.

GOOD

LUCK

A

P

O

L

L

O

11

Montg.

Ala.

The radio was playing in the car. Fred Something-or-other from the Titusville Chamber of Commerce was talking fast. 'And when the folks who were visiting this launch here go home, I want them to tell everybody how beautiful it was from Titusville.'

'Folks,' said an announcer, 'get in on the Apollo 11 Blast-off Sale.' The radio had lost no time.

America – his country. An empty country filled with wonders.

Aquarius did not know how he felt. He was happy all afternoon and went surfboarding for the first time, not even displeased that it was harder than he thought to stand up.

In the evening he left Cocoa Beach to fly back to Houston where he would cover the trip to the moon and back. On the flight, everybody was drunk, and the hostesses were flip and hippy and could have been drinking themselves. The Southern businessmen were beaming.

In the late edition he brought with him, Aquarius read that the Reverend Abernathy together with a few poor families had watched the launching from the VIP area, after making a request of Dr Thomas O. Paine, Administrator of NASA, for special badges. 'If it were possible for us not to push the button tomorrow and solve the problems with which you are concerned, we would not push the button,' Dr Paine said.

Answered the Reverend Abernathy after the launch, 'This is really holy ground. And it will be more holy once we feed the hungry, care for the sick, and provide for those who do not have houses.'

Aquarius thought more than once of how powerful the vision of Apollo-Saturn must have been for the leader of the Poor People's Crusade. Doubtless he too had discovered that his feet were forced to shake. However, Aquarius was not yet ready to call this hallowed ground. For all he knew, Apollo-Saturn was still a child of the Devil. Yet if it was, then all philosophers flaming in orbit, the Devil was beautiful indeed. Or rather, was the Devil so beautiful because all of them, Johnsons, Goldwaters, Paines, Abernathys, press grubs and grubby Aquarius, were nothing but devils themselves. For the notion that man voyaged out to fulfill the desire of God was either the heart of the vision, or anathema to that true angel in Heaven they would violate by the fires of their ascent. A ship of flames was on its way to the moon.

4

The Greatest Week

If lift-off had just provided him with sensations not unlike the very mania of apocalypse, the ensuing flat heat of Houston on his return, the oppression of Nassau Bay off NASA Highway 1 this night in July came into his lungs with the smell of a burnt-out tire. He plummeted into a profound depression. Everything was wrong. Having sneered at the red velvet king-sized bed in the snappy motel off the highway, he now discovered that he was not to have it again. In the avalanche of reporters and corporation representatives who descended on the Texas plain near Manned Spacecraft Center to be near the collective brain which would pilot the flight, he, miserable sophisticate from the East, had somehow suffered a missed accommodation – as agreeable under the circumstances as a slipped disc – and was forced to camp out in the boondocks: in this case, a buried little basement apartment, one-room-with-kitchenette, furnished in convalescent spaced-out colors of dead gray, dull brown, dishwater white, and bargain-furniture green. You may be certain he missed the motel! The apartment had dingy Venetian blinds, plastic of course. Each time he tilted the slats to a new angle of admission of light, the blinds gave an *oooong* of sound which went without pause into the Graves Registration Department of his depressed psyche. Still he kept working the slats. At least it gave him power over some dimension of his environment.

It was one thing to be without ego; quite another to indulge this new modesty in mean quarters. Like many men who lived comfortably for years, he had always taken it for granted that he was superior to his surroundings and could dwell anywhere. Well, maybe one could still dwell anywhere with love, but loveless this week he was obliged to recognize that his basement apartment installed in an interlocking layout of ranch-style apartments, inner patios and underground garages was no place for him to thrive. Not on this job. If he had become a little obsessed with the meanings of a trip to the moon – going on now full attraction into its first night and second day and second night while he languished in his dun coop – if he had come to recognize that the more one brooded on this trip, the more fantastic it became, there was still the thundering and most depressing fact that it was a cancer bud for a journalist to cover. There were assignments which could make a reporter happy – he sometimes thought it would be impossible for a good quick-working novelist to be unable to write a decent piece about a political convention or a well-organized anarchy of the modern young. Give Aquarius a great heavyweight championship fight, and he would give you a two-volume work. There was so much to say. One's senses threatened to sear one's brain with excess of perception. The people at the center of such events nourished you with the tragicomedy of the traps they entered and sometimes escaped. But in NASA-land, the only thing open was the technology – the participants were so overcome by the magnitude of their venture they seemed to consider personal motivation as somewhat obscene. He had never before encountered as many people whose modest purr of efficiency apparently derived from being cogs in a machine – was this the perspective of the century to come or was this the end of a long and insane road?

He did not know, and the fact that he did not know depressed him further. Usually when one did a journalistic piece, the events fit in advance into some part of one's picture of the world. Most interesting about such events was the way they obliged you to make modest or delightful adjustments in the picture. Or even grim adjustments. But you did not have to contemplate throwing the picture away. Aquarius, now plucked up from the circus bonanzas and flaming cathedrals of lift-off, was in Houston dropped smack into the fact that the best way to do the rest of this damnable story was probably to go home and cover the works by television. He simply did not feel himself coming closer in Houston to the value or horror of the oncoming achievement; he did not see that there was any way to come closer. Occasionally, which is to say, five or six times a day, he would drive over to the Press Center at MSC on the other side of the highway, and skulk around the movie theater with the marcelled ceiling. When there was no press conference on – and usually there wasn't – he would look at a blank screen and listen to talk on the squawk box go back and forth between the Capcom and the spacecraft, the astronauts' voices wiped as clean of emotion as a corncob shucked of kernels. In the interim, distances increased. When he got back to Houston that first night, they were fourteen hours out, so their journey had already covered sixty-six thousand nautical miles! All that while he had been surf-boarding, celebrating, and then flying back from Melbourne, Florida, to Intercontinental Airport at Houston. Indeed the astronauts even covered another five thousand miles in the hour it took Aquarius to drive his rented car the fifty-odd miles from the airport through Houston, to Nassau Bay on the other side of town. Next afternoon, thirty hours into their flight, they were over one hundred and twenty thousand land miles out, and their speed had reduced to thirty-five hundred miles an hour. Their voices were of course the same. It did not matter whether they spoke from three miles away or one hundred and thirty-three thousand miles away, the hard peasant facts upon which Aquarius' education had been built, the consciousness that numbers were real units, hard as hours of work and miles one walked, now had to be discarded into some waste-nexus of the mind, some stink of the unusable like the Jersey flats. The real fact was that distance was now an abstract concept; men performing brave and heroic acts were communing over radio whatever the distance. The absence of simple human witness was the fact, not the distance. Sitting in the movie theater, he realized he would find out nothing he desired to know. Yet back of the movie theater was the Newsroom, now jammed with men, rows of desks fifty feet long, hundreds of typewriters, hundreds of phones, hundreds of soft-links in shirt-sleeves transmitting the information which came in at one end of the communication belt from all the publicity pipes in NASA and would soon go out on the other end after all the news-transmitters (human) had retyped the words. Milling through this matrix were forty or fifty men with portable tape recorders and microphones with radio call letters which looked like branding irons. They were always shoving these branding irons in your face, Australian blokes, Swiss blokes, Italian jokers, Japanese gentlemen, Norwegian asthma sufferers, French dudes, Swedish students, even Texas local radio station apprentices. They wanted to interview Aquarius. Aquarius, three weeks habituated to his new uncomfortable racket! Aquarius, Doctor of Rocketry! He said no.

He who had once thought he had only to get on all the radio and television available and he would be able to change the world, now wished only to flee this room with its hundreds of journalists, some so bored and aimless they even wished to interview him, he who now had nothing to say. The latest in the quintessential ironies of his life is that he had become a celebrity at precisely the hour when he ceased to desire it. Oh, what a depression!

ii

Events kept passing him by. A number of feature stories had been written in anticipation of the moment when the astronauts would pass behind the moon. Having turned the spacecraft around to go tailfirst, they would then fire the propulsion motor on the Service Module. That would brake the speed of the flight, and put them in orbit. Since they would be out of radio communication when it happened, no one would know for the next half hour if the burn had been successful, not until they came around the invisible side of the lunar sphere and their antennae were in unobstructed line with the earth again. It promised to offer excitement. Would the motor start? Or were there lunar emanations no physicist had ever conceived?

Actually, Aquarius was bored. Sitting in the Manned Spacecraft Center movie theater, he noticed that the Press was also bored, for few were listening to the squawk box. They all knew the burn would succeed and Apollo 11 would go into proper orbit. There seemed no question of failure, and indeed the burn and the re-acquisition of radio communication went on schedule. Aquarius could detect surly traces in himself, as if he were annoyed with the moon – it should not be so simple to trespass her zones. He was, of course, no longer thinking in any real way – what passed for thought were the dull whirrings of his depression, about as functional to real intellectual motion as the turning over of a starter when the battery is almost dead. In fact he could not forgive the astronauts their resolute avoidance of a heroic posture. It was somehow improper for a hero to be without flamboyance as if such modesty deprived his supporters of any large pleasure in his victories. What joy might be found in a world which would have no hope of a Hemingway?

Or nearest matters first, of a Joe Namath, or Cassius Clay, Jimmy Dean, Dominguin? – it was as if the astronauts were there to demonstrate that heroism's previous relation to romance had been highly improper – it was technology and the absence of emotion which were the only fit mates for the brave. Yesterday, or one of the days which had already become interleaved in the passage of time at Dun Cove, he had read a newspaper story where Armstrong's wife, Jan, had been quoted: 'What we can't understand, we fear.' Even the ladies of brave men spoke like corporation executives on this job. His heart went dull at the thought of the total take-over implicit in the remark, so neat, so ambitious, so world-vaulting in its assumption that sooner or later everything would be understood – 'I paid a trip to death, and death is a pleasant place and ready for us to come in and renovate it.' Abruptly Aquarius realized that for years he had thought of death as located in the milieu of the moon, as if our souls, those of us who died with one, might lift and rise, be free of the law of gravity and on trajectory to the satellite of the craters. Yes, wouldn't it be in the purview of the Wasp, damn corporate Wasp, to disturb the purlieus of the dead? He did not know. His thoughts were always furthest out when he was most depressed, as though like a bird half drowned, the only way to lift was by the wildest beating of wings.

The real heroism, he thought, was to understand, and because one understood, be even more full of fear at the enormity of what one understood, yet at that moment continue to be ready for the feat one had decided it was essential to perform. So Julien Sorel had been brave when he kissed Madame de Renal, and Jimmy Dean been brave in *Rebel Without a Cause*, and Namath when he mocked the Baltimore Colts knowing the only visions he would arouse in his enemy were visions of murder. So had Cassius Clay been brave – to dare to be rude to Liston – and Floyd Patterson brave to come back to boxing after terrible humiliation, and Hemingway conceivably brave to continue to write in short sentences after being exposed to the lividities of the literary world.

But the astronauts, brave men, proceeded on the paradoxical principle that fear once deposed by knowledge would make bravery redundant. It was in the complacent assumption that the universe was no majestic mansion of architectonics out there between evil and nobility, or strife on a darkling plain, but rather an ultimately benign field of investigation which left Aquarius in the worst of his temper.

Next morning came the news of Teddy Kennedy's accident at Chappaquiddick. Dead was the young lady who had been driving with him. How subtle was the voice of the moon. Aquarius remembered a speech Kennedy had given two months earlier at Clark University in Worcester. Mrs Robert H. Goddard, widow of the father of American rocketry, had been there, and Buzz Aldrin as well. Kennedy had urged that future space funds be moved over to such problems as poverty, hunger, pollution and housing. The chill which came back from NASA was as cold as the architecture at the Manned Spacecraft Center. 'We won't be including this item in the daily news reports we send up to the Apollo 10 astronauts on their voyage to the moon,' Thomas O. Paine, administrator of NASA had said. Now the reverberations of this accident at Chappaquiddick went off in Aquarius' brain. As happens so often when a motive is buried, Aquarius felt excitement around the hollows of his depression. For if the blow to the fortunes of the Kennedys was also a blow to one hundred interesting possibilities in American life, if the accident was of such benefit to Richard Nixon that the Devil himself if he had designed the mishap (which is what every liberal Democrat must secretly believe) could have awarded himself a medal for the artistry, yet there was at least a suggestion that the moon had thought to speak. Perhaps that was why there was still a trace of stimulation in the gloom – magic might not be altogether dead.

The day went by, a cloudy day in southeastern Texas. From time to time, Aquarius checked in at the Press Center. Excitement was now divided between Kennedy and the moon. Or was Kennedy even more interesting? The separate phases of the preparation for landing were certainly without high tension. Indeed the Lem even undocked from the Command Module while both were behind the moon. When they came around and signal was picked up again, the voice of Armstrong came over the squawk box. 'The Eagle has wings,' he said, or was it Aldrin who said it? – there was discussion on this for the remark was universally quotable. Yet the happy buzz of conversation among the reporters at the thought of an oncoming climax was dampened considerably by the dialogue which followed:

CAPCOM: . . . *Coming at you with a DOI pad. 101361407981 minus 00758 plus all balls plus 00098 plus corrections 00572 perigee plus 00085 00764 030000293 986 minus*

00759 plus all balls plus 00090 rest of the pad is NA. Stand by on your read-back.
If you are ready to copy the PDI data, I have it for you. Over.

ALDRIN: . . . *Go ahead with the PDI.*

CAPCOM: *Roger. PDI pad, PIG 102330436 0950 minus 00021 182287000 plus 56919 –*

So one got ready for the climax of the greatest week since Christ was born. An hour and twenty minutes later, the Lem having flown around the moon and gone behind it again, the braking burn for the Descent Orbit Initiation would be begun in radio silence. An hour later the final ignition for the final descent would commence. Aquarius, bereft of personal radar or gyroscope, bereft even of the sniff-sensors of his poor journalistic nose, wandered from point to point in the Press Center, rushed back to Dun Cove to look at color television – there were no color sets in the Press Room – then, bored with listening to commentators, and finally incapable of witnessing the event alone, went back to the movie theater and settled in with about a hundred other reporters for the last half hour.

Phrases came through the general static of the public address system. 'Eagle looking great. You're go,' came through, and statements of altitude. 'You're go for landing, over!' 'Roger, understand. Go for landing. 3000 feet.' 'We're go, hang tight, we're go. 2000 feet.' So the voices came out of the box. Somewhere a quarter of a million miles away, ten years of engineering and training, a thousand processes and a million parts, a huge swatch out of twenty-five billion dollars and a hovering of machinery were preparing to go through the funnel of a historical event whose significance might yet be next to death itself, and the reporters who would interpret this information for the newsprint readers of the world were now stirring in polite if mounting absorption with the calm cryptic technological voices which came droning out of the box. Was it like that as one was waiting to be born? Did one wait in a modern room with strangers while numbers were announced – 'Soul 77-48-16 – you are on call. Proceed to Staging Area CX – at 16:04 you will be conceived.'

So the words came. And the moon came nearer. '3½ down, 220 feet, 13 forward, 11 forward, coming down nicely. 200 feet, 4½ down. 5½ down. 160, 6½ down. 5½ down. 9 forward. 5 percent. Quantity light. 75 feet. Things looking good. Down a half. 6 forward.'

'Sixty seconds,' said another voice.

Was that a reference to fuel? Had that been the Capcom? Or was it Aldrin or Armstrong? Who was speaking now? The static was a presence. The voice was almost dreamy. Only the thinnest reed of excitement quivered in the voice.

'Lights on. Down 2½. Forward. Forward. Good. 40 feet. Down 2½. Picking up some dust. 30 feet, 2½ down. Faint shadow. 4 forward. 4 forward. Drifting to the right a little. 6 . . . down a half.'

Another voice said, 'Thirty seconds.' Was that thirty seconds of fuel? A modest stirring of anticipation came up from the audience.

'Drifting right. Contact light. Okay,' said the voice as even as before, 'engine stop. ACA out of detente. Modes control both auto, descent engine command override, off. Engine arm, off. 413 is in.'

A cry went up, half jubilant, half confused. Had they actually landed?

The Capcom spoke: 'We copy you down, Eagle.' But it was a question.

'Houston, Tranquility Base here. The Eagle has landed.' It was Armstrong's voice, the quiet voice of the best boy in town, the one who pulls you drowning from the sea and walks off before you can offer a reward. The Eagle has landed – it reached the Press. They burst into applause. It was the kind of applause you used to hear in the packed film houses of the Thirties when the movie came over the hill of the last reel, and you heard the doctor say the star would live after the operation. Now, a small bedlam of actions began, some of the Press sprinting from the room – could they pretend it was necessary to phone the City Desk? – others talking to each other in babble, others still listening to the squawk box as technology took up again. A few minutes later: 'Eagle. Houston. You loaded R2 wrong. We want 10254.'

'Roger – that is V horizontal 5515.2.'

'That's affirmative.'

Aquarius discovered he was happy. There was a man on the moon. There were two men on the moon. It was a new feeling, absolutely without focus for him. If he felt a faint graveling on the surface of this sentiment, a curdle of emotional skin which formed from his effort to advance heroes he could not find altogether admirable, still he knew he had been dislocated as profoundly by the experience as the moment he learned in the fathers' waiting room at the hospital that his first child had indeed and actually just been born. 'Well, think of that,' he had said. What a new fact! Real as the presence of immanence and yet not located at all, not yet, not in the comfortable quarters one afforded for the true and real facts of the life of the brain.

'Let's go interview the wives,' someone suggested. And Aquarius, grateful for an opportunity to try a few journalistic tools, was happily off and away from the reign of the Capcom, the squawk box, and the abstract incubated existence of brand-new Tranquility Base.

iii

The Armstrong house was modest, with a high-pitched roof of brown shingles. It was a house like half a million other houses in suburbs combining modern and brand-new traditional style. It had hints of an English country inn, for it was a dark-colored warren with small windows and long eaves. Yet the house was situated on a street whose curve had come from no meandering cow but from favorable indices on graphs which showed the relation of income to cost for planned curved-development streets as opposed to planned straight-development streets. El Lago – the name of this suburb – like those others named Kingston and Timber Cove and Nassau Bay – was a soft checkerboard of carefully bent little avenues which ran at reasonable approximations of right angles into other paved prospects, a street occasionally dead-ending, a street just as occasionally completing a full circle. The realty layout-computer in its wisdom for random play in home-road curvature had designed the layout logic so comprehensively, so ready to take into account the variety of desire-factors expressed by consumer dweller-groups oriented in at these precise income-purchase levels, that the effect – what a blow to the goodwill of the progressive designer who had doubtless opted for just once let's have something better! – was as agreeable and sterile to the eye as a model department store living room for brides on a medium-high budget layaway.

What a curious scene then! This moderately undulated street of angled plots and recently constructed private homes, this curved asphalt bordered by trees of the same year of growth, with vista of cars parked in garages, air conditioners, lawn mowers, sprinklers and bicycles, a street in which five children and two adults on foot might be seen on any average hour of the day, what a shock down the block to the habituated eye to see that gaggle of Press all straining at a rope, TV men and still photographers in their customary war for position with word-men – the crush now fortified by the more curious neighbors, Texas grandmothers in the main with gray-greenish eyes, tight not unintelligent mouths, small-town grits in the anticipation, and a real Texas bone-and-leather use of elbows to fight their leading toe in for position against this invasion of hippie journalist gogglers and foreigners with cameras, beards, sideburns, Nehru jackets, turtlenecks, love beads, medallions, shades, tape recorders and foreign tongues, Japs talking to Japs, Germans to Germans, an Italian to other Italians and all the Scandinavians come to NASA-land in El Lago. 'I'm an American!' cried out one full vigor of a gray-haired Texas female being pushed off the frontier by one ham-hock and handlebar of an English ruddy.

There would have been small mayhem, but Jan Armstrong came out then, escorted by an official who treated her with the kind of gravity reserved for the Pope's sister. She was laughing and smiling, she was obviously very happy as she stepped on a raised platform behind the rope. The platform was about a yard square and a foot high, set out for her probably by the television men, and it may have been the color of its candy-pink carpet reflecting pink and violet and delicate rose hues to her healthy skin, but she was an attractive sight on this overcast Sunday afternoon. She was a woman one would not normally have thought of as beautiful, her hair was almost gray and close-cropped, her mouth while full and strong was without coquetry or that hint of duplicity so attractive in a woman for the implicit suggestion that only a real man could ever set her straight. No, her mouth spoke of the stubborn and the steadfast, and she was dressed in no remarkable fashion, she had on a white blouse buttoned at the neck and an orange-red skirt. She had a Scotch nose, strong, not small, not delicate in nostril, the wings cut with determination, the tip showing a hint of two lobes, she had in all the sort of face one sees in the best of small-town schoolteachers or librarians, that complete statement in the features of deep and dedicated strength, so she possessed in this hour a beautiful face, it was radiant – the word could finally be employed – the face was utterly separated from the planned street and the media men surrounding her with their microphones, the face lifted up to that moment in the past when she had directed her strength and her will to one goal, and the goal was now fulfilled – few faces are more beautiful than the dedicated when their deepest hour is in, when the plan utters its first word aloud, and the word is 'yes.' So no question could bother her at this instant. If she had a reputation for being shy, grim, a swimming instructor, a phys ed teacher, a churchgoer, a conformist, a humorless embodiment of the space frontier in female form, she was now on top of it all, she was in rollicking good humor from the release of that safe landing.

'Will you let the children stay up and watch the moon walk tonight?' asked a journalist.

A devil came into her eye. 'I don't care for what they do,' she said with happy idiom, a grammar from the universal interior of the nation, not Midwestern, not Texan, by now from anywhere in there.

The questions went on. She laughed happily at their absurdity. She could not make a mistake. 'Is this the greatest moment of your life?' asked one of the voices.

'No, sir,' said her shining face, 'when I was married, it was the greatest moment of my life.'

They had known each other for three years before they had their first date, they had lived soon after their marriage in a cabin in the mountains with no plumbing, he would wave the wings of the test planes he flew as he burned through the sky over their house. It was a marriage of piano duets and long solo flights in gliders; the death of their only daughter, age three, came from a tumor of the brain. Aquarius hated to comprehend marriages by such contradictory details picked up on the fly, but these details promised to fit, they spoke of the serious physician's daughter from Illinois, and the dream-absorbed boy with the lonely face of early photographs, the boy whose family moved a dozen times when he was young in Ohio. He had worked for a pharmacy at forty cents an hour to earn the nine dollars an hour for flying lessons. Would the relation of husband and wife to each other be so very different from Armstrong's relation to the sky? Everything about Armstrong suggested that he would be happiest in the sky, that surf of space where intimations of a language few could speak might hover on the changing of a cloud.

'Are you pleased with the Sea of Tranquility as a place to land?' asked a questioner.

She liked Tranquility Base. It was obvious the word was agreeable. For those who have been living in dread, tranquility is grace, the very decency of ecstasy.

'What are you having for dinner tonight? Space food?'

She threw back her head like a mare so happy with the day she can support any rider, 'No, sir,' she said and left it there, proper for NASA and the team, but the glint in her eye had its own look, 'We ain't quite so square as you think us, Mister Reporter,' said the unvoiced look.

Another few questions and the NASA representative at her elbow returned her to the house. The Press was off to cars and TV vans, the ride was on to find the Collins home and the Aldrin house, each a block from the other, but back in Nassau Bay itself, back in another soft checkerboard of curves at right angles to the intersection of other curves, a suburb built even closer to men's occupations in buildings with windowless walls.

Pat Collins was another woman in another state. She was conventionally good-looking with the sort of attractive and competent features one finds in secretaries to important executives or in the woman who is supervisor of hostesses for an airline. She had black hair pinned up high and green-blue eyes bright as the lights in a valuable stone. Her arms and legs were very slim, and she smiled a lot as she talked to the Press. She seemed flustered, and not unravaged by the tension of the last few days. Her remarks were polite, enthusiastic, hardworking. She said that by the time they touched down on the moon, she was cheering, but in fact she had the glitter of an actress who is loyal to the company and loyal to the production, and so will mouth the lines clearly even if the theater is half-empty, the play is falling apart, and the cast will be given their notice in another night. She was loyal to what was demanded of her, but the strain was showing – of the three wives she had had the most difficult relation to the event. She was obliged to suffer with the other wives as an equal among equals, share the agonies and the jubilation of the landing with them, yet in fact her husband was not landing, he was up in the air all alone, he would be alone for another day or more, sometimes alone on the back side of the moon while the others explored ground no living man had stepped upon. She had a face which was obviously not without ambition, one of those faces which exhibit no outrageous vice but nonetheless want the best for themselves, not ruthless so much as ready to commit oneself to the partner and work all the way to go all the way. Now her husband had gone 99.9 percent of the way. If it was secretly hard on him, it would have to be twice as hard on her. What a role to play! The interviewers asked her again and again, 'Do you mind that your husband is not landing with the others?' And again and again in a voice which was using the reserves of her good looks she kept replying in a tone determinedly bright (and so bumped over to the edge of the haggard) that she didn't mind a bit and knew Mike didn't either. The laws of propriety at NASA went as deep as the regulations at a hospital – woe to any astronaut or wife who uttered in public any sentiment which would fail to bore the expectations of fifty million viewers. There was a true and proper standard of behavior for every public situation in which they might find themselves. A clear rule of measure: do not under any circumstance say anything more interesting than Richard Nixon would say in the same situation. That was the clubhouse rule; Pat Collins was obeying it. Since she also had the look of a woman who must have real flashes of Irish beauty when relaxed, the expostulations of complete happiness made her remarks so ordinary that Aquarius discovered afterward he had not worked to take a note.

Down the block was the Aldrin house, a structure of pale orange brick with another steep roof. There was a wait for Mrs Aldrin. It had begun to rain out of an uneven lead-colored hesitant sky – a few drops would tattoo on the big Texas leaves of the modest suburban trees and then halt, then rain a little harder, stop again. It was thought at first that the delay was due to uncertainty whether to begin the interview under such conditions – after a while a rumor circulated that Mrs Aldrin was primping for the Press. It made sense to Aquarius. She had wanted to be an actress – she had worked at having a career for a period, had made the rounds against the objections of her father. Having been married to three women who were actresses at one time or another, one of them even a modest movie star who had her career much interrupted for such marrying, Aquarius was able to make broken-hearted jokes about the woes of any man so foolish as to smash into the devotions of an acting career by an act of marriage. As a consequence he was naturally interested in Mrs Aldrin. She, like Jan Armstrong, had had a long courtship. He was obliged to recognize on looking at a newspaper photo that she looked much like her husband, as indeed Jan Armstrong looked like her mate, and Pat Collins like Mike Collins. He did not understand these marriages of people who looked alike and courted each other for years – he did not know if the delay of proper people who looked alike came from deep respect for marriage, or was rather excessive caution to make certain one's narcissism would find the cleanest mirror. He did not know. It seemed to him he had always gotten married in a hurry to women who were remarkably different, except for his final inability to get along with them. Gloomy as the weather was Aquarius. From time to time, like the memory of a telegram whose news was so awful one kept circling the fact of it, came back the simple unalterable fact that Teddy Kennedy had been in a bad accident and a girl had drowned and he had not reported it until morning.

Joan Aldrin had blond hair, she was a big woman with generous features, nose, teeth, mouth, there, all there, but finally she was all eyes. They were the large expressive soulful instruments of a woman who had a real and intense awareness of her stage, which is to say a sense of the air she offered, the way it was received, the space between. She made a good entrance beneath an umbrella thoughtfully held over her by the man who served as guide from the house to a roped-off space between the trees, she was a lady who transmitted palpable gratitudes for courtesies rendered, she had the ability to exercise the air, but not as a flirt or a sexual provocation, rather as a tragedienne one instant, a comedienne the next, she had the quick-changing vital bounce of a woman who might have made a reasonable career in musical comedy – she had as she spoke the slightly slow withheld timing, the meaningless but tasty syncopation that women who belt out a song give to a dull line. So, as she spoke, it was fun to listen.

'What were you doing when they landed?'

'Well, I was holding onto the wall. I was praying,' she said in a loud and syncopated whisper.

She was at once utterly serious, and camping it up. A part of her had been in agonies of suspense which went right into the agonies and deepest attachments of her marriage – another part of her, droll as the humor of her full nose had been obliged to see herself – 'Here you are, big girl, holding onto the wall at a time like this!'

The interview went on. Aquarius monitored it with a mild part of his brain. There was talk about one activity, then another – what the plans were for watching the moon walk – would the children stay up, so forth. The mood was sluggish. The Press had interviewed two wives already.

She was too much of a performer to come in third in a three-horse race. 'Listen,' she cried out suddenly in a big voice, waking up the Press. 'Aren't you all excited?' She looked around coquettishly, carefully, as if to measure what employment could be made of an audience as super-sophisticated and sodden as this. Then it came over her, as it had come over everyone else from time to time. There were men on the moon.

'They did it!' she shrieked happily. 'They did it!'

After she had gone back inside, there was a vacuum. Her vitality was gone. One wouldn't have minded more of her.

Aquarius was left with his gloom. It was finally a dubious male occu-pation to interview the wives of men who were heroes for the day. He was depressed as he walked away thinking of his own wife, and his own marriage, now deteriorating – what work to be obliged to look in on other marriages in their hour of triumph!

iv

That night, the walk on the moon had been scheduled to begin long after midnight, so plans had been laid for late moon-watching parties. But the astronauts, to no one's surprise, were in no mood to sleep, and the moon walk was rescheduled for eight in the evening. Yet, this once, the astronauts were not on time.

Waiting in the movie theater, the Press was in a curious state of min-gled celebration and irritation. It was hard not to feel like a fool. They were journalists, not movie critics, and tonight they would be taking notes on the events which transpired upon a video screen. Of course, the climax of days of the most difficult kind of reporting was finally at hand, but it was a little as if one's nervous system had been appropri-ated and the final shake would take place in somebody else's room.

The psychology of journalists is not easy to comprehend – they scurry around like peons, they have the confidence of God. Over the years they develop an extraordinary sense of where the next victory is located. If a man gives a press conference and is not surrounded by reporters when it is over, he need not wonder how his fortunes are moving – the reporters have already told him. It is for this reason journalists pick up the confidence that they shape events – in fact they are only sensors in the currents of the churn, Venturi tubes to give you the speed of the history which passes. Nonetheless, there is no psychological reality like a man's idea of himself. Even if a writer has lost the best reaches of his talent by putting out facts for years which have been stripped of their nuance – writing newspaper stories in short – still he retains an idea of himself: it is that his eye on an event may be critical to correct reportage of it. Now put five hundred reporters in a room to report on the climax of an event 'equal in importance to that moment in evolution when aquatic life came crawling up on the land,' and put a movie screen in front of them, and a television transmission on that screen which is not only a pioneer effort in communication from a satellite one good quarter of a million miles away, but is also, you may be sure, wildly out of focus. Reporters wear eyeglasses in order not to miss the small print – bad focus on the screen puts a new injury right inside the wound of the previous injury. Something in them reverted. Watching the mooncast, they were like college kids on Friday night in the town movie house – one never knew what would make their laughter stir next, but their sense of the absurd was quick and furious. Like college students who roar with disgust because by God they were being trained to run a supposedly reasonable world with highly reasonable skills, and yet the fools who made this movie had the real power, so the Press took the mooncast on its own literal terms of spectacle – where it was good as spectacle they loved it, where it was poor they mocked.

But let us take it from the start. The screen was dark when the voices began, and since it stayed without image for many minutes while one heard the voices of the astronauts working to get ready, a strain developed in the audience. Would the picture ever come on tonight, or had something gone wrong?

Then one learned from the Public Affairs Officer that the Portable Life Support Systems were working – the astronauts were now connected by umbilical tubes to the big white box on their back, that box which could cool them, clear the fog from their helmets, give them oxygen to breathe, and absorb the wastes of their exhalation. But the minutes went by. There was no image on the screen. Oxygen was being used. They had only a few hours of Life Support in the system – would they be obliged to use it overcoming the difficulties of opening the hatch? Hoots and a hum of restlessness worked through the theater. The journalists were nervous. That rare hysteria which is generated by an inability to distinguish between the apocalyptic and the absurd was generating. What if – assuming they could actually see something – what if Armstrong were to take a step on the moon and simply disappear? Whatever would one do in this theater? The event would be a horror to watch if tragedy occurred; yet it would be a humiliation if it all went on schedule.

A cheer not unmixed with mockery came at the announcement at 9:40 in the evening that the hatch was open. Still no image on the screen. Now followed long incomprehensible instructions back and forth, talk of window clanks and water valves, high-gain antenna and glycol pumps. Out of all this, quiet exhortations from Aldrin to Armstrong. Through the words emerged the realization that Armstrong, made twice bulky by his space suit and the Portable Life Support System on his back, was trying to push through the open hatch of the Lem out onto the small metal porch which led to the ladder which in turn he could descend to the moon ground. It was obviously a very tight fit to get through the hatch. As Aldrin gave instructions there was an inevitable suggestion of the kind of dialogue one hears between an obstetrician and a patient in the last minutes before birth.

ALDRIN: *Your back is up against the (garbled.) All right, now it's on top of the DSKY. Forward and up, now you've got them, over toward me, straight down, relax a little bit.*

ARMSTRONG: *(Garbled)*

ALDRIN: *Neil, you're lined up nicely. Toward me a little bit, okay down, okay, made it clear.*

ARMSTRONG: *To what edge?*

ALDRIN: *Move. Here roll to the left. Okay, now you're clear. You're lined up on the platform. Put your left foot to the right a little bit. Okay that's good. Roll left.*

The Press was giggling. Sanctimony at NASA was a tight seal. A new church, it had been born as a high church. No one took liberties.

Now, two of the heroes of NASA were engaged in an inevitably comic dialogue – one big man giving minute adjustments of position to another. The Press giggled.

Armstrong spoke out suddenly. 'Okay, Houston, I'm on the porch.'

The audience broke into applause. There was mockery, as if the cavalry had just come galloping down the ridge.

A few minutes went by. Impatience hung in the air. Then a loud bright cheer as a picture came on the screen. It was a picture upside-down, blinding in contrast, and incomprehensible, perhaps just such a kaleidoscope of shadow and light as a baby might see in the first instants before silver nitrate goes into its eyes. Then, twists and turns of image followed, a huge black cloud resolved itself into the bulk of Armstrong descending the ladder, a view of confusions of objects, some rough-hewn vision of a troglodyte with a huge hump on his back and voices – Armstrong, Aldrin and Capcom – details were being offered of the descent down the ladder. Armstrong stepped off the pad. No one quite heard him say, 'That's one small step for a man, one giant leap for mankind,' nor did anyone quite see him take the step – the TV image on the movie screen was beautiful, but still as marvelously abstract as the branches of a tree, or a painting by Franz Kline of black beams on a white background. Nonetheless, a cheer went up, and a ripple of extraordinary awareness. It was as if the audience felt an unexpected empathy with the sepulchral, as if a man were descending step by step, heartbeat by diminishing heartbeat into the reign of the kingdom of death itself and he was reporting, inch by inch, what his senses disclosed. Everybody listened in profound silence. Irritation was now gone as Armstrong described the fine and powdery substance of the surface: 'I can see the footprints of my boots and the treads in the fine sandy particles.' Every disclosure for these first few minutes would be a wonder. If it would have been more extraordinary to hear that the moon had taken no imprint in soft powder, or the powder was phosphorescent, still it was also a wonder that the powder of the moon reacted like powder on earth. A question was at least being answered. If the answer was ordinary, still there was one less question in the lonely spaces of the human mind. Aquarius had an instant when he glimpsed space expanding like the widening pool of an unanswered question. Was that the power behind the force which made technology triumphant in this century? – that technology was at least a force which attempted to bring back answers from questions which had been considered to be without answers?

The image was becoming more decipherable. As Armstrong moved away from the ladder in a hesitant loping gait, not unlike the first staggering steps of a just-born calf, he called back to Mission Control, 'No trouble to walk around,' but as if that were too great a liberty to take with the feelings of the moon, he came loping back to the ladder.

Activities went on. There were photographs to take, descriptions of the appearance of the rocks, of the character of the sun glare. One of Armstrong's first jobs was to pick up a sample of rock and put it in his pocket. Thus if something unforeseen were to occur, if the unmentionable yak or the Abominable Snowman were to emerge from a crater, if the ground began to rumble, if for any reason they had to reenter the Lem and take off abruptly, they would then have the chance to return to earth with at least one rock. This first scoop of moon stone and moon dust was called the contingency sample, and it was one of Armstrong's first tasks, but he seemed to have forgotten it. The Capcom reminded him subtly, so did Aldrin. The Capcom came back again: 'Neil, this is Houston. Did you copy about the contingency sample? Over?'

'Rog,' said Armstrong, 'I'm going to get to that just as soon as I finish this picture series.'

Aldrin had probably not heard. 'Okay,' he asked, 'going to get the contingency sample now, Neil?'

'Right!' Armstrong snapped. The irritability was so evident that the audience roared with laughter – don't we laugh when we glimpse a fine truth and immediately conceal it? What a truth! Nagging was nagging, even on the moon.

The television image was improving. It was never clear, never did it look any better in quality than a print of the earliest silent movies, but it was eloquent. Ghosts beckoned to ghosts, and the surface of the moon looked like a ski slope at night. Fields of a dazzling pale ran into caverns of black, and through this field moved the ghost of Armstrong. There were moments when one had the impression it was possible to see through him. His image was transparent.

Aldrin descended the ladder, then jumped back on the lowest rung to test his ability to return to the Lem. The abruptness of the action broke the audience into guffaws again, the superior guffaw a sophisticate gives to a chair creaking too crudely in a horror movie. Now two ghosts paraded about, jogging forward and back, exchanging happy comments on the new nature of hopping and walking, moving faster than a walk but like much-padded toddlers, or overswathed beginners on skis. Sometimes they looked like heavy elderly gentlemen dancing with verve, sometimes the sight of their boots or their gloves, the bend of their backs setting up equipment or reaching for more rocks gave them the look of beasts on hindquarters learning to think, sometimes the image went over into negative so that they looked black in their suits on a black moon with white hollows, sometimes the image was solarized and became positive and negative at once, images yawing in and out of focus, so the figures seemed to squirt about like one-celled animals beneath a slide – all the while, images of the Lem would appear in the background, an odd battered object like some Tartar cooking pot left on a trivet in a Siberian field. It all had the look of the oldest photographs of expeditions to the North Pole – there was something bizarre, touching, splendid, and ridiculous all at once, for the feat was immense, but the astronauts looked silly, and their functional conversations seemed farcical in the circumstances.

'What did you say, Buzz?'

'I say the rocks are rather slippery.'

Huge guffaws from the audience. When the flag was set up on the moon, the Press applauded. The applause continued, grew larger – soon they would be giving the image of the flag a standing ovation. It was perhaps a way of apologizing for the laughter before, and the laughter they knew would come again, but the experience was still out of register. A reductive society was witnessing the irreducible. But the irreducible was being presented with faulty technique. At that they could laugh. And did again and again. There were moments when Armstrong and Aldrin might just as well have been Laurel and Hardy in space suits.

The voice of Collins came into the public address system. He had been out of radio contact for almost an hour during his trip around the back of the moon, so he did not know how the Extra Vehicular Activity was proceeding. He had left communication before Armstrong had reached the lunar surface. Now he asked, 'How's it going?'

CAPCOM: *Roger. The EVA is progressing beautifully. I believe they are setting up the*
flag now.
COLLINS: *Great.*

The audience laughed at this hard pea of envy beneath twenty mattresses of NASA manners.

CAPCOM: *I guess you're about the only person around that doesn't have TV coverage*
of the scene.
COLLINS: *That's right. That's all right. I don't mind a bit.*

Now, the Press roared.

COLLINS: *How is the quality of the TV?*
CAPCOM: *Oh, it's beautiful, Mike. Really is.*
COLLINS: *Oh, gee, that's great.*

The video continued, the astronauts worked on styles of gait, ordinary walking, half-run, kangaroo hops. There was a sense of the
astronauts' happiness as they loped about, and now a delicate envy,
almost tender in its sensibility, went to them from the crowd. There was
finally something marvelous. This old-fashioned indistinct movie of
comedians in old-fashioned suits was in fact but a cover upon the curious happiness everyone was feeling. It was the happiness which comes
from a wound. For with the pain, and there was pain in the thought of
the moon – so private a body to the poet buried in every poke of a
head – the moon being now invaded, there was also the happiness that
accompanies the pain, for the landing was a straight-out wound to every
stable disposition of the mind. Yet a wound in that period when we do
not know which flesh is severed forever and what is recuperable is an
hour of curious happiness. Change may give life. So the world was
watching the loping bumbling skittering low-gravity movements of
these men with the kind of concentration we offer to the study of our
own wound. Something in the firmament was being operated upon.

Well, the flag was up. The Capcom spoke. He asked the astronauts
to stand in view of the camera, then announced that the President of
the United States wanted to say a few words.

ARMSTRONG: *That would be an honor.*

CAPCOM: *Go ahead, Mr President, this is Houston. Out.*

It had been announced in advance that the President would speak to the astronauts, but the liberal portion of the Press groaned, to be answered by a pattering of stiff hands from the patriots in the room.

PRESIDENT NIXON: *Neil and Buzz, I am talking to you by telephone from the Oval Room at the White House. And this certainly has to be the most historic telephone call ever made.*

Large jeers from the audience. The most expensive telephone call ever made! Stentorian hand clapping.

PRESIDENT NIXON: *I just can't tell you how proud we all are of you. For every American this has to be the proudest day of our lives. And for people all over the world, I am sure they too join with Americans in recognizing what a feat this is. Because of what you have done, the heavens have become a part of man's world. And as you talk to us from the Sea of Tranquility, it inspires us to double our efforts to bring peace and tranquility to earth. For one priceless moment in the whole history of man, all the people on this earth are truly one. One in their pride in what you have done. And one in our prayers that you will return safely to earth.*

Every word had its function. It could be said that the psychology of machines begins where humans are more machinelike in their actions than the machines they employ.

'Thank you, Mr President,' answered Armstrong in a voice not altogether in control. What a moment for Richard Nixon if the first tears shed on the moon flowed on the consequence of his words! 'It's a great honor and a privilege,' Armstrong went on, 'to be representing not only the United States, but men of peace of all nations.' When he finished, he saluted.

Some of the crowd jeered again. The image of Nixon faded on home TV screens, his voice was gone from the theater. The moon walk continued. In fact, it was not half done, but the early excitement had ebbed in this last play of rhetoric – the minds of the Press had gone on to the question of whether Nixon was considering it politically advantageous to support a future program of space. As the astronauts continued to walk, to hop, to flit and to skip from one vale of moon ground to another, as the experiments were set out and the rocks picked up, so the temper of the audience shifted. It was a Twentieth Century audience when all was said, and quick in its sense of fashion. By an hour and a half of the moon walk they were bored – some were actually slipping out. All over the room was felt the ubiquitous desire of journalists for the rescue of a drink. Boredom deepened. Now the mood was equal to the fourth quarter of a much anticipated football game whose result had proved lopsided. Now it looked as if rookies were out on the chill field running fumbles back and forth. More and more reporters departed. Even Aquarius left before the end.

V

It was the event of his lifetime, and yet it had been a dull event. The language which now would sing of this extraordinary vault promised to be as flat as an unstrung harp. The century had unstrung any melody of words. Besides – the event was obdurate on the surface and a mystery beneath. It was not at all easy to comprehend. Like an adolescent married before he could vote, the congratulation, 'You're a married man,' had no reality to the brand-new groom. So America and the world would be in a round of congratulations – we had landed a man on the moon. The event was so removed, however, so unreal, that no objective correlative existed to prove it had not conceivably been an event staged in a television studio – the greatest con of the century – and indeed a good mind, product of the iniquities, treacheries, gold, passion, invention, deception, and rich worldly stink of the Renaissance could hardly deny that the event if bogus was as great a creation in mass hoodwinking, deception, and legerdemain as the true ascent was in discipline and technology. Indeed, conceive of the genius of such a conspiracy. It would take criminals and confidence men mightier, more trustworthy and more resourceful than anything in this century or the ones before. Merely to conceive of such men was the surest way to know the event was not staged. Yes, the century was a giant and a cretin. Man had become a Herculean embodiment of the Vision, but the brain on top of the head was as small as a transistorized fist, and the chambers of the heart had shrunk to the dry hard seeds of some hybrid future.

To make sense of Apollo 11 on the moon, to rise above the verbiage (like extinguishers of foam) which covered the event, was to embark on a project which could not satisfy his own eye unless it could reduce a conceptual city of technologese to one simplicity – was the venture worthwhile or unappeased in its evil?

If Marx had done his best to gut the past of every attachment to the primitive, the sacramental, and the magical, if the Marxian formula that history was a reflection of the state of productive relations had thereby elevated reason to that vertiginous even insane eminence out of which technology had been born, then the task now appeared in reverse: one was obliged to make a first reconnaissance into the possibility of restoring magic, psyche, and the spirits of the underworld to the spookiest venture in history, a landing on the moon, an event whose technologese had been so complete that the word 'spook' probably did not appear in twenty million words of NASA prose.

5

A Dream of the Future's Face

Early on the afternoon of July 21, the Lunar Module fired its ascent motor, lifted off Tranquility Base, and in a few hours docked with Columbia. Shortly after, the astronauts passed back into the Command Module and Eagle was jettisoned. It would drift off on a trajectory to the sun. A little before midnight, out of communication for the last time with Mission Control, traveling for the final orbit around the back of the moon, Apollo 11 ignited the Service Module engine and accelerated its speed from 3,600 miles to 5,900 miles per hour. Its momentum was now great enough to lift it out of the moon's pull of gravity and back into the attractions of the earth – the spacecraft was therefore on its way home. Since the trip would take sixty hours, a quiet two and a half days were in store and Aquarius decided to get out of Nassau Bay and visit some friends.

His host and hostess were wealthy Europeans with activities which kept them very much of the time in Texas. Since they were art collectors of distinction, invariably served a good meal, and had always been kind to him, the invitation was welcome. To go from the arid tablelands of NASA Highway 1 to these forested grounds now damp after the rain of a summer evening was like encountering a taste of French ice in the flats of the desert. Even the trees about the house were very high, taller than the tallest elms he had seen in New England – 'Wild pigs used to forage in this part of Houston,' said his host, as if in explanation, and on the lawn, now twice-green in the luminous golden green of a murky twilight, smaller tropical trees with rubbery trunks twisted about a large sculpture by Jean Tinguely which waved metal scarecrow arms when a switch was thrown and blew spinning faucets of water through wild stuttering sweeps, a piece of sculpture reminiscent of the flying machines of La Belle Epoque, a hybrid of dragon and hornet which offered a shade of the time when technology had been belts and clanking gears, and culture was a fruit to be picked from a favored tree.

The mansion was modern, it had been one of the first modern homes in Houston and was designed by one of the more ascetic modern architects. With the best will, how could Aquarius like it? But the severity of the design was concealed by the variety of the furniture, the intensity of the art, the presence of the sculpture, and the happy design in fact of a portion of the house: the living room shared a wall with a glassed-in atrium of exotics in bloom. So the surgical intent of the architect was partially overcome by the wealth of the art and by the tropical pressure of the garden whose plants and interior tree, illumined with spotlights, possessed something of that same silence which comes over audience and cast when there is a moment of theater and everything ceases, everything depends on – one cannot say – it is just that no one thinks to cough.

There had been another such moment when he entered the house. In the foyer was a painting by Magritte, a startling image of a room with an immense rock situated in the center of the floor. The instant of time suggested by the canvas was comparable to the mood of a landscape in the instant just before something awful is about to happen, or just after, one could not tell. The silences of the canvas spoke of Apollo 11 still circling the moon: the painting could have been photographed for the front page – it hung from the wall like a severed head. As Aquarius met the other guests, gave greetings, took a drink, his thoughts were not free of the painting. He did not know when it had been done – he assumed it was finished many years ago – he was certain without even thinking about it that there had been no intention by the artist to talk of the moon or projects in space, no, Aquarius would assume the painter had awakened with a vision of the canvas and that vision had he delineated. Something in the acrid breath of the city he inhabited, some avidity emitted by a passing machine, some tar in the residue of a nightmare, some ash from the memory of a cremation had gone into the painting of that gray stone – it was as if Magritte had listened to the ending of one world with its comfortable chairs in the parlor, and heard the intrusion of a new world, silent as the windowless stone which grew in the room, and knowing not quite what he had painted, had painted his warning nonetheless. Now the world of the future was a dead rock, and the rock was in the room.

There was also a Negro in his host's living room, a man perhaps thirty-five, a big and handsome Black man with an Afro haircut of short length, the moderation of the cut there to hint that he still lived in a White man's clearing, even if it was on the very edge of the clearing. He was not undistinguished, this Negro, he was a professor at an Ivy League college; Aquarius had met him one night the previous year after visiting the campus. The Negro had been much admired in the college. He had an impressive voice and the deliberate manner of a leader. How could the admiration of faculty wives be restrained? But this Black professor was also a focus of definition for Black students in the college – they took some of the measure of their militancy from his advice. It was a responsible position. The students were in the college on one of those specific programs which had begun in many a university that year – students from slum backgrounds, students without full qualification were being accepted on the reasonable if much embattled assumption that boys from slums were easily bright enough to be salvaged for academic life if special pains were taken. Aquarius had met enough of such students to think the program was modest. The education of the streets gave substantial polish in Black ghettos – some of the boys had knowledge at seventeen Aquarius would not be certain of acquiring by seventy. They had the toughness of fiber of the twenty-times tested. This night on the campus, having a simple discussion back and forth, needling back and forth, even to even – so Aquarius had thought – a Black student suddenly said to him, 'You're an old man. Your hair is gray. An old man like you wants to keep talking like that, you may have to go outside with me.' The student gave an evil smile. 'You're too old to keep up with me. I'll whomp your ass.'

It had been a glum moment for Aquarius. It was late at night, he was tired, he had been drinking with students for hours. As usual he was overweight. The boy was smaller than him, but not at all overweight, fast. Over the years Aquarius had lost more standards than he cared to remember. But he still held on to the medieval stricture that one should never back out of a direct invitation to fight. So he said with no happiness, 'Well, there are so many waiting on line, it might as well be you,' and he stood up.

The Black boy had been playing with him. The Black boy grinned. He assured Aquarius there was no need to go outside. They could talk now. And did. But what actors were the Blacks! What a sense of honor! What a sense of the gulch! Seeing the Black professor in this living room in Houston brought back the memory of the student who had decided to run a simulation through the character of Aquarius' nerve. It was in the handshake of both men as they looked at each other now, Aquarius still feeling the rash of the encounter, the other still amused at the memory. God knows how the student had imitated his rise from the chair. There had been a sly curl in the Black man's voice whenever they came across each other at a New York party.

Tonight, however, was different. He almost did not recognize the professor. The large eyes were bloodshot, and his slow deliberate speech had become twice-heavy, almost sluggish. Aquarius realized the man had been drinking. It was a matter of a few shots before this evening, no, there was a sense of somebody pickling himself through three days of booze, four days of booze, five, not even drunk, just the heavy taking of the heaviest medicine, a direct search for thickening, as if he were looking to coagulate some floor between the pit of his feelings at boil and the grave courtesies of his heavy Black manner. By now it showed. He was normally so elegant a man that it was impossible to conceive of how he would make a crude move – now, you could know. Something raucous and jeering was still withheld, but the sourness of his stomach had gotten into the sourness of his face. His collar was a hint wilted.

He had a woman with him, a sweet and wispy blond, half plain, still half attractive, for she emitted a distant echo of Marilyn Monroe long gone. But she was not his equal, not in size, presence, qualifications – by the cruel European measure of this richly endowed room, she was simply not an adequate woman for a man of his ambitions. At least that was the measure Aquarius took. It was hard not to recognize that whatever had brought them together, very little was now sustaining the project. The Black man was obviously tired of her, and she was still obviously in love with him. Since they were here enforcedly together, that was enough to keep a man drinking for more than a day. Besides – if he was a comfortable house guest of these fine Europeans, he might nonetheless wish to leave the grounds. Being seen with her on Houston streets would not calm his nerves.

But there were other reasons for drinking as well. America had put two White men on the moon, and lifted them off. A triumph of White men was being celebrated in the streets of this city. It was even worse than that. For the developed abilities of these White men, their production, their flight skills, their engineering feats, were the most successful part of that White superstructure which had been strangling the possibilities of his own Black people for years. The professor was an academic with no mean knowledge of colonial struggles of colored peoples. He was also a militant. If the degree of his militancy was not precisely defined, still its presence was not denied. His skin was dark. If he were to say, 'Black is beautiful' with a cultivated smile, nonetheless he was still saying it. Aquarius had never been invited to enter this Black man's vision, but it was no great mystery the Black believed his people were possessed of a potential genius which was greater than Whites. Kept in incubation for two millennia, they would be all the more powerful when they prevailed. It was nothing less than a great civilization they were prepared to create. Aquarius could not picture the details of that civilization in the Black professor's mind, but they had talked enough to know they agreed that this potential greatness of the Black people was not to be found in technology. Whites might need the radio to become tribal but Blacks would have another communion. From the depth of one consciousness they could be ready to speak to the depth of another; by telepathy might they send their word. That was the logic implicit in CPT. If CPT was one of the jokes by which Blacks admitted Whites to the threshold of their view, it was a relief to learn that CPT stood for Colored People's Time. When a Black friend said he would arrive at 8 p.m. and came after midnight, there was still logic in his move. He was traveling on CPT. The vibrations he received at 8 p.m. were not sufficiently interesting to make him travel toward you – all that was hurt were the host's undue expectations. The real logic of CPT was that when there was trouble or happiness the brothers would come on the wave.

Well, White technology was not built on telepathy, it was built on electromagnetic circuits of transmission and reception, it was built on factory workers pressing their button or monitoring their function according to firm and bound stations of the clock. The time of a rocket mission was Ground Elasped Time, GET. Every sequence of the flight was tied into the pure numbers of the time-line. So the flight to the moon was a victory for GET, and the first heats of the triumph suggested that the fundamental notion of Black superiority might be incorrect: in this hour, it would no longer be as easy for a militant Black to say that Whitey had built a palace on numbers, and numbers killed a man, and numbers would kill Whitey's civilization before all was through. Yesterday, Whitey with his numbers had taken a first step to the stars, taken it ahead of Black men. How that had to burn in the ducts of this Black man's stomach, in the vats of his liver. Aquarius thought again of the lunar air of technologists. Like the moon, they traveled without a personal atmosphere. No wonder Blacks had distaste for numbers, and found trouble studying. It was not because they came – as liberals necessarily would have it – from wrecked homes and slum conditions, from drug-pushing streets, no, that kind of violence and disruption could be the pain of a people so rich in awareness they could not bear the deadening jolts of civilization on each of their senses. Blacks had distaste for numbers not because they were stupid or deprived, but because numbers were abstracted from the senses, numbers made you ignore the taste of the apple for the amount in the box, and so the use of numbers shrunk the protective envelope of human atmosphere, eroded that extrasensory aura which gave awareness, grace, the ability to move one's body and excel at sports and dance and war, or be able to travel on an inner space of sound. Blacks were not the only ones who hated numbers – how many attractive women could not bear to add a column or calculate a cost? Numbers were a pestilence to beauty.

Of course this particular Black man, this professor, was in torture, for he lived half in the world of numbers, and half in the wrappings of the aura. So did Aquarius. It was just that Aquarius was White and the other Black – so Aquarius could not conceal altogether his pleasure in the feat. A little part of him, indefatigably White, felt as mean as a Wasp. There was something to be said after all for arriving on time. CPT was excellent for the nervous system if you were the one to amble in at midnight, but Aquarius had played the host too often.

'You know,' said the professor, 'there are no Black astronauts.'

'Of course not.'

'Any Jewish astronauts?'

'I doubt it.'

The Black man grunted. They would not need to mention Mexicans or Puerto Ricans. Say, there might not even be any Italians.

'Did you want them,' asked Aquarius, 'to send a Protestant, a Catholic, and a Jew to the moon?'

'Look,' said the Black professor, 'do they have any awareness of how the money they spent could have been used?'

'They have a very good argument: they say if you stopped space tomorrow, only a token of the funds would go to poverty.'

'I'd like to be in a position to argue about that,' said the Black. He sipped at his drink. It trickled into his system like the inching of glucose from a bottle down a rubber tube. 'Damn,' he said, 'are they still on the moon?'

'They took off already,' said Aquarius.

'No trouble?'

'None.'

If the Blacks yet built a civilization, magic would be at its heart. For they lived with the wonders of magic as the Whites lived with technology. How many Blacks had made a move or inhibited it because the emanations of the full moon might affect their cause. Now Whitey had walked the moon, put his feet on it. The moon presumably had not spoken. Or had it, and Richard Nixon received the favor and Teddy Kennedy the curse? Was there no magic to combat technology? Then the strength of Black culture was stricken. There would not be a future Black civilization, merely an adjunct to the White. What lava in the raw membranes of the belly. The Black professor had cause to drink. The moon shot had smashed more than one oncoming superiority of the Black.

ii

That night Aquarius had trouble falling asleep, as if the unrest of the Black professor at the passage of men's steps on the moon had now passed over to him. Nothing in the future might ever be the same – that was cause for unrest – nor could the future even be seen until one could answer the obsessive question: was our venture into space noble or insane, was it part of a search for the good, or the agent of diabolisms yet unglimpsed? It was as if we had begun to turn the pocket of the universe inside out.

He had had at the end a curious discussion with the Black professor.

'It's all in the remission of sin,' the Black man had said. 'Technology begins when men are ready to believe that the sins of the fathers are not visited on the sons. Remission of sin – that's what it's all about,' he said in his Black slow voice.

Yes, if the sons were not punished, then the father might dare, as no primitive father had dared, to smash through a taboo. If the father was in error, or if he failed, the sons would be spared. Only the father would suffer. So men were thereby more ready to dare the gods. So that love on the cross which had requested that the sons not pay for the sins of the fathers had opened a hairline split which would finally crack the walls of taboo. And the windowless walls of technology came through the gap. Back to Sören the Dane. You could not know if you were a monster or a saint of the deep.

In the Nineteenth Century, they had ignored Kierkegaard. A middle-class White man, living on the rise of Nineteenth Century technology was able to feel his society as an eminence from which he could make expeditions, if he wished, into the depths. He would know all the while that his security was still up on the surface, a ship – if you will – to which he was attached by a line. In the Twentieth Century, the White man had suddenly learned what the Black man might have told him – that there was no ship unless it was a slave ship. There was no security. Everybody was underwater, and even the good sons of the middle class could panic in those depths, for if there were no surface, there was no guide. Anyone could lose his soul. That recognition offered a sensation best described as bottomless. So the Twentieth Century was a century which looked to explain the psychology of the dream, and instead entered the topography of the dream. The real had become more fantastic than the imagined. And might yet possess more of the nightmare.

Lying there, unable to sleep, lost in the caverns of questions whose answers never came (Mr Answer Man, what is the existential equivalent of infinity? – Why insomnia, Sandy, good old insomnia) Aquarius knew for the first time in years that he no longer had the remotest idea of what he knew. It was the end of the decade, and the fashion was rising in New York literary lakes to inquire after the nature of the decade to come. He had been a poor prophet of the Sixties, but it was not a century for prophets – poor as he had been, he had still been one of the few who had some sense of what was coming. He had known that marijuana was on its way, and Hip, and the Kennedys, and a time of upheaval, and in the center of the Establishment: loss of belief. Now they asked him what he thought of the Seventies. He did not know. He thought of the Seventies and a blank like the windowless walls of the computer city came over his vision. When he conducted interviews with himself on the subject, it was not despair he felt, or fear – it was anesthesia. He had no intimations of what was to come and that was conceivably worse than any sentiment of dread, for a sense of the future, no matter how melancholy, was preferable to none – it spoke of some sense of continuation in the projects of one's life. He was adrift. If he tried to conceive of a likely perspective in the decade before him, he saw not one structure to society but two: if the social world did not break down into revolutions and counter-revolutions, into police and military rules of order with sabotage, guerrilla war and enclaves of resistance, if none of this occurred, then there would certainly be a society of reason, but its reason would be the logic of the computer. In that society, legally accepted drugs would become a necessity for accelerated cerebration, there would be inchings toward nuclear installation, a monotony of architectures, a pollution of nature which would arouse technologies of decontamination odious as deodorants, and transplanted hearts monitored like spaceships – the patients might be obliged to live in a compound reminiscent of a Mission Control Center where technicians could monitor on consoles the beatings of a thousand transplanted hearts. But in the society of computer-logic, the atmosphere would obviously be plastic, air-conditioned, sealed in bubble-domes below the smog, a prelude to living in space stations. People would die in such societies like fish expiring on a vinyl floor. So of course there would be another society, an irrational society of the dropouts, the saintly, the mad, the militant and the young. There the art of the absurd would reign in defiance against the computer.

In the society of the irrational would be found the weather of the whirlpool. Accelerations and torpor would ride over one another with eyes burned out by visions no longer recalled, motorcycles would climb the trees, a night of freakings when all the hair would be burned for the bonfire of the goat, and bald as the moon would be the skins of the scalp. Hare Krishna! A part of the American world, gassed by the smog of computer-logic, would live like gurus, babas and yogas in the smallest towns, the small towns of America would be repopulated with the poets of the city, and mysticism would live next to murder, for murder was love in freak newspeak, and the orgy was the family. Because the computer was the essence of narcissism (the computer could not conceive of its inability to correct its own mistakes) a view of the Seventies suggested a technological narcissism so great that freak newspeak was its only cure – only the threat of a murderous society without could keep computer society from withering within. How those societies would mingle! Acid and pot had opened the way.

Yet even this model of the future was too simple. For the society of the rational and the world of the irrational would be without boundaries. Computersville had no cure for skin disease but filth in the wound, and the guru had no remedy for insomnia but a trip to the moon, so people would be forever migrating between the societies. Sex would be a new form of currency in both worlds – on that you could count. The planner and the swinger were the necessary extremes of the computer city, and both would meet in the orgies of the suburbs. But was this a vision of the future or the vertigo of the early hours?

Aquarius got out of bed. He was a disciplinarian about insomnia. Having suffered from it years before, he had learned how to live with an occasional bad night. He took no pill, he took no drink, he looked to ride it out. Sometimes he indulged in a game of formal optimism, carrying over from artillery training the injunction to bracket a target. So now if his sense of the future was too pessimistic – he could only hope it was too dark! – he would look for the formal opposite: try to regard science as reasonable, religion as rewarding. He could see – sitting in a kitchen chair, reading by a lamp – how new religions might crystallize in the Seventies, they could give life, for their view of God might be new. And science . . . But he could not regard science apart from technology. Aquarius began to think of Dr George Mueller.

iii

He had had an interview with the head of the Manned Space Program when he was back at Kennedy. Once a professor of electronics, Mueller had become the second highest man in NASA, indeed nobody was higher than him but Dr Thomas O. Paine, yet Mueller was in appearance a very modest man, an archetype to Aquarius of the technician. Mueller – whose name was pronounced Miller – was not tall, certainly he was not short, he was slim in weight, in fact without excess weight, he looked forty-five although he had to be older, and he had a long thin face, a high forehead, straight black hair which he combed straight back – so his black horn-rimmed glasses jutted forth, so did the blade of his nose, the blade of his jaw. He spoke with mild icky-dicky Midwest expletives like 'Golly gee, gee whiz, gosh!' that modest but central sense of presence one might find in a YMCA secretary on the night desk – the manner friendly, impersonal, and on an astral plane, a manner to indicate that of course one is used to talking to all sorts of people – golly gee, you ought to see some of the characters who come in here.

Mueller had the reputation of being tremendously determined when he wanted to get something done, and one could believe it, for he emitted the gentle but total impersonality of a man for whom obstacles if irrational were unforgivable. Perhaps for this reason he was curiously reminiscent of Hugh Hefner. The publisher of *Playboy* was a little in relation to ordinary men like a guy who had been to the moon and back and Dr Mueller could have been his older brother.

Aquarius met him in Mueller's motel at Cocoa Beach, a room as modest as his pretensions. There had been a photographer taking pictures. Mueller apologized, explaining that he had been so busy for so many years that he had hardly had a picture taken – now they had discovered there weren't enough pictures of him for the NASA files. So he posed for a few more, in some degree as pleased and flustered as a man walking into a room, and crack! flashbulbs! they are giving him a birthday party.

Yet once the interview began, Mueller was sensitive to every change about him. Did Aquarius, searching for his next question, feel some intensity of motive or charge of energy, then Mueller was there to respond as quickly as the needle on any of the measuring instruments he had used in all the electrical labs of his youth and academic career. Dart! would go his head; up! would fly a finger; sway! would swing his torso. How alert he must have been to signs of overload or impedance in all the human circuits about his field position in the room. Yes, Dr George Mueller was certainly one full academic counterpart to the

Black professor. Having seen Apollo-Saturn rise from the drawing board to an orbit of the moon, it was as natural for him to live with comfort in the future as it was flesh and drink for the Black to brood upon the past. So Mueller was looking beyond this landing to the uses of space in the future. He talked of rocket shuttles which could be fired up to rendezvous with space stations in orbit and yet be able to return through the heats of reentry to land at spaceports in order to be used again; he spoke of lowering the cost of transport in space from $100,000 a pound to $200 a pound; he outlined future projects for nuclear power plants on the moon whose heats would melt the permafrost and so make available a supply of water. The electrical output of the power plant could then separate the water into hydrogen and oxygen – some of that product would be used to make Lox and LH_2 for rocket fuels; the rest would be mixed with nitrogen extracted from the rocks so they would have the elements necessary to create a livable atmosphere within an enclosed space. Perhaps they would even grow plants. Would it be Aristotle Onassis or Richard Burton who would be first to spring for a bouquet of moon roses?

The space station orbiting the earth could go in for a 'total earth sensing program,' doubtless as comprehensive for the earth, Aquarius decided, as a thorough physical examination conducted daily for a man. The space station would also set up a high-energy physics laboratory. It would certainly accelerate every technique in the manufacture of cameras, telescopes, radars, lasers, and that was just the beginning. Mueller went on in full sentences. He had spoken of these matters a hundred times, but like all high bureaucrats he was equal to a professional actor in his ability to repeat the same dialogue with verve another hundred times. If any reporter had brought a tape recorder, he could by transcribing Mueller's remarks directly have had a printable feature story for a Sunday section. As though unwinding a scroll, Mueller indicated the possibility of a new wonder with each paragraph. Since the space stations would have a weightless environment at their command, it would be possible to grow crystals which would be molecularly perfect. As a corollary, one would be able to build up flawless diamonds of any size. Before Aquarius could ask if the diamond would be as big as the Ritz, Mueller assured him in his cheerful small-town voice that the diamond could be as big as a basketball anyhow – such a cargo might be worth much more than two hundred dollars a pound. Thus the perspective of space factories returning the new imperialists of space a profit was now near to the reach of technology. Forget about diamonds! The value of crystals grown in space was incalculable: gravity would not be pulling on the crystal structure as it grew, so the molecules would line up in lattices free of shift or sheer. Such a perfect latticework could serve to carry messages for a perfect computer. Computers the size of a package of cigarettes would then be able to do the work of present computers the size of a trunk. So the mind could race ahead to see computers programming go-to-school routes in the nose of every kiddie car – the paranoid mind could see crystal transmitters sewn into the rump of every juvenile delinquent – doubtless, everybody would be easier to monitor. In the Systematized Detection Systems of the future, Big Brother could get superseded by Moon Brother – the major monitor of them all might yet be sunk in a shaft on the back face of the lunar sphere.

The possibilities of the new technology glowed in the enthusiasms of Mueller's voice. 'Ball bearings,' he said holding up a finger like an antenna to focus all scattered waves of random thought, 'it's fascinating to consider what possibilities are opened in the manufacture of ball bearings.' He went on to explain in his careful considerate feature-story paragraphs that ball bearings which were cast in a weightless environment would come out as perfect spheres; the deviation on their skin need be no greater than the thickness of a molecule. Earth ball bearings were of course imprecise. In the instant it took the shot to cool, gravity was pulling on the molten ball. So to obtain precision they must be polished, a relatively imprecise technique.

Aquarius was to think again of the ball bearings after he said good-by to Dr Mueller. Such creations of a weightless environment could yet prove monumental for the manufacturer, since ball bearings were as crucial to every load-bearing or load-transporting machine as the valves of the human heart to the flow of blood. Out of the imperfections of ball bearings (which were located after all around the center of every high-speed moving part) came the multiplication of all the other imperfections, since each moving part added the scope of its imprecision to the next moving part. Once perfect ball bearings could be installed, the action of the machines might become a whole order of efficiency closer to the laws of physics, rather than to the adjustments and counterbalances of engineering. That meant a world of future machines whose view of present-day machines might be equal to nothing less than our view of Piltdown man. Or would it merely mean that plastic could now be employed for the ball bearings in order to maintain the built-in obsolescence of machines? Indeed the center of the problem of capitalism's morale was in the perfect ball bearing. Machines built on perfect ball bearings would have a life duration so much greater than present machines that modern capitalism living with the vice of built-in obsolescence as the poison-stimulant to its blood, would be face to face with problems greater and more inescapable than automation. For once space explorers, seeking economic justification, would be forced to develop perfect ball bearings, their use would be bound to explode the sustaining fevers and indulgences of the economy. What then would they all do? Then, capitalism would be as much at war with itself over the continuing nature of the economy as world communism was at war with itself over the direction of its ideology.

So the mood of space which remained with Aquarius after talking to Mueller, that mood elegant and austere as the perfect laws of physical principle, was still a force for disruption. Sitting in the spaced-out colors of Dun Cove, inhabiting the shank end of ruminations like this, was the thought that the moon shot was conceivably the first voyage of the very cancer of the world, for indeed the first journeys of the cancer cell in a body, taken from the point of view of the cancer cell, were certainly bold and dangerous. Not by little effort did a cell leave its own organ and learn how to survive in another. Cancer cells, seen in relation to ordinary cells, were often extraordinary in the variety of their form, as different as a view of Las Vegas at night is different from a village in the Bluegrass, or as different as the internal works of Apollo were in comparison to the works of the family car. Did that account for the curious depression, the sobriety mixed in so many faces with the pride of the achievement? Aquarius did not know.

That was still another reason why he did not perceive the decade to come with any clear picture of events. A dull sense of disaster pushed at the compass of the picture. He was not so certain the decade would have a life like other decades. If space was benign, then on we would continue into space, and the artists would yet be voyaging with the astronauts – think of that happy day when he would nominate himself to be first writer to visit the moon. (Not a chance! NASA would opt for Updike!)

But what if space were not so benign? What if we did not act upon space, explore into space, but space rather acted upon us, drew us toward her dispositions, her plans for us, her intent upon human life, what if we thought we moved up but were drawn up, what if the moon was as quiet as the fisherman when he lays the fly on the water . . .

Having journeyed to the center of his gloom, Aquarius went to sleep. In his dreams a country doctor he had known for years murmured, 'I don't know about all of this. Recognize that the moon could be some kind of catchall simple as the tonsils to protect us here on earth. Maybe those craters come from catching all the cess.' In his dream Aquarius answered back, 'It depends on your idea of God, that's what it must depend on.' Out into sleep he went again, ringings of ether in his ear.

iv

In the morning after breakfast, he found himself rereading a transcript of the postlaunch briefing, a curious activity, but he was like a man on the cusp of a clue. To fall asleep in pursuit of the answer to a mystery was to awaken with the fire in a new place. It had burned beneath the ground while he was sleeping. So he dallied over the substance of the transcript as if some hint of smoke could linger here in the words of men taken down for posterity four hours after the launch.

PUBLIC AFFAIRS OFFICER: *I'd like to introduce Mr Rocco A. Petrone, Director of Launch Operations for the Kennedy Space Center and Launch Director for the Apollo 11 flight. Rocco?*

PETRONE: *Thank you. Well, this is our sixth Saturn V, the Apollo aboard, to go up in a row on time. But I'm not saying, of course – this meant a lot more to us. This is the big one. This is the one we've been working for eight years. The mission is just getting started. I'm sure you all know. But the first step in this historic mission has been just the step we've wanted to take.*

Every detail of the launch, Aquarius recollected, had been Petrone's responsibility. Four hours after lift-off, matters now comfortably out of his hands, he was by the evidence of the transcript as tired as a boxer in his dressing room after a fifteen-round fight.

> PETRONE: *From the moment of truth here, from the moment of ignition and lift-off, lots and lots of equipment have to work for a number of starts to come on our side. We had a few difficulties in the count. I'm sure they already fed it to you. I'd be glad to answer questions. I'm pleased to say that the team was able to handle the problems, keeping the count rolling, and very obviously start this historic mission off. I say on the right first step, which I can assure you is most pleasing to me. And for the team that's worked so hard to get to this point . . .*
>
> PUBLIC AFFAIRS OFFICER: *Okay; thanks, Rock.*

At NASA, the elegance was in the design of the engineering systems rather than in the manners of the men. Which future student of language, unfamiliar with Saturn V, Apollo II or lift-off, would have any idea, encountering this scrap of Petrone's transcript, that the man was describing the emotions he felt after having led thousands of men in Launch Control through the nine hundred hours, the ninety and the nine hours of the preparations and countdowns which put his ship into the air. Who was to know by such a speech – thought Aquarius sipping his breakfast tea – that Petrone had been midwife to the most momentous week, and the mightiest hour. Yes, and that was not so ridiculous if one recognized that to believe in progress and believe in God as well might make it necessary to conceive of our Lord as a vision of existence who conceivably was obliged to compete with other visions of existence in the universe, other conceptions of how life should be. But this had brought him to the heart of the question. It was as if his mind, knowing the style of his thoughts, had directed him to the transcript in the confidence that his sense of irony once aroused, his sense of apocalypse could never be far behind. He was, after all, quick to hunt for reason in absurdity. So to read the language of men who were not devoid of mechanical genius and yet spoke in language not even fit for a computer, of events which might yet dislocate eternity, was a fine irony – the banality of the verbal reaction was an indication of the disease of our time, so advanced in one lobe, so underdeveloped in the other, a fine irony! Unless one was encountering the very desperation of the Lord – there might not be time to develop men to speak like Shakespeare as they departed on heavenly ships.

To believe in God and to believe in progress – what could that mean but that the desire for progress existed in the very creation of man, as if man were designed from the outset to labor as God's agent, to carry God's vision of existence across the stars. If this were true then the intent of the Lord could hardly be to reveal His goodness to us; rather He must employ us to reveal His vision of existence *out there,* somewhere out there where His hegemony came to an end and other divine conceptions began to exist, or indeed were opposed to us. If God finally was the embodiment of a vision which might cease to exist in the hostilities of the larger universe, a vision which indeed might be *obliged* to prevail or would certainly cease to exist, then it was legitimate to see all of human history as a cradle which had nurtured a baby which had now taken its first step. Intended by divine will to travel across the heavens, we were now at least on our way to the moon, and who could know if we were ahead or behind of some schedule the Lord had presented us, a schedule which presumably each man and woman alive would keep in the depths of their unconscious along with everything else most vital for the preservation of life. A large and uncomfortable thought, for if it were so, then the flight of Apollo 11 was a first revelation of the real intent of History. So this much, anyway, had been revealed: one could not make a judgment on the value or absurdity of devoting such effort to go to the moon unless one was ready to recognize that eschatology had conceivably been turned on its head. For if eschatology, that science of 'the four last things: death, judgment, heaven and hell,' was now to be considered in the light of God's need for supermen to negotiate His passage quickly through the heavens, then how much more value might He give to courage than to charity, how much harsh judgment to justice itself if the act to be considered was not expeditious but merely just, yes if speed were of the essence then Hell's Angels were possibly nearer to God than the war against poverty.

This last suggested a step Aquarius was not prepared to take: the idea was as disruptive to a liberal philosophical system as tartar emetic and mustard to a glutton. For it offered a reason why the heroes of the time were technologists, not poets, and the art was obliged to be in the exceptional engineering, while human communication had become the routine function. It was because the Power guiding us had desired nothing less. He was looking to the day when all of mankind would yet be part of one machine, with mechanical circuits, social flesh circuits, and combined electromagnetic and thought-transponder circuits, an instrument of divine endeavor put together by a Father to whom one might no longer be able to pray since the ardors of His embattled voyage could have driven Him mad.

Sweet thoughts for Aquarius to have as a sequel to the ascent, but the questions were grand at least, they could occupy the consciousness of the century. It was somehow superior to see the astronauts and the flight of Apollo 11 as the instrument of such celestial or satanic endeavors, than as a species of sublimation for the profoundly unmanageable violence of man, a meaningless journey to a dead arena in order that men could engage in the irrational activity of designing machines which would give birth to other machines which would travel to meaningless places as if they were engaged in these collective acts of hugely organized but ultimately pointless activity because they had not the wit, goodness, or charity to solve their real problems, and so would certainly destroy themselves if they did not have a game of gargantuan dimensions for diversion, a devilish entertainment, a spend-spree of resources, a sublimation, yes, the very word, a sublimation of aggressive and intolerably inhuman desires, as if like a beast enraged with the passion of gorging nature, we looked now to make incisions into the platinum satellite of our lunacy, our love, and our dreams.

Aquarius would have given much to find a truly revealing face at NASA, for that could have given a clue to these questions, but it was in the logic of such endeavor that no answers be apparent on the surface. If it would take the rest of the century to begin to disclose the real intent of the act, no lightning raid on the evidence, no single happy disclosure, could possibly offer a reply.

Still, Aquarius preferred the first assumption, that we were the indispensable instruments of a monumental vision with whom we had begun a trip. On that conclusion he would rest his thoughts. Having come back at last to earth from the orbits of the dream with such a hypothesis in his pocket, Aquarius was a little more ready to head for home, the writing of a book and conceivably the pouring of a drink. The study of more than one technical manual awaited him.

PART II

Apollo

I

The Psychology of Machines

In the study of literature, much usually depends on direct confrontation with a work. Who would dare to approach *A Farewell to Arms* by a synopsis? It is only natural to distrust a literary experience if we have been guided too carefully through it, for the act of reading must provide by itself that literary experience upon which our senses will later work.

But the study of science is different. Much like the study of history, it begins with legends and oversimplifications. Then the same ground is revisited, details are added, complexities are engaged, unanswerable questions begin to be posed. A scientific account is a story which can always be retold, for the line of the narrative in scientific writing is to be found in the deepening of the concept.

So if we embark once again on the trip of Apollo 11, if once again we proceed to ascend from the launching tower and fire the astronauts into orbit, even take them again to the moon, if in fact we will launch them again and then yet still again, let us recognize that we are enlisted in scientific methods of instruction where our pleasure can be found only by returning over the same ground in order to discover that the story steeped in further detail has become something like another account even as a day recaptured in a dream has acquired the reality of a more extraordinary day. It seems we are off on a journey with mysterious routes, for the implication is unmistakable that a study of the trip technology took to the moon may as well commence in the inner space of the dream.

ii

To speak of the unconscious is to call up the set designer of one's imagination. He bows. Caverns and grottos appear, underwater palaces, the circles and amphitheaters of hell, the rites of barbarians in all the dark forest; we think of witches, and sniff the communions of mood in a church. Yes, the wealth of the theater is here, but it is not all that is here. Something also suggests a domain as closed from light and spectacle as the somber shelves of a library at night – it must be that part of the unconscious which serves as the servant of everyday life. From the depth of its files will it produce on demand, at a rate daily established, the varieties of practical information the conscious mind has requested. It will solve problems, offer analysis of new situations presented by the difficulties of life without, even provide working estimates of how equipped we are to encounter some contingency or meet a crisis and, if the matter is novel enough or dangerous enough, or involves preparation for the most artful kind of work, this instrument of the psyche may even send appeals for information to the mansions, theaters, and dungeons of the deepest unconscious where knowledge of a more poetic and dread-filled nature may reside. So this dark room, this functioning appendage of consciousness, is obviously the living organic embodiment of a computer: one cannot even begin to contemplate the function of the dream until one recognizes what complexity such a computer must possess in its nightly dialogue with the deep.

For years, Aquarius had had the concept of just such a guide somewhere in the human head who – on the basis of what information was available – was usually capable of piloting a man's life through everything from the small decisions of a day to the critical dilemmas of the age. He had given to this pilot a name: the Navigator. The Navigator had varieties of conscious and unconscious information available on request. He was, if one were to think on him crudely, a memory bank of reference to everything which had been learned and was still available to recall, as well as a cerebral library of the opinions and judgments of respected authorities outside oneself. On the basis of one's acquired experience and those worldly guidelines one was ready to accept from without, a future course of action could be estimated, and large and little decisions could be taken. The Navigator was thus the agent of the ego in the unconscious, the dispatcher at the switch.

Over the years, it proved to be too simple a model, and Aquarius was obliged to add some conceptual accessories. To cope with that large variety of daily experiences which were not easy to anticipate or to comprehend, yet were not terrifying so much as confusing, Aquarius now added to the Navigator the services of a Novelist. It seemed to him that everybody, literate and illiterate alike, had in the privacy of their unconscious worked out a vast social novel by which they could make sense of society. Obviously, each novel was different. Obviously, some were better than others. But whether each unwritten novel was a comprehensive work of art, or an unhappy one, the psychic fact was that as life presented new evidence, the book was altered in its details. When such large events as births, deaths, marriages, divorces, successes, failures, social cataclysms and social revelations were sufficiently unexpected to indicate the conception one had of society – that conception so often forged by inferior art and entertainment – was faulty, then the outlines of the novel would be drastically revised; in effect the Novelist was forever drawing up new social charts upon which the Navigator could make his calculations.

In its turn, the dream provided another sort of information for the Navigator. It ran simulations. Perhaps they were not unlike the simulations put into the computers in Mission Control at the Manned Spacecraft Center. Indeed Aquarius began to think the dream might be some psychic equivalent of those equations of celestial mechanics which find it impossible to plot the trajectory of a moving rocket precisely because there are too many unknowns. The earth is in movement, and the moon. Their gravitational effect upon the rocket is forever shifting. Besides earth and moon have special eccentricities in their motion which forbid precise prediction of position at any instant. In turn, the spaceship is not a precise object either. While it is built within narrowly defined specifications, still within such limits, it will always prove to be a minuscule bit smaller or larger than the specified size of the rocket. These variations, while minor, still have effects on the trajectory. In such a field of imprecisions (where everything is known approximately but nothing exactly) the problem is best attacked by inserting thousands of imaginary or simulated trajectories into the computer, each with slight variations in position and velocity assumed for the earth, sun, moon, and rocket. From these thousands of simulated trajectories come a series of imaginary but definitely plotted routes from earth to moon. By such calculation, and by the aid of data which track the path the rocket is actually taking it is possible to estimate the direction in which it will next be moving, and so determine whether it is likely to stray out of the limits of those many thousands of arbitrarily calculated trajectories. It is as if an artist drawing the curve of the arm had chosen in preference to one line, a thousand light strokes none in itself the outline, but taken all together a clear picture of an arm was present.

That was one kind of simulation. Another category of simulation was to present flight conditions to the astronauts. They worked in specially built cockpits outfitted exactly like the Command Module and the Lem, all dials and instruments in functional replica. Thousands of possible problems, dilemmas, and unforeseen breakdowns were played off before them in the simulators over the months of training. We can listen with profit to Collins. 'People think we're baked in heat chambers and whirled in centrifuges until our eyeballs fall out, and there is a little of that, but essentially we are learning an incredibly complex array of machines – all the nuts and bolts and wires – and learning what to do if some of it doesn't work as advertised. We go through the intended missions tiny tedious bit by bit to make sure that we understand it and that the equipment is right and what its malfunction modes are and what our alternate plans are.'

Aquarius had long ago decided, long before he had been introduced to celestial mechanics or the workings of mission simulators, that dreams, yes all that mighty symbolic and theatrical equipment of the dream could not possibly be limited to so meager a function when all was said as wish fulfillment. If there were desires which were near to unmanageable, and the imprisonment of those desires left the body charged with psychic wastes which had to be eliminated, then certainly the dream would serve in the way Freud had declared – certainly most dreams, indeed some part of every dream, must be a wish fulfillment, but surely all that panoply, horror, and pleasure, that intimation of worlds beyond worlds, that explosion of ogres in every corner of endeavor, could hardly exist in the paraphernalia of the dream merely to serve as some sort of small and large intestine of the psyche. Indeed there had been intellectual knots in all those long careful arguments through which Freud labored in order to transpose nightmares over into scenarios of wish fulfillment. No, Aquarius thought, there was a statement in the nightmare all direct, a clap of psychic thunder, a vibration from the deep. There had to be more to the dream than Freud had ever given it – the dream was like the third eye of the Navigator, it looked into many a situation the eyes of reality could hardly assess. Perhaps the dream was indeed a simulation chamber where the possible malfunctions of life tomorrow and life next year could be tested, where the alternate plans could be tried. That at least must be one essential function of the dream. For as one moved through the situations of the day, reality kept giving intimations to the senses that reality was not what it appeared to be, not altogether. A crack in a man's voice might give a clue to an oncoming disease, an odd laugh in a friend could leave its echo of possible treacheries and deceits. Sometimes in the middle of racing across the street in front of a car, there might be curious hesitations in one's gait as if unvoiced but large areas of the psyche were ready for an accident. A thousand such intimations of a reality subtly beneath reality, yet ready – it was possible – to become outer reality before very long, was apparent in everyone's waking day. And this reality was more exciting, more threatening, more demanding and more rewarding than the easier reality of the working-day surface. So as these pieces of extra-real information were noted, so might they be stored, so might they be marinated in the vats of the imagination; later that night they could be served in the dream as the ingredients of a scenario which would look to test and explore a hundred possible avenues of that subterranean future which had been offered so many curious and sometimes threatening intimations in the previous day. It was possible that in the dream, one traveled through a scenario where one was his own hero, and in the dream one might learn how one would react to the death of the man with a crack in his voice, and conceivably have glimpses of reaction to one's own death as well? Reliving the joke with the untrustworthy friend, his laugh was now, yes, overtly treacherous. So did one face up to him, dare him, cow him? Or was one more afraid of him than ever conceived? Was there murder at his base? Running in front of that car, one was now running in a nightmare. Whole schisms of potential suicide were revealed in a more glaring light: the Navigator could recognize that matters internal were worse than his previous estimate. So next day, the charts of the Novelist would be redrawn for the trip through the social world, new reefs to avoid laid in, new channels discovered and marked. Now subtle changes in the person might be evident in his relations with the sick man, the friend, or in his own actions as he crossed the street.

That had been Aquarius' measure of a new approach to the dream, and he liked it. It explained much. It gave dignity to the dream and to the dreamer. The dreamer was no longer consoling himself. Rather he was exploring the depths of his own ability to perceive crisis and react to it; he was exploring ultimate modes of existence in sex and in violence, in catastrophe and in death. So the real substance of a dream was a submersion into dread. One tested the ability of the psyche to bear anxiety as one submerged into deeper and deeper plumbings of the unknowable until one reached a point where the adventurer in oneself could descend no longer, panic was present – one was exploded out of the dream. But a dangerous shoal had at least been located.

iii

Yet even to think of the dream as a set of simulations which explore into dread is to open some obvious comparisons with the trip of Apollo 11 to the moon. For if it is in the nature of our lives to explore for meaning not only in the duties and surprises of a working day, but at night in the alleys of the unconscious (where revelations of terror beyond the terror we already know suggest the very perils of our soul), what force resides then in the parallel thought that our voyage to the moon was finally an exploration by the century itself into the possible consequences of its worship of technology, as if, indeed, the literal moon trip was a giant species of simulation to reveal some secret in the buried tendencies of our history. It was as if technology had determined to invoke the god of magic it had already slain, even as a priest might step via his nightmare into the powerful passions of sexual instinct so primitive he had once cast it out, and wished to see if he were powerful enough to cast it out again.

Such remarks are large, they are grand, they roll off into the murk of metaphysical storm. Still there are quick clues to be sniffed and landmarks in the murk. If the title of our chapter is The Psychology of Machines, the bewilderment of the reader at the notion is a hint direct to the anxieties of technology. For if machines have psychology, then technology is not quits with magic – technology is founded on the confidence that magic does not exist and so machines may be designed to perform the most extraordinary acts. It is the premise of magic that if the same act is repeated ceremoniously enough times, it will invoke a spirit. Or at least it will if the conditions are appropriate, the servants are possessed of no unruly forces, the gods are sympathetic, the animals and maidens to be sacrificed are virgin, and the equipment is unpolluted. It is the premise of technology that spirits do not exist, and the same act repeated in obedience to a system of procedure and well-oiled machinery will produce not a spirit, but in fact, the same result as the preceding occasion. Whether the gods are well- or ill-disposed, the car will start, the rifle will fire, the stereo will play. Actually, there are any number of occasions when the car won't start, the rifle jams, the record-changer on the stereo develops a mind of its own. A mind of its own! That is the threshold of the psychology of machines. For such a psychology exists, or it does not exist, and technology is founded on the implicit belief that machines are not possessed of psychology; the rifle jammed because of a speck of dirt in the breech, the car engine was flooded by the nervous foot of the driver, and the record-changer, far from having a mind of its own, rather had its record-changing procedure altered by careless handling. For every malfunction there is a clear cause technology must argue, a nonpsychological cause: psychology assumes free will. A human being totally determined is a machine. Psychology is then a study of the style of choice provided there is freedom to choose. Even a title like The Psychology of Machines assumes that the engine under study, no matter how completely fitted into the world of cause and effect, still has some all but undetectable horizon between twilight and evening where it is free to express itself, free to act in contradiction to its logic and its gears, free to jump out of the track of cause and effect. Since such events take place, if they do take place, on those unexpected occasions when no instruments are ready to examine the malfunction, the question is moot. No one alive can state to a certainty that a psychology of machines exists or does not exist – indeed it would take a theoretician of the dimensions of Einstein to prove the presence of such a psychology or, indeed, what would be even more dazzling, prove definitively that such a psychology could not conceivably exist. If it is the passion of technology to live as if such a proof were already here, there is a primitive residue in man which is far from convinced, face to face with the presence of a machine, that the engine is not possessed of a variety of spirits benign and wicked. Indeed the practical experience of everyday life is forever suggesting that complex machines behave in more extraordinary fashion with complex and highly charged operators than with calm and easygoing mechanics. If rational arguments sweep in immediately to speak of the lack of science in such observation (indeed even the lack of simple organized observation in the observation itself) still the enormous anxiety of technology remains. Either it has extirpated magic, or it has not. And if it has not, if magic still exists amid machines, then the reign of technology could be ended at a stroke, for where there is a little magic, there can be a mighty magic, even as the first fission of the atom inspired the terror among physicists that a chain reaction might occur which would destroy the earth. If one machine became sufficiently magical to set out on a life of its own, who could be certain that a resonance producing similar activity might not appear in all similar machines everywhere?

To the technician this fear is no more outlandish than the private terror of doctors that the whole human race may abruptly debouch into cancer. For so far as the human body is a machine, a cancer cell exhibits evidence of magic (or what we have been calling The Psychology of Machines), since the cancer cell has left the organization of human flesh and is exhibiting a mind of its own. If the doctor knows fear of a cancer plague because of the guilty departure of Twentieth Century medicine from the controls and safeties of the past – we prescribe pills before their side-effects are even detected – so the technologist has another kind of dread. He knows that technology, having occupied the domain of magic, now has a tendency to invade every last social taboo. Indeed, what is technology if it is not an ability to photograph the act and put it on television so that we may study our own creation? 'Son, that's how Ma and me were moving when you were conceived.' Yessir, that's technology, that's where the box office is – the century is so full of dread at the godlike proportions man has assumed, that the only cure for dread is to extirpate every taboo and see which explosions fail to come. Yet all the while we root out the taboos, everything primitive in us which still gives credence to the taboo, all the unspoken and conceivably tribal experience in the ducts of our dream rush up primitive, even primeval findings into our profoundest simulations. So the century feels a profound anxiety. That anxiety lives like the respirations of a clam in the clammy handshakes of all too many technologists and technicians. They know their work is either sufficiently liberating to free man from the dread of his superstition-ridden past, or their work smashes real and valuable taboos, and so becomes sacrilegious acts upon a real religious fundament. Could this not yet destroy the earth as it has already disrupted every natural economy of nature? That is the primary source of the great anxiety of the technologist as he stands before the idea that a machine may have a psychology.

So it is to be expected that a discussion of the psychology of machines has more fascination for a physicist or a serious engineer than for the average student of humanities. Aquarius, armed by his four years as an engineering student, would sometimes try to tease intelligent ladies and literary critics out of their supposition that science was an exact study with certain knowledge. On the contrary, he would assure them, there was no final knowledge whatsoever in science. We knew that gravity was an attraction between bodies, and we could measure that attraction, but why they chose to be attracted to one another was nicely out of our measure. We knew that energy was required for all work and that it sometimes took the form of light or heat or electricity, we knew it was stored in fuels and in explosives, we could measure energy in its various forms, measure it as majestically as the tides of an ocean, or gauge it so finely as the drops in the sea, but we did not have the remotest notion of what energy might look like. Nor did we know what electricity was. We knew how to use electricity, we could use it with precision in uncounted different instruments, but finally we knew no more about its nature than to say it was usually present as a flowing stream of electrons in a wire, and we did not know what an electron looked like. Certainly we did not know why passing a wire between the poles of a horseshoe magnet would have so magical a result as to cause an electrical current to run through the wire, anymore than we understood why any electrical current in a wire would in its turn produce a new field of magnetism. The entire world of communications was built on electricity, yet we did not even have the right to say that if magnetism was mood, and someone interrupted your mood, the quick impulse of irritation you might feel was electricity in your nerves. We did not know. We did not understand the ultimate nature of electricity. Nor did we comprehend time. There were numerous theories of time, but time remained as fundamentally mysterious as the notion that space was infinite, or matter consisted of individual atoms whose makeup was as complex as solar systems, and then proved more complex. Each year the number of subatomic particles discovered was greater. To laymen who had grown up on electrons, protons, neutrons and positrons, there were now mesons and photons and mu mesons, still more names and concepts no layman could follow, and science had less certainty today about the periodic table of the elements and the structure of the atom than at the turn of the century. The list could continue. If the nature of time was not comprehended, nor gravity, nor magnetism, nor the final meaning of the word electric, what did we know finally about sound? We could say that sound was caused by waves, but we could not explain why the waves made sound in the ear. That was still a mystery. So, too, was touch if one thought about it, and smell, and taste, and pain, and death, but these were all beyond physics, yes, somewhere between sound and smell the domain of physics ended. Small matter. Even in the very center of physics, before the phenomenon of light, darkness abounded. Nobody could be certain whether light was composed of little pellets, or traveled like sound in a wave, or was both. Both! When it came to ultimate scientific knowledge we were no further along than the primitive who thought light came from God. Perhaps it did. No physicist could begin to prove it didn't. So we didn't even know what a flame was. We had forgotten the majesty of fire, the impenetrable mystery. What indeed was a flame? A burning gas we were told. But why did a gas burn with a light? And from where did a gas come? Savages had looked once at fire and knew. God was in the wood of the trees and in the core of everything which burned, but now one could hardly remember that to look into a fire hot as the manifest of immanence might be equal to staring into the fires of Apollo 11 as the ship of flames began its way to the moon. What confidence was in that fire.

No, nothing was known at the root of science. To worship science was like being married to a beautiful woman who furnished your castle, bore your children, decorated and illumined your life, filled your days, was indispensable. Yet all the while you did not know the first thing about her true nature. Was she in love with you or a masterpiece of hate? There, in the center of the dream, was not an answer but an enigma. Was light corpuscular or a wave? Or both. Both!

These ultimate matters, before which the most sophisticated scientists could feel consummate ignorance, were subtly troubling to the technologist because he above all men was married to the beautiful woman and had relations with her every day, near to predictable relations, he was ready to devote his life to her predictability, and yet anxiety remained profound for he knew enough to know that the psychology of machines was not necessarily a fiction since even the function of an ordinary radio was not without its elements of savage awe. If an electromagnetic wave could travel for hundreds, or thousands, or millions of miles at the speed of light and all invisible, and be detected by an apparatus of no great complexity when all was said, being occasionally so simple as a length of wire whose end was attached to an ordinary crystal, and in any case no matter how elaborate an apparatus, one which consisted finally of endless refinements of that wire and that crystal, well, who was to say that the incredibly complex and artful networks of the brain fed by all the electricity of the nerves could not send electromagnetic waves as well (or some not yet detectable variant of them) and receive extraordinary communications in return. One did not have to be a champion of telepathy to recognize that if wireless communication was so simple, nay, so basic to the physical order of things, then telepathy might be more simple to conceive in its presence than in its absence; if one added to this thought a recognition of all the metal conductors in the rocks and in the earth beneath the magnetic fields of the atmosphere, and the unalterable fact that any metal form which passed through a magnetic field would generate electromagnetic waves which might be received by another form somewhere else sympathetic to it – well, the existence of natural receivers, of organic radios in every square mile of earth and God knows what unknown forms of psychic electricities still undetected by any instrument of measure but radiated in messages between every insect, blade of grass, and tree, the possible existence then of the earth as one giant communications bank of invisible transmitters and receptors! – it was enough to knock a technologist on his ear if he began to think too much of the psychology of machines. Indeed the electrical phenomena of electrical machines were so often so incomprehensible in small but painfully inexplicable ways that the astronauts and Capcoms and commo engineers at NASA used the word 'glitch' to account for an unaccountable electrical phenomenon like the light on an instrument panel suddenly turning on when the machine it serviced was most definitely off. That was sometimes not merely hard to explain, but impossible to explain. So they called it a glitch. God's own luminescence was in the switch! Give a better explanation! 'I just threw a glitch into the light when I was turning my warning lights off and on,' said Gordon Cooper during the flight of Mercury-Atlas 9 when a gravity signal showed on his switchboard during an orbital free fall. Cooper was renowned for his phlegm, but one butterfly of the night must have beat its wings in his throat when he looked at a dial which showed the force of gravity was present at a time when he knew he must be without weight. Yet note: whether it is with vanity, woe, or awe, he still takes credit for throwing the glitch. Who indeed has not felt the force of his own personality before a sensitive machine?

iv

If a covert belief in the existence of a psychology of machines is the sepulchral drama of many a physicist and engineer, there can be no branch of technology where the force of this unspoken possibility is felt more acutely than in the design of space vehicles. The early history of rocketry reads like an account of the burning of witches. One sorceress did not burn at all, another died with horrible shrieks, a third left nothing but a circle of ash and it rained for eight days. At the *Raketenflugplatz* outside Berlin in the early Thirties, rocket engines exploded on their stands, refused to fire, or when tested in flight lifted in one direction, then all but refired at right angles and took off along the ground. Indeed the early history of rocket design could be read as the simple desire to get the rocket to function long enough to give an opportunity to discover where the failure occurred. Most early debacles were so benighted that rocket engineers could have been forgiven for daubing the blood of a virgin goat on the orifice of the firing chamber.

Even at the end of the Fifties and into the beginning of the Sixties, the malfunctions of rockets were legend, and the first of the large launch vehicles like Thor and Jupiter, Juno and Vanguard and the early Atlas were all failures, misfiring on the launching pad, or failing to behave under control once in the air. The flat mournful moors of Cape Kennedy with their abandoned launching towers and gantries could remind the tourist of earlier stricken hours when Echo had failed and Mercury 1, Explorer, and Ranger I and Ranger II and Ranger III, Mariner I and Beacon Explorer A, Explorer XXI and Mariner III, Atlas-Centaur in its vehicle test and OSO-C Delta. Even as late as 1965, Gemini 6 had an engine abort on the Launch Pad.

The basic principle of a rocket was simple enough, it was as simple as lighting a cannon. Inflammables were brought together, ignited, fired through a chamber open only at one end – the gases of the flame pushing out pushed back against the rocket, and the rocket accelerated from the continuing push. That was simple enough, simple as Newton's Third Law of Motion: for every action there is an equal and opposite reaction – but physics was divine and engineering was sweat. The theory had all the elegance of natural economy, the engineering had to deal in the beginning with the almost intolerable problem that the heat of the flames melted the engine which contained them – so rockets never were able to function with any predictability until their cooling systems were perfected – metallurgy, ceramics and plumbing were all to take a mighty leap through mighty pipes and valves before Saturn V would lift. Then had come a host of problems on variations of trajectory: a rocket was not unlike a ball inhabited by smaller not quite symmetrical balls which rolled around within – so deviations were present in every trajectory. Rockets with solid fuel had a firing chamber which grew larger as the fuel burned away – therefore, the thrust altered; rockets with liquid fuel were obliged to react to the fact that the fuel sloshed around in the tanks. A world of instruments, of gyroscopes, radios, telemetric devices, computers and various electric monitors and controls moved into position on the rocket, each instrument to exhibit its own peculiarities, working difficulties, tendency to malfunction, and subtle hint of private psychology.

So dread inhabited the technology of rockets. Two of the most primitive and mysterious actions of nature, the force of fire and the transmission of thought, had been harnessed in machines which sat within other machines – the fire was controlled to aim and hurl a ship to the moon; the thoughts of men on the guidance and preservation of that ship were directed into electromagnetic circuits which propelled these thoughts and observations into other electromagnetic instruments thousands and hundreds of thousands of miles away. If the mysteries of physics were still unplumbed, if men were able to perform these actions without knowing altogether why they worked, no ordinary dread had been engaged, particularly if we consider that the management of fire and the management of thought are two of the most perilous activities for primitive man. Speak next of what it means to invade the heavens, then occupy the moon, the moon! that pale sister of Creation. Dread sat in the personality of NASA technicians, sat in the cold loveless air of concentration they brought to their work, sat like the guilt which squats in the vaults of all those banks where the ill-gotten gold of the world is stored. Like a capitalist who risks all the moral future of his soul on the gamble that God believes in capitalism and wants each man to seek to enrich himself as part of God's design, so the engineers at NASA lived in that ice-chamber of the moral heart where they could not know if their actions were divinely approved or abhorred, but dread showed in the chill dank air of air-conditioning and human relations at the Manned Spacecraft Center south of Houston.

And indeed the technicians had much to concern themselves with. The huge size of the launch vehicles, and the delicacy of their instruments, the length of the trip and the fields of magnetism and radiation through which they must pass, the virtues and talents of the astronauts, and their fatigue and occasional lapses had resulted hitherto in all the eight years of manned space flights in America in some astounding triumphs, and an unbroken list of mishaps, errors, and deviations in function in every one of the flights. Errors had cropped up in the ultra-complex machine of the spaceship, among the astronauts in their space suits (also machines) and in the network of communication on the ground. All of the Mercury flights had trouble: Alan Shepard had leaks in his thrusters – the language is not without its humor, but rocket terminology is not without its sexual reference – the hatch on the Liberty Bell 7 blew before the capsule had been secured – Grissom almost drowned; John Glenn had endless delays in lift-off, weeks went by. In flight the automatic steering did not work too well, the left thruster failed, then the right, the gyroscope indicators were at odds with what

Glenn could see for himself. Ground control received signals that the lock holding the heat shield in place for reentry had opened. If that was true, then Glenn would be dead on return. Carpenter in Mercury-Atlas 7 also had trouble with his control system; his deployment switch failed to release the landing parachute. He had to throw the switch by hand. With these malfunctions he overshot the recovery area. Schirra in Mercury-Atlas 8 had a flight suit which overheated, and the launch vehicle did a clock-wise roll after it left the pad – the booster engines were misaligned, Cooper flying Mercury-Atlas 9 had a host of malfunctions, the suit, the gravity light, the carbon dioxide level, finally, the automatic pilot system failed. He had to fly the ship into reentry himself. Not for nothing had they called their ships Freedom and Faith and Friendship. If the experience of being weightless was new even for a test pilot, and the sight of earth from a hundred miles up might have taxed the language of a literary giant ('The earth is real beautiful from up here, babe, I wish you could see how beautiful the earth is from here,' said the astronauts) certainly the recognition that one might soon be dead gave flesh to Freedom and Friendship. Such words may have arrived late to certain of the astronauts, have come to them indeed with the force of new concepts. Just as a Marxist who had abruptly discovered religion might have found whole echelons of thought in words like charity or grace, so a pilot, hard-bitten, self-centered, stingy, and anchored in military systems of authority might have found words like Freedom and Friendship not only strange and new, but not without comfort as they stood before these spooky technological ventures into the vacuums of space knowing that malfunctions in their equipment were bound to appear. So it had been with every Mercury flight, so was it to be with each of the ten Gemini missions, from Gemini 3 to Gemini 12, thrusters failing, short circuits, electrical problems, control systems affected, cooling systems failing to cool, overshot landings, missed rendezvous, computer failures, fading fuel cell batteries, transponders running out of power, engine abort because of a dropped plug, Armstrong in direct peril in Gemini 8 when the capsule began to whirl around frantically at sixty revolutions a minute, face-plates on EVA suits fogged up, eyes irritated by antifogging mixture, defective turbopumps – yes, mixed into all the triumphs of the Mercury and Gemini programs were the flaws whose reason could be found and the flaws which defied reason, and then the total of all these triumphs and all these ruptured taboos, all the flaws which had been corrected and all the luck which had been gathered, blew up into one vast moment when a machine to everyone's horror might even have developed a psychology of its own, for Spacecraft 012 mounted on top of Uprated Saturn I booster Number 204 during a pilot-manned check-out had a fire in the cockpit at 6:31 in the evening on January 27, 1967, there out on Launch Complex 34 at Kennedy Space Center and before the hatch was opened, the crew, Gus Grissom, Ed White and Roger Chaffee were dead. Of course, one did not have to look for a psychology of machines. There was a psychology of politicians as well, there was that complacency which finds its way into the most serious tests of the most critical materials when a large corporation has the highest influence in the highest places, as indeed which large corporation does not. One could look for the villain everywhere, even in the White House. There were more than a few to whisper maliciously that the bag if ever opened would have blasted Lyndon Johnson to the moon. And Bobby Baker could have been his launch vehicle. The investigation which followed was obliged to consider such possibilities as careless exposed wires, pure oxygen atmosphere, untested materials which had been termed fire-resistant but actually burned in pure oxygen, then had to look at discrepancies between what North American Aviation had delivered and what NASA had specified. The investigating board had been selected by NASA, there were some who accused it of whitewash, yet even so, how much was revealed! It was announced that the quality of work fell below standards and specifications. There were manufacturing discrepancies, unestablished performance goals, high rates of rejection of product, a lack of means to measure work done against work planned, 113 Engineering Orders considered 'significant' which were not accomplished by the time the Command Module in which the astronauts were killed was delivered to NASA, and 623 other Engineering Orders only finished subsequent to delivery. The ensuing cries of congressmen demanding to see an old and most critical report by General Phillips on work at North American in 1965 now resulted in a replay of stories about the investigation of Bobby Baker by the Senate Committee on Rules and Administration in 1963, for this long-gone inquiry had uncovered the fact that Baker and two partners had majority control of a vending machine corporation called Serv-U which received most of its income, close to a million dollars a year, from a contract with North American Aviation. That good contract had been signed three months after North American Aviation succeeded in obtaining the contract to build Apollo, a deal which would yet be worth over three billion dollars. Baker was described as being 'like a son' to Senator Robert S. Kerr, who was Chairman of the Senate Committee on Aeronautical and Space Sciences. Kerr had a company called Kerr-McGee Oil. One of its directors was James E. Webb. After a recommendation from Senator Kerr, James E. Webb became NASA Administrator, the highest job in NASA. Of course Webb was well-regarded by Lyndon Johnson. 'More than any other individual, James E. Webb deserves credit for United States success in space,' Johnson had later been bound to say. Yet the contract for Apollo had been granted after one hundred and ninety panels of experts in NASA had ranked not North American but the Martin Company highest in technical approach, technical qualifications, and business qualifications for the project. Martin received an overall rating on the scale employed of 6.9; North American, which came in second, had 6.6. Webb however and two of his deputies in NASA overruled the Source Evaluation Board of NASA, and gave the contract to North American after all. Three months later Bobby Baker had his Serv-U contract with North American. Bobby Baker – Senator Kerr – James E. Webb. This was of course the purest kind of guilt by association. But what association! North American was to weather the examinations and condemnations of the Apollo 204 Review Board after the fire. North American would keep the contract to make Apollo.

Yes, one could look for the villain everywhere, even in the White House, one could chase the death of those three men down all the lobbying corridors of the Capitol, one could make a political case, yet no engineer could be certain it would not have happened just as well if Martin had made the machine, for a wrench left behind by a workman was found near wires whose insulation may have been bruised by the opening and closing of the hatch. It needed only one minor short circuit in that capsule full of oxygen, one evil spark.

Besides, Grissom had a reputation as well. He was – bear the word! – a sport, and sports were known to attract witches late at night, and witches might have a curse. Grissom had almost died once in the Liberty

Bell, and White, his crewmate, now also dead, had been the first to walk in space on Gemini 4, and John Glenn, first American to orbit in space, had fallen in his bathtub and been harshly incapacitated. Was a curse building like the curse of the Pharaohs on the explorers who would open their tomb? Onto such Sunday feature stories may the dread of engineers be attached.

Whatever the reason for the fire, it was certain that everyone at NASA and every corporation attached to the project of Apollo 11 had worked in a different way since. Space was no longer on corporate effort so much as on a coordinated effort to produce the finest achievement in every detail. A massive holy wedding of the entrepreneur and the engineer had produced a massive vehicle tested and retested, checked and rechecked, triumphant in the four mighty flights of Apollo 7, 8, 9, and 10, and now ready to demonstrate before the world that the psychology of the machine did not exist and man could land on the moon. What an enormous subterranean fear that indeed it did.

V

In the Project Engineering Facility Building at MSC in Houston, a white building with neutral cream walls, red tile floor in the lobby, and a plastic ceiling with inset lights and metal baffles for the air conditioning, there was a room just off the entrance furnished with nothing but soft-drink machines and automatic food vendors. A man could buy his lunch off those walls. A choice was presented of preheated cans of Vienna sausage, corned beef, beef and macaroni, Dagwood sandwiches, pastry, chili, chiliburgers. One could have been back at the food trailer at Cape Kennedy still waiting angrily for a cold drink and the launch, but there was a difference. In this NASA building for space engineers, the automatic food vendors were a natural feature of the interior landscape: here men walked around with gnawing technical problems on their brain, that very brain they must accelerate each day with coffee and cigarettes in order to pull up from the gut a little more juice than had been allotted to the head by the body's requisitions on the energy pool. NASA technicians were men obsessed with the vertiginous interface between physics and engineering, and so they were likely to stay locked in their heads for hours, working and worrying at their small piece of the continent of techniques and inched-up improvements which lay over their work. So they did not want to go to lunch and talk. It was sometimes important not to have to meet people while taking on food, important not to have to go through the barrage of interpersonal relations and random Brownian collisions of personality in the company cafeteria – that was the worst interface of all – saying hello to a mixed set of friends, acquaintances and strangers. So they slouched in alone into the automatic vending machine room, deposited their dimes and quarters with all the small libidinal benefits which accrue from small insertions, and rose thoughtfully in the elevator back to their office and the hot can of lamb stew. Yes, people at NASA tended to have what Aquarius had come to call a lunar air. If there was a human aura that surrounded the body, an invisible spiritual envelope which if large enough, spiritual enough, dense enough, menacing enough or glamorous enough could be felt as a presence, an emanation, a – God save the mystique! – a charisma, then of all the people who had it and some had it like Sonny Liston and Muhammad Ali and Bobby Kennedy and Teddy Kennedy and Pearl Bailey and Jimmy Breslin and Brigitte Bardot and Charles Laughton, there were others who did not have it at all, hordes of hippies, for example, livers and aura consumed in forgotten vertigos of LSD. One had the impression it was possible to go up to such people and lay a finger on their skin. No notice of presence or unseen halo, no sense of a forced passage through some invisible envelope about the person, no welcome, no hostility would cause the finger to hesitate, not even for an instant, not even a quarter of an inch from the skin. Such hippies were moonmen: like the moon they had no atmosphere surrounding them. So they possessed not an aura but a lunar air – their envelope was gone. As FBI men used to be able to detect Communists anywhere (by a gift of the nose, by a gift of the nose!) so Aquarius always thought he had a sense of anyone who had ever taken too much LSD. Some salt of the libido was washed out; the aura was gone.

Now he felt that same absence, that same bemusement, same lunar air in all so many of the NASA technicians. It was not that they took LSD – he was confident most of them had not even gone near marijuana – it was rather a reflection of that same intoxication which thinned the blood and sucked the life from the complexion of the hippies: it was that rush of thoughts extorted by the iron of the will, that liege paid by the body to provide the brain with intellectual ecstasies.

Now, just as hippies might owe their lunar air to the consumption of the soul (and worse! the soul of futures unborn) in the fires of mystical states they had not earned but seized by a flight of drugs, so could the technicians derive their lunar air from an accumulation through the days of the little dread which never left, that particular dread which inhabited their insomnia like an incubus until the query over coffee in the morning break on how well did you sleep was no longer equal to a remark on the weather. There was a psychology to machines or there was not. Who could forget that eighty minutes after Apollo 7 had lifted off there had been a power failure in the Mission Control Center at Houston. In the Mission Control Center! One could not get nearer to the center of the technical brain. If the mission computer was kept operating by emergency power, by special backup power, still forty or fifty mission controllers stood before dark consoles for two minutes, unable to follow the flight, nothing to see in the room but the dull nightmarish red of emergency lamps. How their faces must have looked in that deep boudoir light! Then the power came back. The rest of the flight went more or less on schedule, except for fifty specified mishaps and the counterbalance of an announcement by good General Phillips, '. . . a perfect mission . . . we have accomplished 101 percent of our intended objectives.'

But give all power to the psychology of machines, there was as much to worry about in the very real interface between physics and engineering. Interface! Perhaps it was the biggest word at NASA. Interface was that no-man's-land where you joined the mouth of one bag to the mouth of a very different bag. Kissing, for example, was an interface. Indeed all primitive, which is to say, unclarified relations, were interface. The place where the first stage of a Saturn V was joined to the second stage was an interface because they would be first together then apart, and when they sundered there had to be a way to separate them, blow them apart, cut all the communicating wires at the same time. If the human leg could be used and immediately detached in such a way that no bones would groan and no blood flow we might speak of an interface at the groin. Interpersonal relations were interface – you bought your food in the automatic vending machine room to avoid interface, the waitress coming up to your dinner table for the first time was inter-face, the taxicab driver to the new passenger, the cop to the kid he has just arrested, the boy making love to his girl – interface was at every joint in Apollo-Saturn, the joint where the Command Module holding the astronauts met the Service Module holding the motor and the fuel to reach to the moon. Interface was where the Service Module met the SLA which held the Lem. Where the SLA met the Instrument Unit which directed the three stages of Saturn V. Interface was between each of those three stages. Think of the engineering which had gone into the ability to separate objects weighing hundreds of thousands of pounds and traveling as fast as twenty or twenty-five thousand miles an hour, yes, interface was the step from physics to engineering, from the state-ment that such a separation was possible by the laws of physics into the mechanical arrangements which made it work, and work 100 percent of the time (unless there was a psychology of machines to assert itself on the thousandth or ten thousandth time). Physics was a study of the order and courtliness and splendor and bewildering mystery of the rules of action in nature, a contemplation of its forces; engineering was immersion into the slippage coefficient of the adhesive applied to the nut which held the bolt of one ten-millionth of the total conception fleshed into a machine, that conception which first was drawn by a physi-cist on a blackboard. At a stroke! 'Here, we will have the interface. The stages will separate.' So physics was love and engineering was marriage. Physics was sex, conception and the communion of the family – engineering was getting the eggs out on time. Physics was the quiet remark, 'Give an object an escape velocity of 36,000 feet per second and it will be able to leave the gravitational field of the earth.' Engineering was the fifty years of rockets digging furrows in cornfields and catching fire on the pad from leaky valves. Engineering was the five hundred thousand men who burned up libido and slaved for years in the effort to put together enough collective effort to get a spaceship weighing six and a half million pounds to lift itself and acquire velocity sufficient to escape the six and a half million pounds of attraction the gravitational field of the earth would put on that ship.

Then, of course, not all questions could be solved by physics. Some-times engineering was obliged to work on its side of the interface with no certainties whatsoever. Think of a marriage created by computer, a couple found instrumentally suitable to one another: the computer, having studied the questionnaires they filled out, had joined them. But the computer, being new at this variety of human engineering, did not necessarily conceive of every requisite detail and property. Some characteristics are buried. So the couple were perfectly suited except that the husband had a body odor which was repugnant to the wife. Across the interface, physics was immediately consulted.

'Unfortunately,' said Physics, 'there is no acceptable science of smell.'

'Shouldn't a simultaneity of all other compatible attributes tend to suggest a compatible odor?' asks Engineering.

'That is the diabolical aspect of all these stinks,' confesses the doctor of Physics. 'Sometimes a smell is in concert with the collective attribute, sometimes it is out of phase. You will have to solve your dilemma with no help from viable theory.'

Engineering has its means. A deodorant is used. Should the problem be immediately solved, engineering has staked out another habitable territory free of physics. But wait. The solution proves temporary. The deodorant inspires a genital rash in the man. Solution B: other deodorants. Answer B: other eczemas. Solution C: inhibit the wife's olfactory nerve by surgery. Solution C is temporarily satisfactory. Wife, however, loses a portion of her ability to cook as well as she did before, and the marriage rolls down a slump vector in another direction. That is engineering without help from physics. For every flaw there is a compensatory move which can kick off new malfunctions which in turn will request other solutions with still other difficulties. Occasionally an elegant move arises – a male perfume is concocted to alter the husband's armpit: it blends, alters, nay *diverts* his natural odor over to one end of the wife's spectrum of acceptance. Now the function is relatively normal. That's engineering when it's happy. When it is miserable, the tests go on and on, and across the interface, physics turns out its pockets, speaks of the unplumbable depths of its mysteries, and the tests themselves drip with breakdown and malfunction – the machines to test the machines are even exhibiting a psychology of machines. Leave dread alone – out of the sheer exhaustion of the intellectual search is enough libido burned to give the average engineer his lunar air. No wonder there were feature stories now and again in local papers about the high rate of divorce in NASA circles. 'He has no time left over from his work to give to me,' was the cry of the wives across the interface. They could as well have screamed: 'The machines have sucked up his libido.'

vi

Now it is the product of all this anxiety, all this hard work, dread, political lobbying, corporate corruption, and messianic teamwork, this overburn of libido, and sense of mission in a new monastery of men, a sense of mission indeed unique in the moral junkyards of the century, that, yes, as a product finally of the twenty-four billion dollars spent, and ten years of concerted effort, and mishaps beyond number, and the tragedy of a fire in which no high executive associated with NASA could be certain he was without guilt, yes, as that product, Apollo 11 stood at last on its Launch Pad on top of Saturn V, and was ready on the morning of July 16 to blast-off for the moon. And as the astronauts lay on their backs, at the top of that enormous stack, eyes pointed up to Heaven (except that they could not see the sky for the nose cone of Apollo 11 was covered by a heat shield) as the astronauts lay waiting, Neil Armstrong in the left seat, Buzz Aldrin in the center, and Mike Collins on the right, the view above them was a bank of instruments within reach of their hands, and from that short distance near as numerous in appearance as the stars, some six hundred and fifty switches, dials, meters, circuit breakers, controls and displays in formation after formation of instruments overhead and to their sides, black buttons, gray buttons, striped yellow and black latches with hooded covers for controls so critical they must not be touched by accident, squares of glass for warning lights, computer windows, abort lights, master alarms, velocity indicators, flight controls, controls for stabilization, propulsion, attitude indicators, altitude indicators, just some of the banks of buttons and switches set on the gray panel overhead for the commander, a mounted variety of toggle switches, rotary switches with click stops, thumbwheels and push buttons, even controls with locks which must be released before they could be operated. Now, add the controls of the pilot in the center seat with his commands over reaction control propellant management, over environment control, over fuel storage subsystems, and the pilot on the right now in charge of his two-hundred-odd pieces of selection and monitoring over communications, electrical control, data storage and fuel cell components, service propulsion subsystems, the three pilots all capable of performing each others' jobs in emergency, all of them aware, indeed immersed for months in the practice and malpractice, the meaning and occasional ultrameaning of the six-hundred-odd varieties of warning and choice personified in these controls, these switches, these breakers, these buttons, these instrumented schools of technological fish so ready to detect every portrayal of emergency, and watch over the sequences of their mission, the changes of their velocity, the measure of their fuels, the control of their environment in that gray conical home and chamber and ship, not twelve feet wide inside, not ten feet high, the Command Module with its several systems of radio, its tanks and batteries and fuel cells, its caution and entry systems, and below these overhead dials, there out in front of their extended feet another host of cabinets and spigots and new dials for food, for medical monitoring instruments, for film and sanitation equipment, hot water, navigation instruments, guidance-system optics, rendezvous radar transponders, sextant and telescope, waste-water tank, oxygen regulators, pressure regulators, suit test valves, temperature controls, cabin-pressure controls, mission timer, fan motors, purge valves, radiators, exterior lights, interior lights, the Ordeal panel (Orbit-rate-drive-earth-and-lunar panel) and two hand controls on the armrests of the commander's seat, the Thrust-Translater Controller, which enabled him to accelerate on any one of three axes, forward or back, up or down, right or left, and holding the same handle to be able to begin the abort sequence, which could cut the mission short merely by turning the T-shaped handle in a counter-clockwise direction, as well as the Attitude Controller Assembly on the right armrest, which was used to maneuver the spaceship via its small rocket thruster engines into any combination of pitching nose-up or nose-down, pointing the craft into a yaw left or right, or rolling to either side – yes this whole constellation of systems and subsystems of control over power, and control over the beaming out of thought translated into electromagnetic waves, and control even over the disposal of simple human waste sits over, above, around, beneath, yes and even behind the three astronauts in whole congeries of Twentieth Century concepts and forces which have come to focus that this effort may fly to the moon. And so Armstrong, sitting in the commander's seat, space suit on, helmet on, plugged into electrical and environmental umbilicals, is a man who is not only a machine himself in the links of these networks, but is also a man sitting in (what Collins is later to call) a 'mini-cathedral,' a man somewhat more than a pilot, somewhat more indeed than a superpilot, is in fact a veritable high priest of the forces of society and scientific history concentrated in that mini-cathedral, a general of the church of the forces of technology, for think not of the fifteen miles of wire in that small capsule, but of the vast multibillion dollar technological bands which belted the very economy of the nation, bands which gathered together the technological solutions to problems intricate, explosive, and sometimes not comprehended by physics (merely manipulated by engineering) and ran them up and down and in and out of every dial, switch and circuit breaker in the plastic interior of that capsule, and even belted the plastic surface of that gray Armalon fireproof fiber-glass cloth which made the bed of the couch on which the astronaut in his space suit sat, no, lay under the roof of the rocket, flat on his back as a baby in his crib, and talked into the plastic band of black microphone at his lip, yes, Armstrong was a general of the church of all these forces holy or most uncomfortably unholy and now soon to be on the loose.

It was American capitalism most undeniably in the saddle, not that capitalism which had come out of the hard Scotch-Irish fiber, the very wings of the Wasp, not that capitalism in which greed and piety had done their wild Celtic jig, some blessing or curse from the Druids still stinging the toes, not that good old Midwestern Scotch-Irish, German, Swedish, English, Welsh (and who knows what other good cold bloods?) which had gone into the making of that old-fashioned profit and propriety, that greed and service, itch for exploitation and nose for growth, that quickness to cheat and that sense of solid procedure, gambling man's love of destiny and workman's love of tools, pillar of moral rectitude, friend of high hypocrisy – no, not that particular capitalism which had come round the bend into the Twentieth Century full of its wild dreams and orderly homes, that turn-of-the-century American capitalism, the only prosperous middle-class power ever built on a series of outrageous gambles, no, that capitalism had passed over to its son, corporate capitalism, and corporate capitalism was the marriage of huge profit with huge service, of teamwork – the methods of the hospital mixed with the methods of the football team – and of detestation of contradiction. Where the old capitalist had a rock glint, 'I'm a crazy old bastard,' he would confide to any reporter, thinking of his ability to water the stock of widows and head the drive to distribute Christmas packages to the poor, proud of every paradox in him, as if in the boil of his contradictions were the soups and nutrients of his strength, so his son was a dull-eyed presence, a servant of reason – contradictions as odious to him as words of filth before a table of the immaculate. Therefore, the son was never about to work alone. There is no contradiction in a committee, not at least after it has delivered its collective measure of finding. Like the father, the corporate son sought to make profit for the corporation, but neither he nor the corporation did it for profit, they did it for reason, did it to remove contradiction from the earth, remove problem and heartache from human interface; subtract germs, viruses, pests and bugs from the bounty of nature; subdue contradictory ideologies in foreign affairs; extirpate irrationality from conduct and inefficiencies from machines; and sometimes Aquarius would even suspect they wished to remove human activity from divine punishment. For the more quickly that technology demands new certitudes from physics, and receives mysteries instead (being obliged thereby to call a halt or advance blindly on the engineering side of the interface) the more does technology secretly suspect that God is irrational in His judgment, and best kicked upstairs where He will not interfere with human disposal of life after death. Such at least is Aquarius' view of the American corporation. He thinks it is not sublimely suited to deal with heartache on the male or female side of the interface – in fact he sometimes thinks the reason the American family is in so much danger is because life in ranch houses is so deadly. He knows every bug the corporation slays by reason creates two by the stimulations of mutation, he knows the bounty of nature is gagged, in fact plastic wastes are stuffed into every one of her mouths; he knows Communism thrives on the opposition of the American corporation, and degenerates into civil war when left alone; he suspects rational solutions to poverty will end in delinquency, and finally he knows that the corporate son of the old capitalist does not often get the inefficiencies out of his machines, for he is more interested in communicating his love of reason than in applying it.

So he designs new machines before the old are tested and vitiates the intestinal fortitude of the old mechanical insides by using inferior materials which break down too early. The money which used to be spent for good materials is now spent for advertising the product. He speaks of a society of reason and tells lies every time he opens his mouth. The more simple and ubiquitous is his commodity, the greater are his lies. No surprise then if the lunar air of engineers at NASA was reminiscent of the hearty hollow manner of many executives in many corporations, that look which suggests one is trapped on the inside of a drum. Every corporation executive told lies because it was in the nature of his work to be expected to make positive statements, but he was obliged to make them about matters which were larger than a man could comprehend. How indeed could one know what the result would be of five million people all brushing their gums with your toothpaste each morning, and who knew if your toothpaste was better, and if not, why bother? Why consume one's life in this nasty nugatory activity of pushing at products, and indeed what did it matter if one toothpaste was better than another – let each consumer find his own. So a quiet sense of woe lay over all the antiseptic aisles of all the corporations, mixed in with the office jokes and the good Wasp regard for positive living and healthy cheer, a subtle woe fine in its strain for most, but intense in its concoction when rendered, if any corporation executive cared to render such a bitter drop, for it was a death-heavy guilty stone-heavy woe that man exhausted his working hours on earth in corporate activities so ridiculously petty that the only answer could be that he did not dare to push further and faster into real ideas of nature, manufacture, joy, dread, death, and human drama because his own ideas had gone dead before his sense of shock at the taboos he had smashed in the century.

Consider that sense of life programmed and wasted, of reason so overapplied to life that all contradictions having been killed, the light of reason had finally left the eye, yes that was also in the bands of force and process and network which flowed across the country and came to focus in the bank of instruments eighteen inches over each astronaut's head as he lay in his plastic suit on a plastic couch – lay indeed in a Teflon coated Beta-cloth (laid on Kapton, laid on next to Mylar, next to Dacron, next to neoprene-coated nylon) space suit on his Armalon couch – plastic, that triumph of reason over nature! – and the astronauts, whether feeling a focus of such forces or oblivious to them, were nonetheless a true focus of concentration for every corporation executive in the country, even for the man who sold the toothpaste, since aerospace was not all that was in this capsule, the food and drug industries were here as well in their special medicines, extraordinary toilet papers, and exceptionally processed foods, so too the construction industries, the gas and fuel industries, metal industries, forget the corporation, occupation, or industry which did not have its relation to one of the factories responsible for one of the buttons in the six-hundred-and-fifty-odd buttons, switches, dials and controls which fanned over the heads of these generals of the church of corporate endeavor. The corporation had replaced the old existential gambles of the small-town capitalist and the lonely tycoon with the security of the organization. The corporation substituted cooperation for competition, and reason for struggle, but because its activities were usually at once immense and petty, like the manufacture of toothpaste, immense and noxious, like the production of cigarettes or poisons for war, or immense and depressing, like the shoddy production of slovenly functioning automobiles, or even immense and scandalizing, like the ways in which aviation contracts were garnered, the corporation not only gave security but engendered loneliness and woe, that same heart-heavy death-heavy stone-guilty woe. Small wonder there was a communion of feeling around the nation for these generals of the forces of a new church which offered a new communion in replacement for the loneliness of meaningless effort: here at last was American capitalism attached to a corporate activity which was momentous, dangerous, awesome, and a palliative to dread, for if the venture succeeded, could the heavens despise the nation? Supported by the memory of tragedy, the three burned and asphyxiated corpses in the hideously charred cockpit, that very recollected sense of pain gave sentiments of nobility to corporation executives looking to find a line of connection between their work and the vault of this endeavor.

It was not that NASA had done a perfect job, in fact there were critics enough to accuse it of timidity on one hand and rash acceleration on the other, scientific critics to decry the lack of real scientific investigation on this trip, critics who called for unmanned flight, critics who cried for an end to explorations in space altogether until our problems on earth had been properly solved, and no one could deny that NASA, once recovered from the twenty-month paralysis which followed the death of Grissom, Chaffee and White, had accelerated its flight schedules at so inhuman a pace the suspicion had to arise that they seemed to think if they did not make the moon in '69, fulfill the dead and martyred President's directive to land a man and return him safely by the end of the decade, that indeed there would be no landing at all, and the end of all space effort. If NASA did not believe in a psychology of machines, it certainly seemed to subscribe to a psychology of events, for from October 11, when Apollo 7 with Schirra, Eisele and Cunningham went up, until nine months and a few days later, to this day of July 16, four flights in all had been made, Apollo 7, 8, 9 and 10, flights on sixty-day centers, flights rotated so quickly that the lessons of the last flight could hardly be catalogued, let alone absorbed and employed before the next flight was ready, and indeed some very large chances had been taken. Before Apollo 7 all sorts of flaws had been found in Apollos 4, 5, and 6, all unmanned. Apollo 4 spilled fuel during the flight, and a computer station suddenly shut down. On Apollo 5, the Lem, being tested for the first time, was able to fire its engine at only ten percent of maximum thrust, and that for only four seconds instead of thirty-eight (it was not to be tested again until McDivitt and Schweickart took the large risk of flying it in Apollo 9) yes, Apollo 5 had been so unsatisfactory a flight that even the windows of the Lem had broken while readying for the launch, and that for no reason engineers could discover. And Apollo 6 had been a near disaster. Two of the five second-stage engines had cut out prematurely. The third-stage engine failed to re-ignite when it should. As a result, the spacecraft went into the wrong orbit. On Apollo 7, to repeat, there had been fifty mishaps, everything from malfunction of the navigational guidance and control equipment, to circuit failures in flight, plus sticky control handles, mysterious surges of the capsule, a nine-minute communications blackout on the seventh day of the mission, three days of medical monitoring lost. Perfectionists could have spent six months recovering from those fifty mishaps, but this was October 1968, this was the year in which the war in Vietnam became a debauch, and Bobby Kennedy and Martin Luther King had been killed, and Lyndon Johnson abdicated, this was the year in which the colleges showed open signs of revolution and corporation executives for munition firms were barred from campuses, it was a year in which industry was still recovering from the shock that General Motors had put private detectives on Ralph Nader. (Didn't General Motors understand how modest and uneventful were the private lives of savage writers?) So NASA had barreled ahead, NASA had pushed for results as if the very nerve of endeavor in the nation depended on their effort, depended on doing it before '69 was out (what a joke for the sophistications of Jack Kennedy that men would first kiss the moon in '69). Ignoring the dubious performance of Apollo 7, they rushed Apollo 8 into readiness in the next sixty days. That was the first flight of men ever to go up in Saturn V, but the trip went all the way to the moon, passed around it ten times in orbit, some orbits so low as seventy miles away, while Frank Borman read the opening words of Genesis on color television to a Christmas Eve audience and brought his crew back before the year was out. After that, it seemed as if NASA could not make a mistake. Gone was the bad publicity of the days when Gemini flights were regarded as stunts, gone were the horrors of the investigation of the fire. The sense of waiting for a blowup had abated. A new sense seemed to have arisen that this congeries of corporations must work in ways they had not worked before, cooperate as they had not yet learned, tell the truth across the board in fast-working practice – it was no longer corporate procedure but mission procedure. The hundreds of thousands of workers in NASA and the corporations who worked for NASA now had a holy task, they had the most serious and honorable work of anyone hired by any corporation in the nation and so, like that Russian people which learned how to wage war and discover efficiency in their factories only when the country was in danger of whole occupation by the Nazis, so did American capitalism finally put together a cooperative effort against all the glut, waste, scandal, corruption, inefficiency, dishonesty, woe, dread, oversecurity and simple sense of boredom which hounded the lives of its corporate workers, and prove finally to react as if the country were being occupied. And indeed it was. For years, the forces of irrationality had been mounting into a protective war against the ravages of corporate rationality run amuck. Now corporate rationality to save itself would commit the grand, stupendous, and irrational act (since no rational reasons of health, security, wisdom, prudence or profit could be given) of sending a ship with three men to the moon. What a mighty congregating, what a streaming of workers and techniques toward the base of the rocket, what a gathering of the hours as the countdown proceeded, what a last week of huge effort to lift Apollo 11 from the ground.

vii

Not all the men who worked for NASA were hearty and hollow or had a lunar air. Of the five thousand who worked on the launch, perhaps half were present at one time or another on Pad A at Launch Complex 39, and they were 'so highly motivated,' said the Launch Operations Manager, Paul Donnelly, at a press conference, 'that even a loser comes in to do a good day's work.' Donnelly was a middle-aged leprechaun of a man with boiling green eyes, pepper-and-salt blond hair, a snubbed Irish nose, a green jacket, a white shirt, a greenspeckled tie. When asked if he was tense about the launch, he snapped, 'Suppose you have to have a brain operation? Would you want the surgeon to be tense?' He was firm in his answers, tough as a bowstring ready to send its arrow. 'We lift off if the ceiling is anywhere over five hundred feet,' he said in reply to one query. 'Maximum allowable wind is twenty-eight to thirty knots,' he gave back for another. He had diagrams, he showed details of the countdown, he had been a semipro baseball player in his youth.

That was one kind of man around launch operations. The Director, Rocco Petrone, already quoted, was another, and he was a big massive ex-football player at West Point with thighs for forearms and eyes which suggested the compressed power of oil about to come in at the bottom of a well, his mind – by reputation – a store of endless detail. Many men who looked like him were attached to the work of the launch. It was heavy work. Over six million pounds were going to go up in the air, and go up by the force of a fire which consumed its fuel so fast – over half a million gallons to be burned in two and a half minutes – that one might as well think of it as a controlled explosion. Where everything at MSC in Houston was drawn, wired, small, neat or locked up, the buildings without character, the technicians lunar – the structures here at Kennedy, the VAB, the Mobile Launcher, the Transporter and the Mobile Service Structure were in contrast gigantic powers gathered for a climax so rapid that months of work would have their fruition in two visible minutes across the sky, and indeed the power of commanding the flight would soon pass from the Control Center at Kennedy to Mission Control at Houston, where it would remain for the next eight days. So it was no accident that men who were strong and definite, men agile as baseball players or as isometrically bottled as guards or fullbacks should be attracted to launch operations. Indeed many of them had the kind of sullen collected hurricane violence you find in the Mexicans and Indians who work in Texas oil fields; a few of them dressed with the sportiness of Las Vegas spenders; if whisky and jazz was to be located anywhere in the ranks of NASA-men, it was in these technicians and workers whose problems revolved around pumping the million gallons of fuel into the bellies of the stages of Saturn V. If their eyes seemed tuned to the depths of fuels ready to roar into hundred-footed flames, if preparations for the launch left worry on their heavy faces, it was not because they had the kind of dread which comes from smashing taboos. No, Cocoa Beach had been wild and raucous in its time, and you could still hear such jokes in the strip bars as: 'I want to commiserate with all you folks from Georgia because Governor Maddox just had a rectum transplant' . . . gulp . . . 'and the rectum rejected *him*!' No, if there was dread here, it was the kind professional athletes have: will we win the game? They know they cannot even think of not winning, for luck comes in strings. If they lose, worse retribution will follow. So they concentrate upon Wernher von Braun's mighty Saturn V, and like good athletes take the effort of raising the rocket into their dreams, even in sleep they go over the function of each pipe and valve.

Apollo 11 had moved out of the VAB building on May 20, almost two months before, through doors advertised to be large enough to receive the UN Building, doors which left an opening forty-five stories high in the fifty-two-story-high wall of the VAB. Through that exit, inching along at about the speed a turtle can trundle down the road, had emerged Apollo-Saturn, attached next to a movable structure larger than itself, the Mobile Launcher, declared to be 'the heaviest portable structure known to the Free World.' The launcher was four hundred and forty-five feet tall, it weighed twelve million pounds. Its base, on which Apollo-Saturn sat, was two stories high and covered half an acre. Its vertical structure rose for forty-four stories, an open pillar of battleship-gray girders and platforms which resembled a skyscraper in early construction. Apollo-Saturn was tall, it was 363 feet high, thirty-six stories long, and it stood straight up, but the Mobile Launcher was larger, it towered above the rocket like a basketball player seven feet tall reaching over a player who is five and a half feet tall – there was even room for a twenty-five-ton-load overhead hammer-head crane to extend comfortably from the top of the Mobile Launcher over the needle-sharp nose of the Launch Escape Tower on the very top of the rocket. Below, nine retractable steel bridges, some sixty feet long, reached out like arms, umbilical cords and phalluses from the Launcher to Apollo-Saturn, offering access back and forth from the space vehicle to seventeen work platforms, some movable, some fixed in the four-hundred-plus feet of rise of the Mobile Launcher, the work platforms providing a base for loading propellant, charging pneumatic systems, checking out the instruments, and overseeing the electrical networks.

Now this huge piece of sculpture composed of shining Apollo-Saturn and the intricate girder-work of the Mobile Launcher, each embracing the other over nine bridges of steel, came crawling out of the doors of the VAB at the pace of a tortoise, or the pace of a caterpillar. The last velocity is not ill-chosen. Apollo-Saturn and its mate were installed on a giant crawler, a six million pound behemoth of a moving fundament with four tracks ten feet high and forty feet long, each track at the corner of a two-story steel structure 131 feet in length and 114 feet wide, an area larger than a baseball diamond. This tractor, the size of three long barges strapped together broadside had a weight in combination with its vertical cargo, forty-five-odd stories high, of over eighteen million pounds, so this caterpillar tractor had individual treads on each crawler track which measured over seven feet in width and weighed a ton. This tank which, in relation to ordinary Army tanks, made the men clustered at the feet of its tracks look like Lilliputians, this mechanical monster and marvel with its modest two diesel engines each but a fraction under three thousand horsepower, had an overall top land speed when loaded of one mile an hour, but generally it took six or eight hours to travel the three and a half miles from the VAB to Pad A on Launch Complex 39, for although the Crawlerway was a special fine highway as broad as an eight-lane turnpike with median divider, and had six feet of various kinds of hydraulic fill, crushed rock, asphalt prime coats, and small river rock for topping to support these ambulatory eighteen million pounds, the road had nonetheless a couple of demanding turns to be negotiated, and a five-degree incline to be ascended at the finish in order to reach the launch base of the pad itself, that last some forty-eight feet above sea level. Before one even begins to conceive of the terror of Apollo-Saturn and the Mobile Launcher keening their forty-five stories over some five degrees (as if to reduce the Tower of Pisa to a tortured sapling!) rest in the comfort that the crawler had a hydraulic leveling system powered by two one-thousand horsepower diesels (1,065 h.p. to be precise) which could adjust the base hydraulically within an inch, no, two inches of the horizontal while climbing the last five percent of the incline up the concrete launch hill. That was done with the aid of a sort of carpenter's level one hundred and thirty feet long from end to end – doubtless the longest carpenter's level in the world – so the top of the space vehicle was kept vertical within ten minutes of arc, not bad, a deviation not much larger than the diameter of a volleyball.

One could assume that the possible presence of high winds, slow turns, and upgrades would be the reason the crawler travels at less than its full speed, travels rather at turtle speed, half a mile an hour, but since the sight of the open skyscraper and the rocket in nine-armed embrace clanking along the Cape Kennedy moors at a rate somewhat less than one foot a second is a sight no man has ever seen before he has seen it, it is indeed a moment in the symbolic pageantry of legend perhaps not unequal to that hour when Birnam Wood came to Dunsinane: perhaps the exquisite sense of caution in Rocco Petrone which is reflected in the speed of half a mile an hour, instead of the possible rush through at twice that rate is due to some secret pleasure taken in the magnified luxury of treating all the workers at the Space Center to the pleasure of watching their mighty moonship edge along the horizon from morning to dusk, or even more spectacularly at night, with lanterns in the rigging, like a ghost galleon of the Caribbean! The beginning of the trip to the moon was as slow as the fall of the fullest flake of snow.

The trip had been made on May 20, just two days after Apollo 10 had taken off – Apollo 11 was in fact on the pad before Apollo 10 had made the first of its thirty-one orbits about the moon. There Apollo 11 was to remain for the next fifty-seven days while finishing touches were put to its works, and a Mobile Service Structure even stouter than the Mobile

Launcher, if not quite so high, was brought in from its parking station a mile and a half off to approach the rocket from the other side. Now two giants of gridwork, one forty-five, one forty stories tall flanked the rocket, their mutual arms encircling it from all directions, the new Service Structure employing five adjustable platforms which could move up and down, and open and close each floor like adjustable jaws. Swathed and swaddled in open air, technicians with access to her at every hatch and every port, the space vehicle was given its Flight Readiness Test. If the components had all been tested before in vacuum chambers and simulators, if, in fact, the assembled rocket had recently been checked in the VAB via a Plugs-In Test, a full mission simulation of every moving part expected to function in flight, every switch, valve, gimbal and sensor having been activated, now again, outside, on the ground of Launch Pad A Complex 39, the Flight Readiness Test was given, the complete vehicle was counted down, the Command Module was put through an imaginary mission to the moon, all connections hydraulic and electric tested between the ground support equipment, the technicians at the base, the technicians in the firing room, and the computers buried in their insulators and shock-proof containers at the base of the Mobile Launcher, all to be coordinated finally in that Firing Room of the Launch Control Center built on a wing of the VAB, three and a half miles away. All this had been done in the intervening weeks, and then there was a last Countdown Demonstration Test, a rehearsal in full of the real countdown. This exercise required loading liquid oxygen and liquid hydrogen aboard the stages, then draining the fuels after the simulated time for ignition had been reached. Yes, Apollo 11 had been filled and unloaded in order to strain every connection and joint, probe every hose and every seam for leak, and since the huge rocket was as sensitive to change as a harp taken from the cellar to the sun room, the pipes and valves shifted their accommodations subtly: launch engineers having learned to live with some sympathetic equivalent of a sense of pitch, the rocket was in effect retuned after the simulation. As the pad leader, Guenter Wendt, would put it, 'The whole thing moves and groans. It has its own noises you haven't heard before.'

It was a Promethean preparation for a Herculean task, and by its end the last week had been reached, the last ninety-three hours of countdown had begun, this count to be spread over five days, with pauses or holds designed to give time to solve unglimpsed problems or relieve crew fatigue; on they pushed now into the final tuning and tightening of preparations on the rocket, every gauge and every dial, every display and every computer on the space vehicle now monitored in the Firing Room at Launch Control Center, the nerves of hundreds of engineers (with the experience of previous launchings to fine those nerves) were now alert for the subtlest signs of any malfunction which might first be revealed by some hesitation of a valve, some fluctuation in pressure, some recalcitrance in an electrical sensor – for five long days the count proceeded, and then with nine hours to go, over six hours before the astronauts would even approach the hatch to the Command Module, the most critical operation began, the loading of the four and a half million pounds of liquid oxygen and liquid hydrogen into the tanks of the three stages.

viii

For the first and largest stage, a kerosene called RP-1 would be ignited with liquid oxygen. That fuel would fire the F-1 engines. There were five of them, each capable of delivering better than one and a half million pounds of thrust, the largest rocket engines ever designed – any one of the five motors was in itself ten times as powerful as the rocket in the Mercury-Atlas which first carried John Glenn into orbit, and all of this force fifty times greater than Mercury-Atlas, was grouped at the base of Apollo-Saturn, one engine at the center, the other four at the corners of a square, each motor eighteen feet high and with a thrust chamber or bell near to twelve feet in diameter at the bottom, thus a huge bell comparable for example to the great bell at Moscow raised by the Emperor Nicholas in 1836, or the bell at Notre Dame, or at St Paul's in London – the bell of the thrust chamber in these rocket motors was in fact larger by far than Big Ben or the Liberty Bell, and flames hundreds of feet long would fire from it when the oxygen and kerosene were ignited. Indeed much more than two thousand gallons would be consumed each second. That was one good reason to use the kerosene called RP-1. Since it was stored at normal temperature, that fuel was relatively easy to handle. In fact, it was even loaded into the first stage before the ninety-three hour countdown had begun. The problems of storing it in the RP-1 tank of the first stage were no greater relatively than holding fuel oil in the heater tank of a house basement for a few weeks. While kerosene was about half as efficient by weight as liquid hydrogen (which was used with liquid oxygen in the second and third stages) the ease with which it could be stored encouraged the designers of the F-1 engine to employ it. There is a tendency in engineering design to take only one large gamble at a time; if you are planning a new machine, sound procedure suggests taking on no more untested techniques than are necessary. So, while kerosene for the first five engines would weigh almost a million and a half pounds, and the use of liquid hydrogen would save three-quarters of a million pounds in weight, still no one had ever tried to transfer that much liquid hydrogen through pipes at a temperature of 423 degrees Fahrenheit below zero. (As Marx's Engels had been the first to point out: Quantity changes quality.) Besides, liquid hydrogen was so light in relation to kerosene that it took up more volume; it would have needed a tank five times as large, which would have taken back some of the saving in load. So the engineers designed the huge F-1 engine to use oxygen and kerosene. The additional weight would not prove so crucial at the base because one could always increase – at least up to a point – the amount of fuel, and therefore the duration, of the first stage and so lift the extra heft of the kerosene. It was in the upper stages, and in the Apollo spaceship itself, that one had no choice but to conserve on weight, for those fuels would be dead load for all of that period the first stage was burning. It was one thing for kerosene to be obliged to lift its own relatively heavy mass, quite another to have to raise kerosene which would be doing no work until later. So RP-1 was employed for the first stage, and was pumped into its tank one hundred and twenty-six hours before lift-off. (The tank, incidentally, held 212,846 gallons.)

The liquid oxygen and the liquid hydrogen, the Lox and the LH_2, were a different matter. They were gases converted to liquid and therefore highly unstable. The physics was basic. In the same way boiling water becomes steam, so every liquid will convert to a gas if it is given enough heat; in turn, every gas will concentrate itself into a liquid if the pressure upon it is high enough, or the temperature sufficiently low. As rocket fuels, oxygen and hydrogen were used in liquid rather than natural form because more oxygen and hydrogen could be kept in the same space once the gas had been compressed to become liquid. But what problems to contain it! What higher powers of hydrodynamics were now demanded to pump it! For it had to go from its storage tanks through a quarter-mile of pipe all the way to the bowels and belly of the rocket, and that at a rate which must be so high at times as ten thousand gallons a minute, more than a hundred-sixty gallons a second of a liquid down 423 degrees below zero had to pass by any given point – what an icy rush of sludge! Yes, hydrodynamics, that sturdy engineering science of plumbing, pipes, valves and pumps, had taken its own leap to create Saturn V – the world of technology had been shifted in its natural evolution by the development of the moonship, whose inbuilt emphasis on hydrodynamics and electronics had altered the balance of technological achievement as much as our conception of man might be staggered by a new species whose capacity to think and pass food through his system had raced far ahead of his ability to walk, run, play, work, or do anything else. The pipes which carried the liquid hydrogen were vacuum-jacketed tubes of stainless steel, pipes fifteen hundred feet long, ten inches in diameter and pushed from storage tank to rocket tank by a pressure so great as sixty pounds per square inch, which once one recognizes how light is liquid hydrogen – hardly more than half a pound a gallon, half a pound a gallon! – is certainly at sixty pounds a square inch, one mighty push on such froth. Each pipe proved to be a pipe within a pipe, a vacuum installed between to maintain the insulation of the extreme cold, and indeed the speed of the pumping was doubtless to reduce the risk of the leak, for vacuum or no, that kind of cold fuel, that intensely low cryogenic liquid energy, was bound to warm if it tarried in the pipes. It would then expand, leaks would spring. 'We can live wiz some leak,' Kurt Debus, the Director of the Space Center, had said in his fine Junker accent, leaks in such high pressure pipes were the order of procedure, indeed every pipe and valve in the whole plumbing combine of Apollo-Saturn, the Launch Pad and the storage tanks was redundant, which is to say that for every pipe there was another unused pipe equal to it and ready to be used if the leaks in the first could not be found and controlled, leak they could live with – if it were modest – but everything in their technology looked to remove the likelihood of such conditions, and so they rushed the fuel through the pipes, through the ingeniously fitted double pipes and valves jacketed to keep the vacuum intact, past the jacketed couplings, through the insulated umbilicals, up into the insulated tanks of the rocket stages, there to sit, and boil, and boil off, a small percentage of this million gallons of varied fuels stirring in their icy containers, turning to gas, and then being vented into the air to form the clouds which clung to Apollo-Saturn through the early morning and on into the dawn of the day of its launching. As the fraction boiled off, they topped it with new fuel, kept up the topping, kept it up indeed until the last of the countdown minutes was ticking.

Now, nine hours earlier, they had cleared the launch area for the loading of propellants, all of that blast area which could go up in conflagration if the rocket blew. But it was beyond the imagination to conceive of a million gallons of fuel suddenly ignited. Such an explosion would bear comparison to the early atom bomb. So no one was in the area while the fuel was transferred, no need, the loading was done by automation, Saturn V sitting silently, ship of space on her pad. The Service Structure having been drawn down the road by the crawler at lift-off minus 10 hours, Apollo-Saturn was alone again with her Mobile Launcher. For four hours and thirty-seven minutes, from lift-off minus 8 hours and 15 minutes to lift-off minus 3 hours and 38 minutes, first the liquid oxygen and then the liquid hydrogen and associated cryogenics (or low-temperature fuels like liquid helium, liquid nitrogen) were loaded into the stages of Saturn V, and into the fuel cells and thrusters of the Service Module, the Command Module and the Lem, floodlights playing on the white skin of the spaceship, hints of the dawn beginning to breathe over the ocean. Back at MSOB, eight miles away, they awakened the astronauts at four-fifteen in the morning and gave them a last quick medical examination before they proceeded to breakfast on orange juice, toast, steak and eggs. Then silver chloride electrodes called biomedical sensors were pasted in place, taped to the skin – at least four had to be worn by each astronaut in order to monitor his heartbeat and respiration, the wires running into a bio-instrumentation harness, a belt each astronaut wore which had little radio transmitters called signal conditioners, each the size of cigarette packages. Equipped with a built-in power converter, the bio-instrumentation harness sent constant signals to an electrocardiograph and an impedance pneumograph in the spaceship. Therefore, the astronauts did not have to be plugged into any socket for a medical record to be kept. Plugged in they would be in other ways, by hoses from their space suit to oxygen supply, by electric cable with a twenty-one-socket connector which ran from earphones and microphones (worn in a black and white helmet on their head) to their garment. But then they even wore film packs on their legs and dosimeters on their arms to measure the belts of radiation through which the spaceship would pass, yes, space suits donned, and thus installed with detectors on every flank, they picked up their life-support systems, which fed oxygen to their helmets, and life now passing into them through the umbilical hose attached to their suit, passed, helmets on, through the aisle of reporters and photographers waiting for a glimpse of them and rode in the crew transfer van to the Launch Complex, where they arrived a little after six-thirty, rose up in the elevator to the 320-foot station, entered the Command Module over the highest of the nine swing-arms, a view below thus vertiginous that they might as well have been walking along the cable of a bridge, and settled in, Armstrong in the left seat, Aldrin in the center, Collins at the right for the last two hours and forty minutes.

It was a time for checking out the last of the tests. A crisis had occurred already and been solved: a liquid hydrogen line serving the third-stage engine of Saturn V had begun to leak, but the line was shut down. Its twin, hitherto redundant, was now opened to serve. The problem had been solved. If there was a psychology to machines, it was not unduly in prominence this morning. Tests with the astronauts proceeded. Pushes of current were sent out to activate certain needles, then checks of secondary dials could be made, for they were obliged to react to the first dials. There was a list of tests to be made by each astronaut: if rocket safety was designed on any principle, it was built on the idea of precautions founded on other precautions. If there were six hundred and thirteen laws to the Talmud which could not be broken, restraints or fences had been built into the traditional customs to avoid getting near a situation where one might be able to break one of those six hundred and thirteen laws. If you were not supposed to engage in adultery with your neighbor's wife, the safest precaution, the fence, was never to look at her. Wherever else this construction of safeguards to inhibit even the approach to taboos had entered the Judeo-Christian heritage, certainly it had entered engineering practice. There safety factors abounded. You calculated the greatest stresses a bridge could ever receive, and then built it five times stronger than it needed to be. So now the Emergency Detection System of Apollo-Saturn heaped attention on the whisper of a clue that a slight deviation from the norm could, if it continued to be unbalanced to the tenth of a hundredth of an acceptable degree . . . the launch was not about to be called off because things were wrong to the tenth of a hundredth, but the dials were there to measure such deviation. The readiness to intercept crises before they began to develop was nowhere richer than in launch systems. Thousands of rockets had not been launched for nothing; the safety precautions were enormous, and they were articulated. There were capacities to abort the mission on the ground and in the air, by automation and by decision, by the crew and by Mission Control. Checks sat upon precautions and points of no return abounded where choices had to be made for GO or NO GO, choices to continue or abort. The limits tolerable to each instrument were laid out, were red-lined on every dial.

ix

Nobody had forgotten the fire. The memory of Grissom, Chaffee and White was always most intense when the crew were in the Command Module and waiting for a lift-off. Virtually trapped at the top of the stack (for even the high-speed elevator would take them thirty seconds to the ground) a gondola had been rigged at Swing-Arm 9 of the Mobile Launcher to slide down a wire to a point on the ground over two thousand feet away. No veteran of roller coasters would necessarily take it in stride, but if fire broke out at the base, and the elevators seemed too slow, the astronauts, lumbering along in heavy space suits, could still work out of their hatch exit, clamber into the gondola, and take the quick ride along the wire to that bunker seven hundred yards away where presumably they would live through the explosion. That was one means of egress in emergency, the elevator was another, but their best means, once the close-out crew had departed in the hour before lift-off, was by means of the Launch Escape Tower, that extra rocket on top of the total stack of Apollo-Saturn. Once the swing-arm could be pulled back, the Launch Escape Tower was ready to be armed. Now if something went wrong, the Command Module would be cut loose instantly from the Service Module by explosive bolts and the Launch Escape Tower, with only the Command Module attached, would zoom off from the rest of Apollo-Saturn and go flying out to sea, there to float down by parachute in the midst, ideally, of a flotilla of Range Recovery ships already waiting, even as the *Hornet* and its attendant ships were already moving into position in the Pacific for splashdown eight days later.

The thirty-one minute mark was passed in the countdown. A test checked the batteries and fuel cells once again. In all the tests of all the systems and subsystems which went on during the countdown, all the tests of propellants and purge systems, of abort and detection and destruct systems, all monitoring of the loading of the liquid gasses, Lox and LH_2, all monitoring of the loading of nitrogen and helium for those purge systems which would put pressure on the fuels to move once inside their rocket tanks, and would trigger some of the valves, and clean out systems already used, yes of all the functions of purge and ullage performed by the helium and nitrogen, functions which had to be constantly tested and monitored through the hours of the countdown, and the tests to measure boil-off of the stored fuels, the check-off of tests for the integrity of the structures, tests for the environmental control of air and the purging of used air in the cabin of the Command Module, the check-out of environmental control in the Lem, the checks of the ability of the motors to swing on their gimbal rings (for Saturn V would direct itself by turning its heavy motors through six degrees of arc), yes and tests of flight control with the small rocket motors called thrusters, tests of instruments already tested, everything proceeding through its place in the countdown, no factor was necessarily more sensitive or more studied or of more concern than the inert non-moving potentialities of the fuel cells and the batteries, for a rocket had no pistons and no propellers, no belly nor hold for men to move about in, no gears and transmission – it was a ship designed for space, to travel through space, and so it was a curious ship, a braincase on the tip of a firecracker: ultimately it was nothing but fire for force, and electricity for thought, for direction and dialogue between itself and the heavenly body it left, and the heavenly body it would explore. Electricity was half its existence. Without fire it could not move; without electricity it could not think. Say, once up, it could not even light every new fire. So, it came as a surprise to recognize that all the electrical power in Apollo-Saturn was derived from fuel cells and batteries only. A ship the size of a destroyer suggested huge generators, but they would have weighed too much, so the power for Saturn came from batteries, and for Apollo from batteries and fuel cells which used Lox and liquid hydrogen as solution for the electrodes, and therefore was able also to employ the water formed as waste product for the astronauts to drink, an elegant conversion.

While the spark to ignite the mighty motors of Saturn V did not come from the batteries, but from a cable on the ground, the batteries were all there was of electrical power to feed current to the instruments in the first stage and the other stages once in flight, all there was to ignite the five J-2 motors of the second stage, the single J-2 motor of the third stage. Yet these stages and Instrument Unit were powered by only eleven batteries, two fifty-six-volt batteries, nine twenty-eight-volt batteries, their combined weight perhaps not a thousand pounds, and of it all, only two twenty-eight-volt batteries, weighing twenty-two pounds and fifty-five pounds, would serve for all the functions of the mammoth first stage, 33 feet in diameter, 138 feet tall, weight when loaded five million pounds and more – just seventy-seven pounds of battery, total of fifty-six-volts, to take care of all that mass, but of course the fuels in the first stage would be consumed at almost three thousand gallons a second, and life in the flight would be only two minutes and forty seconds before the motors would shut down and the giant empty stage cast loose to fall in the sea. So fifty-six volts would doubtless suffice for two and a half minutes of instrument life.

The minutes went by. The closer they came to the moment of lift-off, the less there was to do. Each test was designed to begin at the earliest moment its function could fit into the chain of the process, but everything which could be finished in less than the time allotted was gotten out of the way; this countdown had moved smoothly. At Launch Control Center, in the big firing room three and a half miles across the moors and bogs were rows and rows of consoles with technicians in front of them, television screens, lights, gauges, charts and graphs, TV pictures of Saturn V from sixty possible angles feeding sixty television cameras with different views of the ship and the Launch Pad in operation, and hundreds of technicians before hundreds of gray consoles in a dozen and more rows, key events in the time-line of the launch up on display, and the completion of each event signaled on the screen by a rectangle lit up with the name of that event. All the tests, check-outs, and readings of all the systems and subsystems now funneled into a climax, an apocalypse of communications for the last few minutes. As the tension of the previous weeks burned into the clear life-giving ozone of these critical instants, these superlivid adrenalins, as each of the systems-engineers reported in on the status check their men had made of their banks of instruments, as each chief responsible for engines, for computers, for commo, for guidance, for abort, for stabilization, for propellants, for purge subsystems, for environmental control came in with GO, GO for the first stage, GO for the second stage, GO for third stage, GO for the Instrument Unit, GO for the Lunar Module, GO for the Service Module, GO for the Command Module, GO for the Launch Escape Tower, GO for the astronauts, the system moved over from man to machine. For the last three minutes the countdown would be automatic; the machines would come in with information so quickly that they would monitor their own systems, approve their own systems, give their own signal for GO or NO GO, the computer was supervising the last hundreds and thousands of events in the last three minutes.

'Good luck and Godspeed from the launch crew,' Paul Donnelly called out on his microphone, and Armstrong, from three and a half miles away answered, 'Thank you very much.' Quietly, he added, 'We know it'll be a good flight.' There was confidence between them, the confidence of missionaries, the very air of messianic love – that love which, like Robert Frost's cube of ice, traveled on its melting.

Automatic sequence, and the members of the launch team stood at their dials looking for red-line values, looking for some last crisis or betrayal, some silent scream of the needle across the red line. The sequence was automatic, but the men were still considered more trustworthy than the machines themselves: the last control, the control above all others was a manual abort. Automatic or no, there was still the psychology of machines to be feared; so, man was still entitled to have the final say if some event which was incomprehensible to the machines occurred, and someone human must decide if the mission proceeded or stopped. On went the sequence. The gas generator valves in the base of Saturn V now closed on command, the main fuel valves from the Mobile Launcher shut off, the Emergency Detection System was activated in every circuit, the exhaust igniters came forward, the explosives for a destruct in midair were made potential, the hydraulic pressure in all systems were checked at once – OK; the voltage in all systems – OK; Instrument Unit ready for firing – OK; check-out valves in ground return position – OK . . . OK . . . OK . . . Oxidizer tanks in the upper stages now pressurized. Transfer to internal power on entire spacecraft. Astronauts report in. They are GO. The guidance system for controlling the ship in flight is now on full internal power. Seconds go by. Fifteen seconds to lift-off. Twelve seconds. The swing-arms begin to pull away. Five hundred volts pass through a cable still attached by its umbilical and goes into the bowels of the rocket to ignite the turbo-pump exhaust gases which burn the igniter links which trigger an electrical signal to open a four-way valve which opens the main Lox valves and propellants flow into the combuster. In seventeen separate split-second steps are gases ignited into fires which ignite other gases whose exhaust pressures open giant valves which release the orifice in the main tanks and on the fire of other fires are the rocket engines lit.

Eight and nine-tenth seconds before lift-off, the first flames burst out of the base of the rocket motors and vault down a concrete flame trench on the pad, a trench fifty-eight feet wide and forty-two feet deep. At its center is a cusp of metal concave on both slopes, a flame deflector forty-odd feet high and one million three hundred thousand pounds in weight. It receives all of the fury of the heat and blast as the five engines of the first stage build up in nine seconds to their seven and a half million pounds of thrust. Refractory concrete, volcanic ash, and calcium aluminate are the heat-protective skin for this flame deflector, which proceeds to divide the fires and send them away on each side down the trench to break into open air a hundred feet away on either side. Nozzles in the walls of the flame trench spew thousands of gallons of water a minute to cool the deflector, fifty thousand gallons a minute pour over the Mobile Launcher as the spaceship goes up, steam and smoke worthy of a volcano rise into the sky.

But for the moment the spaceship does not move. Four giant hold-down arms large as flying buttresses hold to a ring at the base of

Saturn V while the thrust of the motors builds up in the nine seconds, reaches a power in thrust equal to the weight of the rocket. Does the rocket weigh six million, four hundred and eighty-four thousand, two hundred and eighty pounds? Now the thrust goes up, the flames pour out, now the thrust is four million, five million, six million pounds, an extra million pounds of thrust each instant as those thousands of gallons of fuel rush every second to the motors, now it balances at six million, four hundred and eighty-four thousand, two hundred and eighty pounds. The bulk of Apollo-Saturn is in balance on the pad. Come, you could now levitate it with a finger, but for the hold-down arms. Now in the next second and the next, the thrust is up to full launch, to seven and a half million pounds, more, more than one million pounds of surplus force is now ready to push upward. And still the rocket is restrained. The hold-down arms, large as buttresses, still retain the ship for two more seconds before lift-off. The last check-outs race through the automatic sequence and GO comes back, and the hold-down arms – what engineering in those giants! – pull back, and Apollo-Saturn rises inch by inch in those first seconds, pulling tapered pins through dies to slow the instant of its release. Inch by inch, then foot by foot, slowly, story by story, swing-arm by swing-arm, the swing-arms pulling back in the last five seconds, the last two seconds, umbilicals snapping back, slowly Apollo-Saturn climbs up the length of the Mobile Launcher, the flames of apocalypse no more than the sparks of its chariot, and spectators cry, 'Go, baby, go.'

X

As Aquarius continued to write in later days and weeks, and then in the months after Apollo 11 lifted off from the pad at Cape Kennedy and began its trip to the moon, as he continued to brood about the chasm between technology and metaphysics, the psychology of machines and the dreams of men, the omens of the future amid the loss of taboo, the horror of the ascent and his fear of the heavens – was the Devil chief engineer of the ship which went to the moon? – he came at last, through the nice agency of a friend, upon a passage in Revelation, Revelation itself! The passage, 8:6 *et sequentia*, reads:

Now the seven angels who had the seven trumpets made ready to blow them.

The first angel blew his trumpet, and there followed hail and fire, mixed with blood which fell on the earth; and a third of the earth was burnt up . . .

The second angel blew his trumpet, and something like a great mountain, burning with fire, was thrown into the sea; and a third of the sea became blood . . .

The third angel blew his trumpet and a great star fell from heaven, blazing like a torch, and it fell on a third of the rivers and on the fountains of water . . .

The fourth angel blew his trumpet and a third of the sun was struck, and a third of the moon, and a third of the stars . . .

Then I looked and I heard an eagle crying with a loud voice, as it flew in mid-heaven, 'Woe, woe, woe to those who dwell on the earth, at the blasts of the other trumpets which the three angels are about to blow!'

And the fifth angel blew his trumpet, and I saw a star fallen from heaven to earth, and he was given the key of the shaft of the bottomless pit; he opened the shaft of the bottomless pit, and from the shaft rose smoke like the smoke of a great furnace, and the sun and air were darkened with the smoke from the shaft. Then from the smoke came locusts on the earth, and they were given power like the power of scorpions of the earth . . .

In appearance the locusts were like horses arrayed for battle; on their heads were what looked like crowns of gold; their faces were like human faces, their hair like women's hair, and their teeth like lion's teeth; they had scales like iron breastplates, and the noise of their wings was like the noise of many chariots with horses rushing into battle. They have tails like scorpions, and stings, and their power of hurting men for five months lies in their tails. They have as king over them the angel of the bottomless pit; his name in Hebrew is Abaddon, and in Greek he is called Apollyon.

Apollyon was not Apollo, no more than Ἀπολλύων is equal to Ἀπόλλων. Apollo was in fact the god of light and Artemis, his twin sister, was goddess of the moon. But Aquarius ignored the fact that Apollyon was not Apollo and took it as a sign. However, he also ignored the last two of the seven angels, and so ceased quoting before the sixth blew his trumpet and by fire, smoke, and sulphur was a third of mankind felled by plague. Nor did Aquarius know 'that in the days . . . sounded by the seventh angel, the mystery of God . . . should be fulfilled.'

2

A Trajectory to the Moon

If there is a crossing in the intellectual cosmos where philosophical notions of God, man, and the machine can come together it is probably to be found in the conceptual swamps which surround every notion of energy. The greatest mystery in the unremitting mysteries of physics must be the nature of energy itself – is it the currency of the universe or the agent of creation? The basic stuff of life or merely the fuel of life? the guard of the heavens, or the heart and blood of time? The mightiest gates of the metaphysician hinge on the incomprehensibility yet human intimacy of that ability to perform work and initiate movement which rides through the activities of men and machines, and powers the cycles of nature.

Still, the laws describing the behavior of energy are sound, they are usually simple – they may be called fundamental for their results do not vary. If, on consideration, it might still be a mystery why a liquid, a solid, or a gas can store energy which is capable of prodigies of work once the forces are released, still the precise results of such liberation of energy have been well studied. In three centuries, physics has moved out of the rough comprehension that shifts in matter from solid to liquid, or liquid to gas, involve discharges of energy, to an application of that knowledge onto half the working technology of the world. So the employment of such principles in the design of rockets has been no great work for physics. The physics is simple.

The burning gases which push out of the throat of the rocket engines push back against the rocket itself. Therefore, the rocket can rise, thereby can it defy gravity once the push produced by the expansions of the burning fuel is greater than the pull of gravity upon the ship. Therefore it does not matter if the rocket is on the ground or in the clouds – it does not lift by pushing against the earth, or in flight by pushing against the air, no, it is rather the simple push of the escaping flames against the ship itself which gives thrust to the voyage. Once entered into that bay of space between the earth and the moon where the effects of gravity are hardly to be noticed, so the weight of the ship and the men in it are hardly to be noticed.

But that will be later. On the ground, the full force of gravity is present: if the ship weighed six and a half million pounds, it would need as much force, and a little more to lift it. In fact, its thrust would be designed to reach up to seven million seven hundred thousand pounds in those nine seconds before the four hold-down arms were released and the rocket began to rise. It could have risen with less force, it could theoretically have drifted upward at the very moment the thrust was minutely greater than the force of gravity, but that was an impractical mode of ascent, for the smallest loss of thrust at such a critical moment would have obliged the rocket to collapse back and topple on its pad. It was the life experience of such rocket engineers as Von Braun, rather than the laws of physics, which decreed that Apollo-Saturn be chained to its base until the thrust upward was a million two hundred thousand pounds greater than its weight. For that reason, it was manacled by four giant metal hold-down arms. You can be certain there had been cracks in the early forgings of test metals of the hold-down arms for they were not easy to design, being massive in size yet required to let go their million-pound grip on the split part of an instant. The unlatching interval for the four arms had to be all but simultaneous – the separation was geared not to exceed one-twentieth of a second for its duration: in fact if any of the four arms had failed to complete their operation in more than a fifth of a second, the liberation would have been effected by properly placed explosives. With one million excess pounds pushing it, Saturn V was hardly to be kept back on one side while being released on the other – it would have begun to pitch over – yet note that even with all four hold-down arms sprung at once, the rocket ship was still restrained for the first few inches of travel. Something exactly so simple as eight tapered pins had each to be drawn through its own die – as the vehicle rose through the first six inches of flight, each die was obliged to straighten the taper in its own iron pin – the eight dies to travel up with the ship, the eight shucked pins to be left in their fastenings on the hold-down brackets. If not for such a simple mechanism, Apollo-Saturn might have leaped off its pad fast enough to set up a resonance, then a vibration strong enough to shake the ship and some thousands of its instruments too critically. For consider: if when empty, the space vessel weighed less than half a million pounds, it was now carrying a weight of fuel twelve times greater than itself. But there were no bones or muscles in this fuel, nothing in the fuel to hold the ship together, just liquids to slosh and shake and seek to distort the rigidity of the structure. Most of the spaceship was nothing but its own fuel tanks, and there were few places where the hide of the rocket was more than a quarter of an inch in thickness and sometimes so thin as one-twenty-fifth of an inch, and aluminum alloy at that, places where the fuel tank was literally the skin. Of course the ship had corrugations in its surface for stiffening and bulkheads for bracing, which also served neatly as baffle plates to reduce the sloshing of the fuels. Even so, one would look to reduce every quiver in so delicate a structure – the restraining pins performed just such a function for the first half-foot of ascent.

In the course of this act, at an instant when the spaceship was not yet three-quarters of an inch off the ground, specific switches on the hold-down arms tripped loose a pneumatic system which gave power to surges of compressed gas which ran in pipes up the great height of the launching tower: the gas tripped the couplings of the five service bridges still connected to the rocket. Their umbilicals now detached, these arms pulled away as the ship began to rise. Six inches up, and loose from the pins, the stages of Apollo-Saturn climbed up the stories of the Mobile Launcher, climbed up on its self-created base of flame, up past the flying withdrawal of its bridges and its umbilicals. To clear the tower, to be free of any sudden gust of wind which might lash it sideways, a yaw maneuver, programmed into the rocket, was initiated one second after lift-off, and turned the nose a few degrees from the vertical further away from the tower. For the onlookers three and one-half miles away, the rocket appeared to waver, then stagger. In fact, it did. There was wind blowing, and the rocket had been designed not to fight wind (it was not stressed for that) but to give way to wind, to relinquish the trajectory it was on, and compute a new trajectory from the slightly different position where the gust had just left it. So separate commands kept issuing from the Instrument Unit at the top of Saturn, sometimes every half-second, and the motors kept responding with little spurts and sags of speed. The result was a series of lurches and bumps in the first few seconds. 'Very rough,' said Collins afterward, 'very busy . . . It was steering like crazy. It was like a woman driving her car down a very narrow alleyway . . . She keeps jerking the wheel back and forth . . . a nervous, very nervous lady . . . I was glad when they called Tower Clear because it was nice to know there was no structure around when the thing was going through its little hiccups and jerks.' That was after eight seconds. At close to twelve seconds, the four outboard engines were swiveled through a few degrees, a pitch maneuver was initiated, and a roll. The roll would end in twenty seconds, the shift in pitch would continue for two minutes and twenty-five seconds by which time the rocket would be climbing no longer straight up, but rather at a reasonable angle close to twenty-five degrees from the horizontal, and would have already passed through the severest structural strains of its trip. The astronauts, lying on their backs, unable to see out the Command Module for the first few minutes with the heat shield covering their windows like a blanket, would feel this dynamic pressure at a minute and twenty seconds after lift-off, fifteen seconds after breaking through the sound barrier. Acceleration would continue, and as in a centrifuge or a mechanical whip, their bodyweight would go up to 2 G's, 3 G's, to close to four times their own weight, their eyes would feel a mean pressure, but 4 G's was not intolerable for an astronaut – they were familiar with high-speed dives in test planes and gravity simulators. Besides they were lying on their backs – the blood would not drain from their heads. All this while, the noise of the rocket motors, if sounding like a prelude of apocalypse to the spectators on the ground, was no more than a quiet rumble in the sealed conical volume of the Command Module. If not for the five numbered Launch Vehicle Engine Lights on a dial before them, they could not know if a single motor went out; the volume of sound in the cabin was not high enough to distinguish the difference.

Indeed there was not that much in those moments for the astronauts to do. The first eleven and a half minutes would be spent in reaching up into an earth orbit. The firing of the motors, their cutoff, the guiding commands and the separation of the stages would take place as part of a sequence which had already been programmed into the Instrument Unit of the launch vehicle. The astronauts were effectively in automatic flight. They would have no need to touch a flight control unless something went wrong. Indeed their work in these early minutes was to watch the dials and so be on the alert for the first hint of any malfunction in a system or subsystem vital to this portion of the trip. Listen to how little is said by them through the first three minutes of flight. Of course they are flying on their backs and the weight of their intense acceleration lies like lead on their tongue.

ARMSTRONG: *Rolls complete and a pitch is program. One BRAVO.*

CAPCOM: *All is well at Houston. You are good at one minute . . . Stand by for Mode 1 Charlie MARK Mode 1 Charlie.*

ARMSTRONG: *1 Charlie.*

CAPCOM: *This is Houston, you are GO for staging.*

ARMSTRONG: *Inboard cutoff.*

CAPCOM: *Inboard cutoff.*

ARMSTRONG: *Staging and ignition.*

CAPCOM: *11, Houston. Thrust is GO all engines. You are looking good.*

ARMSTRONG: *Roger. Hear you loud and clear, Houston.*

In these three minutes from lift-off the rocket ship had accelerated from a rate of travel of a few inches a second to almost ten thousand feet a second, it was now forty-three miles high and seventy miles away and traveling at about one hundred miles a minute or six thousand miles an hour. It had burned half a million gallons of fuel weighing almost five million pounds and had already dispensed with its first stage, an object 138 feet long, and 63 feet wide at its fins, a short-lived stage which was more than three-quarters of its weight and half of its volume. Now, it was about to fire the ullage rocket or opening gun of the second stage. What speed, what acceleration, what onrush! Perhaps we must quit the rapidity of the process in order to discover how the speed is achieved.

ii

Just as the Greeks could be confident they had discovered the secret of beauty because the aesthetic of their sculptors permitted no blemish to the skin, because their sculptors said in fact that the surface of marble was equal to the surface of skin, so classical physics remained simple because it did not try to deal with anything less than ideal form. Later, Western aesthetics was sufficiently ambitious to wish to discover the laws of beauty in skins with blemish and bodies with twisted limbs (and indeed would never quite succeed), just indeed as engineering could never prove simple and comprehensible to amateurs. At its best engineering was a judicious mixture of physics and a man's life-experience with machines: one insignificant dial on one bank of instruments was often the product of the acquired wisdom of a good engineer who had put in years of work reducing the deviations of an imperfect instrument of measure.

Any attempt to explain the mechanics of the flight of Apollo 11 in engineering terms is then near to impossible for one would be obliged to rewrite a set of extracts from technical manuals, and each manual would finally prove nothing but an extract from other more detailed manuals, which in turn would be summaries of the verbally transmittable and therefore less instinctive experience of veteran engineers. Yet, the pure physics of the flight was still simple, so simple and pleased with itself as a Greek statue.

The rocket rose because the forces which were pushing it up were larger than the forces which held it down. The thrust of its motors was greater than the heft of its bulk. So it rose upward, even as we can jump in the air for a moment because for just a moment the push in our legs up against our body is greater than our weight. Speak of potency! – the force of our legs immediately ceases; almost immediately we descend. While the rocket had no legs to propel it upward, it had rather a burning gas expelled from its rear, and this force did not cease. So the rocket continued to rise. In the beginning it did not rise very quickly. Seven million seven hundred thousand pounds pushed upward against six million five hundred thousand pounds of weight which pressed downward. The difference was therefore to be calculated at one million two hundred thousand. That was the same as saying that if the rocket had been mounted on wheels in order to travel down a level road (and so did not have to be lifted), one million two hundred thousand pounds would be pushing the same six million five hundred thousand pounds of rocket. It can be remarked in anticipation that as this force continued to push, the rocket would begin to go faster. Its velocity would increase at an even rate if the push remained the same and the weight remained the same. If at the end of a second, its measured speed was about what it should be – five feet a second – it would reach fifty feet a second after ten seconds, and one hundred feet a second after twenty. The reason was not complex. The push did not diminish. Therefore the rocket would go five feet faster every second than the second just before. After two seconds it would be going at ten feet a second because five feet a second would have been added in that second interval of measure to the first five feet a second. After three seconds, fifteen feet a second would be its speed. The velocity would increase five feet a second, every second, so long as the push remained steady on that rocket rolling on wheels down that level road. At the end of seventeen and a half minutes the rocket we have used for an imaginary model would be moving at an imaginary speed of a mile every second.

Yet that hypothetical rocket is still traveling at a much slower rate than Apollo-Saturn. When Apollo-Saturn went into orbit one hundred miles up and fifteen hundred miles out, not twelve minutes were gone, yet it was traveling at five miles a second or eighteen thousand miles an hour.

The explanation is agreeable to a liberal mentality, for it suggests that expenditure is power. The greatest weight in the rocket is fuel, and the fuel is being consumed. The rocket loses weight at a rate as immense as thirty-five hundred gallons of fuel each second. Somewhere about thirty-five thousand pounds of weight vanish in the same interval, which comes out by calculation as close to two million pounds a minute. At the end of a minute, seven million seven hundred thousand pounds are pushing not six and a half million pounds but four and a half million. Thus, the ship is accelerating more rapidly each instant. Its speed of increase now would be not five feet a second but more than twenty. Since the engines, however, also increase their effectiveness as the rocket takes on high altitude and the near-vacuum of the thinning atmosphere offers less resistance to the fires of the exhaust, so at the end of two minutes and fifteen seconds of flight the thrust has actually reached over nine million pounds and is then pushing only a little more than two and a half million pounds. Now the rocket is being propelled by a force almost four times as great as itself: so its acceleration would be not five feet a second as at lift-off nor twenty feet a second at the end of a minute, but more like ninety feet a second.

Apollo-Saturn however does not travel that fast for long. It takes two minutes and fifteen seconds to reach such acceleration, and then the center motor is shut down. The thrust reduces to seven million two hundred thousand pounds from the four continuing engines. Twenty-five seconds later the outboard motors are cut off. A few more seconds, and the first stage is released. The rocket begins to travel on the motors of the second stage, and these next five engines are not nearly so powerful. Never again will Apollo-Saturn pick up speed so quickly.

It hardly matters. The more modest acceleration of the second stage is added onto the high velocities already attained by the first stage. Apollo-Saturn will increase its speed to four and a half miles a second, and will be altogether out of sight when the second stage is discarded after nine minutes and twelve seconds of flight. Stripped of its first stage and its second stage, powered now by but a single motor which develops hardly one part in forty of the force the first engines developed to get off the ground, the ship now weighs only four hundred thousand pounds, or a sixteenth of its original weight. Drastically reduced, it is still in need of a little more speed, and the third stage will give it that, the third stage will take it up to something near five miles a second, or eighteen thousand miles an hour. To reach the moon it will yet have to go faster, it will have to reach twenty-five thousand miles an hour to escape the force of the earth's gravity. But that is a subsequent step. Now the ship is wheeling through the near-heavens. A little bit more than one hundred miles overhead, it proceeds to circumnavigate the earth every hour and twenty-eight minutes. Its weight, fuel of the third stage partially consumed, is now down to three hundred thousand pounds and it is in that magical condition of defiance to gravity which is known as orbit.

iii

But what exactly is this condition? What is orbit? How can a rocket continue to circle around the earth if its motors have been shut down? Why indeed is earth orbit even required? Why don't the astronauts head directly to the moon?

If the last question is the first to be answered, that is probably because the answer is at hand. The excitement of the first twelve minutes of flight has been so great, so much has happened, so many parts of the ship have fallen off, and so many conditions have altered so rapidly that the astronauts, much like the men at Mission Control, need an hour or more to collect themselves, say farewell to earth, check over the conditions of their ship of space after these violent, hurtling, and most abrupt changes, and give us time to learn about the nature of orbit and what in fact is the very appearance of the ship.

So, despite all honors given to physics, the line of explanation may be obliged to pass again through engineering long enough to offer one last description of Apollo-Saturn as she sat on the pad. Already not twelve minutes in the air, she is of different appearance, and soon will be unrecognizable. So it is appropriate to offer a final opportunity to the reader to begin to glimpse the structure of that remarkable mechanical white whale which rose into the air at Cape Kennedy and soon disappeared.

The ship of space was after all not one rocket, but essentially two – a brain on the end of a firecracker is how Aquarius first vulgarly would seize the idea of it – but actually it was composed of two rockets, one on top of the other, separate rockets each possessed of brain and force of propulsion. It was just that the lower rocket, the first rocket, the mighty rocket which would lift it into the air, while not entirely without a brain, had in effect the little skull which sits on the neck of a dinosaur, and the second rocket, which would travel to the moon, while almost unimaginably intelligent (and pregnant as well with a mechanical child who would reach a new land) was also possessed of its own sources of fire. In comparison to the dinosaur called Saturn, the power which propelled Apollo's brain was modest, it was in fact about one four-hundredth of the initial force which would lift the two rockets Apollo and Saturn together but this small motor was still powerful enough to propel Apollo through all of its intricately conceived plans and projects.

Therefore, let us think of it again, not as one rocket, but two joined together, Apollo on top of Saturn, each with an explosive and a brain to direct that explosive. Of course, the first, Saturn V, the dinosaur, the launch vehicle, is not quite one rocket, it is rather one rocket made of three stages called S-IC, S-II, S-IVB, rockets one behind the other to the number of three with one brain and Instrument Unit to service them all, time them all, ignite them properly, separate them, and keep them connected to Apollo, which could in emergency even control them. Yet the launch vehicle, while composed of three rockets, can be thought of as one, because the three were merely a convenience, like successive boxcars of fuel to feed a train. Taken all together with its four parts (the first stage S-IC with the five F-1 motors, the second stage S-II with its five J-2 motors, and the third stage S-IVB with its single J-2 motor plus the Instrument Unit), our launch vehicle made up a total of two hundred and eighty-one feet in length, almost three quarters of the total height; it took up more than ninety-eight percent of the total weight; of the 6,487,354 pounds of Apollo-Saturn on the pad, 6,377, 354 pounds belonged to Saturn.

In its turn, the 110,000 pounds of Apollo spacecraft also had four parts. Its base, which was not even approached until one had risen close to the height of a football field from the ground, was the SLA, the Service Lunar-Module Adapter, a shell to hold the Lunar Module or Lem. Then came the Service Module, and then the Command Module, where the astronauts were installed. Above all this was the Launch Escape Tower, the needle on top of the Command Module. That was not designed to travel through space. Indeed the needle would not even reach into orbit. Shortly after the first stage separated, the escape tower was required no longer, and so it was fired off, pulling away at the same time from the Command Module a heat shield which had protected the astronauts from the high temperatures of friction caused by their rapid ascent through the atmosphere. Therefore, it was only after three minutes and seventeen seconds of flight, at a velocity now near to two miles a second, that the astronauts were at last able to obtain a view. The tower which had just departed would not be seen again – it can be stated for obituary that it was thirty-three feet long, weighed 8,910 pounds, and had three motors. The jettison motor, designed to carry the tower itself away from Apollo, had a thrust of 31,500 pounds; the escape system motor, required to pull the Command Module away from any trouble below it at a very great speed, had almost 150,000 pounds of thrust, and a relatively little motor of 2,400 pounds of force was used as a species of rudder. The bothersome details of these three motors which are not even to be seen again have been mentioned to give an idea of the variations in thrust between different and separate rocket engines on Apollo-Saturn. There were eighty-seven motors all together, some for propulsion, some for steering, some for ullage to keep pressure on the fuels, some were even retro-rockets designed to push the separated stages away from the flying ship. Thirty-seven such motors, as different in size and thrust as the million and a half pounds of F-1 in comparison to ullage rockets of seventy-two pounds, no more! were distributed over the three stages of the launch vehicle, and that count does not include the numerous explosives (also a form of rocket) which were used to rip bolts apart, separate stages, and in case of abort, blow the tanks.

Indeed, an explosive wire wrapped around the circumference of the launch vehicle was used to separate each stage. On signal this wire belt would be detonated; hundreds of aluminum straps attaching the cylinder of one stage to the cylinder of the next would be severed at guillotine speed. Retro-rockets carefully ignited by the first explosion would then fire for a fraction of a second, just enough to push the exhausted stage a few feet back. Next the motors of the newly ignited stage would take off like a relay runner from the velocity handed over by the previous stage. In this manner, the launch vehicle rose, separated stage by stage, was consumed, and left the spacecraft in earth orbit. There, in a great circle of earth one hundred miles up, revolved the second rocket, Apollo, with its Command Module, its Service Module, its Lem, and in some nice defiance of definition the partially used third stage, which would soon be employed to fire Apollo out of earth orbit and toward the moon. That combination of spacecraft and last expendable stage of the launch vehicle, making in combination an object one hundred and ten feet long, now sailed in a circle around the earth.

iv

Orbits are not difficult to comprehend. It is gravity which stirs the depths of insomnia. We can remind ourselves that the idea that what goes up must come down is still a momentous discovery to an infant (who finds every balloon a magical beast). An adult on the contrary builds the bedrock of his common sense on the certain fall of an unsupported object. What goes up *must* come down. Of course the Lem in

Apollo 10 would never come down again – it was off in orbit around the sun, and Mariner would not be back again, nor Ranger, nor Lunar Orbiter. And Surveyor I had landed on the moon and there it had stayed. So the moon apparently had its own kind of gravity, as did Mars, and Venus, and the planets and the sun. Every heavenly body had its domain of gravity, its field of attractive influence. Indeed every physical body did. There was a force which drew bodies together, and physics had calculated it neatly as in reverse proportion to the distance between, which was another way of saying that attraction accelerated as bodies drew nearer, it increased as the inverse function of the square of the distance. Two feet apart, bodies were only attracted to one another with one quarter of the power the same bodies would feel at one foot apart. At three feet, it was one-ninth the power, at five feet one-twenty-fifth. Of course if the bodies were only six inches apart, the force of attraction was four times as great as at one foot. That was the first part of an extraordinary relation. The second was that no matter how attracted, whether by much or little, they were attracted in inverse relation to their mass, which crudely is to say in opposite proportion to their weight. Heavy bodies would attract light bodies toward them faster than light bodies would induce movement in heavy bodies. Very heavy bodies would hardly move at all as light bodies accelerated toward them. It was more than theory: the laws of gravity were the soundest laws of behavior in the kingdom of physics – it required only that one not personify the bodies, not give them private will or curiosity or whim, no independent desire or independent resistance. Obviously, a theory of sexual mechanics could never be based on gravity. Still, gravity was a damnable mystery – why did inanimate or dead objects attract each other? And why did animate objects like human beings act like inanimate objects under certain conditions, such as jumping off a high building? Then the human body entered the laws of physics at a weight of two hundred pounds. Since the earth had a literal weight of six billion trillion tons, there was no visible reciprocity. Dispensing with all sizable calculations of air resistance, the body fell at an increasing rate of thirty-two feet per second every second, which was the acceleration gravity gave to falling objects. From a fall of a hundred feet the two-hundred pound man would be moving about sixty-five feet a second when he hit the ground – the earth, with all of its six billion

trillion tons, would not be quivering at any perceptible rate of acceleration up toward the body. Still by the width of a hair taken from an
electron it had been moving toward the body which approached.
Because we cannot see the movement, we tend to think it does not
exist, that bodies merely fall, but in fact the reason a two-hundred
pound man exhibits much the same behavior as any inanimate object in
a ten-story fall is that the power with which the earth pulls on him is
enormous, and the muscles which express his will are not equipped to
be effective in air. So it is not so much that he falls as that he is *drawn* by
an intense and accelerating introduction to the earth. And indeed anyone who had ever had a high fall has felt such a fierce force pulling on
him. 'The earth came up and socked me,' is the good tough statement
a child gives to witnesses. What a ringing in the ear, what a memory of
the lightning bolt!

Metaphors then arise of a charged and libidinous universe with heavenly bodies which attract each other across the silences of space. If they
do not all gather together in assembly at a point, it is because they do
not act simply upon each other, but are each in relation with many
bodies in many directions, and all are moving as well. If they are moving
fast enough and with enough force, they can resist powers of attraction
calling to them from vastly larger bodies. An airplane need not descend
so long as it has fuel to give its motor strength.

We are preparing, however, to brood upon the astronauts in orbit,
and that is another case. They are now not flying with engines, they are
coasting with motors off. They are a hundred miles up in the air, and
their motors are off, and yet they do not fall. They merely continue to
circle the earth once every hour and twenty-eight minutes.

Of course they are traveling fast, at eighteen thousand miles an hour
we can remember, and while that is nowhere near the speed of light,
186,000 miles a second (a speed which is probably the walls of the vessel
which contains the time of our universe), it is still interesting to note that
eighteen thousand miles an hour is as many times greater than the rate
at which a baby crawls (or Apollo-Saturn began its trip from the VAB to
the pad) as the speed of light is greater than the present velocity of the
spacecraft. If men could move out of infancy at half a mile an hour and
get up to eighteen thousand miles an hour in one lifetime, well, who was
to assume that the walls of the universe were safe from future men?

At any rate, the spacecraft was traveling at eighteen thousand miles an hour, and that speed was just great enough to keep it in orbit a hundred miles above the earth. It was of course falling, it was in fact in free fall and in a virtual vacuum (for the presence of air at one hundred miles of altitude is next to nonexistent) but it was also traveling so fast in a forward direction that it fell forward like a ball thrown into an endless chasm, and as it fell forward it fell around the curve of the earth. The earth pulled on it of course, it pulled on the spaceship with all the force of its gravity, but only succeeded in bending its path around the circumference of the earth.

It is easier if we conceive of gravity as a general condition of which magnetism is a special case. Magnetism only works a force of attraction on iron objects, whereas gravity works on all objects. Think of a cylindrical ring highly magnetized, and a steel bullet fired from a revolver which passes the ring in just such a way that the intense force of the magnetism bends the path of the bullet around the cylinder. The bullet would whip around until air resistance had slowed its speed to the point where it would circle in and touch the ring. Orbit was comparable. Even better, at one hundred miles up, air resistance was next to nonexistent, so the spaceship could stay in orbit for a long time. What peace and communality to drift over the earth from such a height, to see nations appear below on the extended field of a continent, nations no larger to the eye from a hundred miles up than the spread of a city from a passenger plane, and oceans the size of a cove. Clouds covered the earth like the wet white feathers of a bird just born, clouds with lines of logic and reentrant curves of thought, clouds gathered in nodes of spiral to signify a shift of atmospheric being – a storm was blowing across a thousand miles of sea. A planet manifest beneath, filling the window out which the astronauts would look, a planet blue and brown with white trailings of celestial ripple and wake, hints of green and dark gray and silver in the curvings and stuffings and pumpings of thousand-mile odysseys of clouds, caravans of weather, pulsations of weather. What a peace, what a calm, what a silence!

Not a chance! The sound of static filled the module, the crackling of questions and answers from earth to spacecraft, from air to ground. Into the ear of each astronaut came burning searching spitting sounds, the unclassifiable sounds of static, so much like the rush of crackling air, the consumption of something vital in space. If smell, Aquarius long had thought, was related to time, then sound was some current in space. But what was static, what special case of space? Static was a form of speech man did not comprehend. Like the dialogue of the dolphin, the communication was faster than the ear. But what fury of irritabilities was loose in the ether at this harshness perpetrated upon it!

Below the astronauts was the earth. Above them, behind them, around them, was a sky no longer blue but dark, dark as the endless night of space. They would not always see that black sky. Sometimes the shine from the earth, the moon, or the light of the sun would reflect from their module, and the sky – almost impossible to see stars in such curious reflected light – would seem a dark rose, a hellish color with glare and plays of transparency and curtains of dark. They were weightless. As their ship fell in its ever continuing circle about the earth, so they fell too. If not strapped down, or with hands on a grip, or with Velcro-soled shoes hooked to a Velcro mat, they floated through the fall of free fall as a man will fall and turn in the air before he opens his parachute. They could do small dives inside the locked spaces of the module, twists and tucks and ever revolving turns, disport like seals and monkeys and otters. That very weightlessness, that absence of gravity which space doctors worried might produce profound deteriorations of the organs and the flesh in a trip of some months' duration was for now a delight, a languor. Think of reaching to dispose of a full can of beer. To one's surprise it is empty. How languorous is the arm as it comes up. John Glenn experiencing weightlessness in his first orbit had said, 'Contrary to this being a problem, I think I have finally found the element in which I belong.' Of course, years later, Glenn had slipped in a tub and almost been killed. Who knew what price was in the pounds lost to weightlessness – for in every intoxication was some kind of price. Otherwise, nothing divine to economy.

Now, however, they did not have time to play. There was static in their ear, and systems and subsystems to monitor, align, check out, adjust, a Puritan ethic in the whisper of the empty spaces – keep busy in the empyrean was the whisper of those empty spaces. So they kept busy for those two and a half hours before they took off for the moon. Names and duties came their way. Alignments, static, stations.

Ground stations ready to receive their communications and relay them to Houston came into range as they moved around the earth, acquired signal, then lost signal when they moved further on. The tracking ship *Vanguard* in the Caribbean lost contact, the Canary Island Station acquired it, Tananarive in the Indian Ocean was next, then Carnarvon in Australia, Honeysuckle in Australia, Goldstone in California. Somewhat later, starting from an altitude of ten thousand miles, on all the way to the moon, there would be but three stations for radio communication, each one hundred and twenty degrees apart on the circumference of the earth, stations with enormous eighty-five foot dish-shaped antennae located in Madrid; in Goldstone; and Canberra, Australia; but for now the stations were more modest, the antennae were no larger than thirty feet, the communication was close, and the data received was fed by relay into computers at Mission Control at Houston, there to go up on the wall to show in display the position of the ship.

They passed the other tracking stations in their orbit, passed them at Hawaii, at Guaymas in Mexico, Corpus Christi, in the Bermudas, the Bahamas, the tracking ship *Redstone*, in all there were seventeen ground stations, plus six units from ARIA (Apollo Range Instrumented Aircraft, what an acronym was ARIA!), six jets which could be used for voice relay in emergency or on unusual azimuths if Apollo passed over in unconventional orbit. But nothing was out of the ordinary today. The orbit and a half before Trans-Lunar Injection was a time of testing out communications and making their checks. The rocket thrusters which would give their spaceship its attitude, put its nose up or down, left or right, or leave it to roll, were fired in brief maneuvers, and studied by instruments in Mission Control in Houston to pass on the results. The temperature of one rocket thruster was soon discovered to be lower than the others. Queries came up from the ground.

CAPCOM: *Would you confirm that your RCS heater switch for quad BRAVO is in primary? Over.*
ALDRIN: *You're correct. It was not in primary. It was off. It's on now. Thank you.*
CAPCOM: *Roger. Thank you.*

The time-line lengthened. They were in second orbit. At a point somewhere near the Gilbert Islands, about halfway between Australia and Hawaii, they would ignite the engine on the third and remaining stage of the launch vehicle, and proceed into Trans-Lunar Injection, which is to say they would quit their earth orbit and fire up to the moon.

This is the transcript on that mighty event:

CAPCOM: *Apollo 11, this is Houston. We are slightly less than one minute to ignition and everything is GO.*

COLLINS: *Roger.*

ARMSTRONG: *Ignition. (The sound of rocket motors are faintly heard.)*

CAPCOM: *We confirm ignition and the thrust is GO.*

CAPCOM: *Apollo 11, this is Houston at 1 minute. Trajectory and guidance look good and the stage is good. Over.*

COLLINS: *Apollo 11. Roger.*

CAPCOM: *Apollo 11, this is Houston. Thrust is good. Everything is still looking good . . .*

COLLINS: *Roger.*

CAPCOM: *Apollo 11, this is Houston. Around three and a half minutes. You're still looking good. Your predicted cutoff is right on the nominal.*

ARMSTRONG: *Roger. Apollo 11's, GO.*

CAPCOM: *Apollo 11, this is Houston. You are GO at five minutes.*

ARMSTRONG: *Roger, we're GO.*

CAPCOM: *Apollo 11, this is Houston. We show cutoff and we copy the numbers in Noun 62 . . .*

COLLINS: *Roger, Houston. Apollo 11. We're reading the VI 35579 and the EMS was plus 3.3. Over.*

CAPCOM: *Roger. Plus 3.3 on the EMS. And we copy the VI.*

ARMSTRONG: *Hey Houston, Apollo 11. This Saturn gave us a magnificent ride.*

CAPCOM: *Roger, 11, we'll pass that on, and it looks like you are well on your way now.*

V

Conceive of a sinner who is a Catholic and devout. What complexity in his feeling for the Church, what pieties of observance live between his sins. He has to make such intricate shows of concealment to his damned habits. Yet how simple is the Church's relation to him. Extreme Unction will deliver his soul from a journey through hell.

So it is with physics and engineering. Physics is the church, and engineering the most devout sinner. Physics is the domain of beauty, law, order, awe, and mystery of the purest sort; engineering is partial observance of the laws, and puttering with machines which never work quite as they should work: engineering, like acts of sin, is the process of proceeding boldly into complex and often forbidden matters about which one does not know enough – the laws remain to be elucidated – but the experience of the past and hunger for the taste of the new experience attract one forward. So bridges were built long before men could perform the mathematics of the bending moment.

Now, Apollo 11 has had a trip up from earth with bumps and blasts, clockwork and sharp explosions, communications and fires hot as five thousand degrees Fahrenheit in the furnace of the engines and it has been a trip which on the one hand amounted to no more than a passage through the simplest laws of acceleration in classical physics, but represented on the other a fair climax to the best and most complex engineering techniques of the century, yes, Apollo 11 after this voyage from earth to earth orbit, and from orbit (via its last burn) into Trans-Lunar Injection, yes, Apollo 11, much in debt to engineering, and still trailing the ghosts of the earth's atmosphere, now pushed its nose toward the moon (or toward a rendezvous with the circling moon three days out) and the final burn of its third stage completed, passed over at a speed of 25,000 miles an hour from the sweat-yards, crooked contracts, closet Mafia and mud-lanes of engineering to the rare temples of physics where one law at a time was enshrined, and one could observe the pure effects of that law. Now, the spaceship, all motors off, coasted up to the moon, lifting on its own momentum against all the loosening bonds of earth's powers of desire to pull all flying bodies back to it, Apollo 11, shining templar of alloy, ascended in all delight as pure exhibit of Newton's Laws of Motion. How beautiful they were. The First Law interned the mystery of the ages for it stated that 'every body continues in its state of rest, or of uniform motion in a straight line, except when it is compelled by external force to change that state.' That was a way of saying bodies at rest and bodies in uniform motion were in the same state. Indeed they were, once one could recognize that bodies everywhere were in motion, all bodies, the earth, the sun and the stars. Rest did not exist, rest was relative, a special condition of uniform motion – no easier place to grasp that phenomenon than in the fastest travels of man, for the more rapid the vehicle, the less was the sensation of speed. One passed from a sense of rest sitting in a chair on the ground to a cognition that it was only the sense of rest one knows in balance with the movement of larger parts of the universe. Further and further away from the earth coasted Apollo 11 at an initial translunar velocity fifty times greater than the fastest auto ever to scream past five hundred miles an hour on the salt flats at Bonneville; and in the Command Module, freed of the sensation of weight, objects traveled as freely as the men, flashlights switched on, then given a twist, spun like illumined beacons on a tower, free-floating as they revolved, crumbs from the bread of a sandwich hovered for hours like motes of dust the size of flies.

It is a picture of great happiness, of harmony, of souls at rest, and evil matter released from the bondage of its weight, but the laws of motion like the laws of morality invoke every notion of balance. Newton's Second Law was harsh and just, as severe as the Third, which has been already encountered and taught us – if we did not know – that for every action there is an equal and opposite reaction. So the Second Law stated that the rate of change in a motion must be proportional to the force applied. One could not cheat life. One did not accelerate for nothing, nor slow one's speed without braking force and heat. As a body moved away from the earth, so, too, did its speed diminish, for the force of gravity weakening, it was still a force to be applied against the effort to escape. Therefore Apollo 11 moved at a half, a quarter, finally a tenth of its greatest speed as it ascended to escape the gravity of the earth and enter the new field of the gravity of the moon. It was like a ball being rolled up a hill – if it reached the crest it would go over and roll down the other side no matter how slowly it was traveling over the crest. In fact, it was easy to think of the earth as being at the center of a bowl, and objects seeking to escape the earth's gravity would have to be fired out at an initial speed sufficient to travel up the wall of the bowl and roll over the lip – that initial speed was almost seven miles a second for all objects which would escape the earth, and at almost seven miles a second (and a little more) was Apollo 11 fired up – the rest of the trip was given over to the hours and the days of the long coast in space as the speed of the spacecraft diminished from thirty-five thousand feet a second to three thousand feet a second, yes, the ship of space was still moving at the respectable speed of half a mile a second when it passed finally out of the last lingering lulling attraction of the earth and moved over the lip into the bowl of the moon's gravity somewhere at an unmarked point about one hundred and eighty seven thousand nautical miles from the earth. And the computers moved over in the calculations with it. Now the force of the moon's gravity would draw the spacecraft with its three men nearer and nearer to it, now the descent to the moon had begun.

vi

Long before, almost two and a half days before, not a half hour after the third stage had completed its five and a half minute burn and pushed Apollo 11 up to its highest speed, the final separation took place between this last of the launch vehicle and the as-yet unfired spacecraft.

The Command and Service Module, two objects which when isolated together looked like nothing so much as a tin can with a cone on its nose and a motor at its rear, these two objects in combination about thirty-five feet long and thirteen feet wide, now separated themselves from the SLA or Service Lunar-Module Adapter. The SLA was attached to the S-IVB or third stage and in turn these last two objects, S-IVB and SLA, also might have looked like a tin can with a tin funnel attached, for the S-IVB was a cylinder with a motor at its rear, and the SLA was composed of four closed curved aluminum panels, much like the petals of a bud, designed to be sprung free a fraction of a second after an explosive wire blew the joining between the CSM and the SLA just as the small rocket thrusters of the Service Module gave a small push away. On the instant, the four aluminum petals, each twenty-one feet long, and each wide enough to encircle a quarter of the Lem, blew off in four separate directions to go scaling through space and reveal what the SLA had contained – the folded legs, the head, thorax, and sac of a complex mechanical craft or creature which had the look of a particularly nasty insect. It was the astronauts' first glimpse in space of the Lunar Module or Lem. And the Command and Service Module containing them promptly backed away from this sight of the leg-folded Lunar Module mounted on the end of the spent third stage, and at a rate of less than a foot per second, such slow speed offered it by a judicious use of its thrusters, the CSM withdrew about seventy feet, and then did at the end a complete back somersault to point in the opposite direction. The turnaround had taken about two seconds. The nose or probe of the Command Module had now reversed to point toward a small cavern in the Lem called the drogue, named doubtless for its resemblance in shape to one of those conical canvas sea anchors which are set out on lines during storms at sea. Now the CSM chugged forward slowly on its thrusters, moving relative to the Lem (still attached to the S-IVB) about as fast as a tugboat at maneuvering speed, although absolutely through space at a velocity somewhere over twenty-seven thousand feet a second. As the CSM moved, it proceeded by the use of other thrusters to go through a sixty-degree roll in order to align certain catches on its probe with the drogue. Then, at about one foot per second it closed the gap between probe and drogue. At the edge of rendezvous the speed was reduced to some fractions of an inch per second. With not much more impact than the lips of thoughtful lovers coming together, the probe of the Command Module entered the drogue, passed through a hole where a set of metal tongues called captive latches might lock, and the rendezvous was completed. The CSM and the Lem were locked.

The terminology of rocket engineers was no stranger to the act of coupling – the stages of a rocket when brought together were *mated* – the astronauts, aware of the risibilities of passive and active collaboration between probe and drogue, would play with the humor. During a television transmission two days later, when Armstrong and Aldrin had to remove probe and drogue in order to enter the Lem, this following dialogue took place.

CAPCOM: *It looks like you almost got the probe out.*

ALDRIN: *Yeah, it's loose now . . .*

CAPCOM: *Looks like it's a little bit easier than doing that in the chamber.*

ARMSTRONG: *You bet.*

ALDRIN: *It's pretty massive, but it goes where you direct it.*

ARMSTRONG: *Mike must have done a smooth job in the docking. There isn't a dent or a mark on the probe.*

On the flight of Apollo 12, there was heard, much to the quickly concealed shock of the commentator, the remark at the end of rendezvous – 'Move over so I can stick it in your giggie.'* A docking was a docking, mechanical or no, a sweet feat in fact, equal in difficulty to two humans in a weightless environment trying to make love (with no hands) while they floated about, nothing more to direct them forward or away from each other than the wind they could expel at either end – or from their ears! What suggestion of future activities for tourists to the moon.

Actually this particular docking had a few complications. 'When I started to pitch up,' said Collins, 'for some reason it – it stopped its pitch rate and I had to go back . . . and hit . . . an extra proceed on the DSKY. And during the course of that, we drifted slightly further away from the S-IVB than I expected. I expected to be out about sixty-six feet. My guess . . . I was around one hundred or so, and therefore I expect I used a bit more [gas] coming back in.'

Pilots were notorious for being stingy. Folklore concerning them was filled with tales of waitresses who had been stiffed by a thin dime, or the woes of any airline stewardess who shared a cab with the pilot and paid the extra nickel in a bill of $5.75. So Collins now apologized for wasting gas – stinginess was a virtue in the air – birds are not the most conspicuous of the romantic figures. No, one does not waste gas. A useless move in the air burns fuel or loses momentum which may not be available later: so pilots keep clean charts and apologize for the expenditure of consumables they might have been able to save. What gives the flavor here is Collins' generosity of manner. Whether superficial or straight from the heart, Collins had his implicit way of presenting himself as the most generous of the astronauts. Ergo, his apologies and copious remarks on the shade of this faint inefficiency have their hint of comic excess.

It was not serious, no, indeed, and after docking, work went on. The hatch back of the probe on the Command Module was removed, the twelve locking latches on the ring which sealed the Command Module to the Lem were checked, the umbilicals for oxygen and electricity were connected, and the hatch was replaced. All this was accomplished with the astronauts feeling no sensation of motion even if they were coasting through space between four and five miles a second. And from what they could see through their windows the spaceship now looked bizarre. The realignment and docking had left the third stage at one

---

* A Russian cosmonaut just after contact had cried out, 'I'm being raped.'

end, the Command and Service Module at the other, and the Lem installed between. In this curious position, broadside to the direction of their flight, they traveled for a period.

The Lem was pressurized. Other adjustments were made. Then the Command and Service Module with Lem attached separated from the third stage. Four knee bolts holding the Lem's legs to the S-IVB were exploded with enough force to push away the CSM-with-Lem-now-attached at a rate of 1.2 feet a second. Separation completed, the spacecraft was pitched around once more, and its main motor was started. The burning time was but three seconds, just enough to give a speed of twenty feet a second. At that rate, the now liberated moon-craft, consisting of the CSM and Lem, pulled away from the third stage about as fast as a modest powerboat. Then, as a result of radio commands issued from the ground, the unburned fuel still in the third stage was vented. For every action, there was an equal and opposite reaction. From the simple discharge of the fuel there was a shift in the direction and speed of the abandoned third stage. It was a small shift, and a small change in speed, but it would be enough to divert the trajectory of the S-IVB away from the translunar coast of the Command Module. Now the S-IVB would pass far enough away to go past the moon and on to the sun. It would never be seen again. As time elapsed, the S-IVB drifted two miles away, four miles away, was finally lost from sight. The Command and Service Module with the Lem attached in front now sailed toward its rendezvous with the lunar sphere. But for a brief midcourse correction or two, its motor would not have to be fired again until the moment came to brake speed and ease into orbit around the moon. So on went this self-contained universe, born in high motion on the airless silences of space, born with the term of its human life dependent on the continued function of its oxygen and provisions, its human and propellant consumables, on through space it went, a bullet fired up from earth and now free of gravity and no readiness to stop.

vii

To speak of a self-contained universe when one is only dealing with a vehicle which is self-sustaining for a short period is to trespass on the meaning. A man is a universe by that measure, indeed he is more self-contained in his ability to adapt and survive than the ship of Apollo 11. In fact the Command Module is more like the sort of universe complete in itself one glimpses in a flower cut for a vase. Such an ornament receives food, breathes, exudes, molts, can even preside over a fresh development like the opening of a bud, and presumably this cut flower is capable of sending and receiving messages from other flowers and plants (if such communication is one of the functions of a flower) but still! we know the flower will live only a few days. It is a self-contained universe whose continuation is sealed off from itself.

The same was true of the Command Module. The men in it could live no longer than there were supplies of oxygen for them to breathe, and that was for two weeks. Nonetheless, Apollo 11 was more a cosmic expression than an ornament. Its vase was space, and through space it traveled, a ship, a species of man-made comet, a minuscule planet with an ability to steer. We will inhabit this ship for the rest of the journey, we will be taken up with the activities of the men in it, we will learn – if we will not learn precisely how to fly it – still we will pick up a little of what it might be like to endure, even at times to enjoy, the near to two hundred hours of the complete round trip. So it may be best to fix these living quarters in our mind.

The spaceship, now free of the third stage, and five and a half hours out, was a most peculiar-looking object, about sixty-seven feet long, and in a variety of widths, for the top of the Lem (that mechanical spider!) was attached to the nose of the Command Module and in turn the base of the Command Module, we can remember, was faired into the Service Module. What then could be your honest width? The Service Module all complete was close to twenty-five feet long and twelve feet ten inches in diameter, a perfect tin can in appearance but for the bell of its motor at the base which took up perhaps ten feet of that length, a large bell for a modest motor. The Command Module was ten and a half feet high and at its base identical in diameter to the Service Module, but it tapered to the top, where it was no more than three feet across. There its probe locked into the drogue of the Lem, and the Lem had a width which was most confusing. Its body was fourteen feet in diameter, but its four legs were thirty-one feet across their diagonal when extended (somewhat less when closed) and the ship, while ready to coast in every attitude, either Lem first, or bell of the Service Module first, whether with its structure broadside to the line of its flight, or canted at an angle, cocked for example to the left, cocked to the right, upside down and all but inside out (any position was possible in a vacuum), still usually coasted with the Lem forward. The spaceship was therefore a most peculiar construction, not unreminiscent now of a bullet fixed to the knotted body of a bug. Indeed it had to be an object unlike anything seen in any machine on earth for its different components were designed to function in separate compartments of the universe. The Command Module and Service Module were obliged to withstand the friction of the atmosphere at high speeds – so their skin was streamlined to help minimize heat and drag, but the Lem had been designed to function only in a vacuum, and therefore needed streamlining no more than an earth borer would need eyes, indeed its form was the perfect expression of the sum of its curious functions, just as an office skyscraper which looks like a Kleenex box on end is the perfect expression of the innate architecture contained in a one hundred million dollar bill. The Lem looked about the way the organs, arteries and veins of the human body might appear within the human skeleton if the body lived in an environment which required no epidermis or sheath of outer muscles.

There it was – spaceship! An object without grace or unity of design, sixty-seven feet long, anywhere from thirty to twenty feet wide at the legs down to three feet wide at the joining of the Command Module and the Lem, and yet in all of this ship, the astronauts – except for that period when they would inhabit the Lem, were confined to the conical interior of the Command Module. Its interior space was not twelve feet wide, not ten feet high (indeed for purposes of standing not everywhere six feet high) and since its walls were covered with instrument panels, equipment and cupboards, and its floor when all three couches were out was all but unavailable, it had to prove a close cramped near-intolerable capsule for a claustrophobic spirit. Fifteen miles of wiring and two million functioning (which is to say sensing or bearing, protecting or moving parts) had been crammed into a paltry seventy-three cubic feet of space for each man. That is less volume than is offered to each passenger in a comfortable car. Nonetheless, it was a boy's dream of a habitat, for the Command Module was at once a workshop, a submarine, a pilothouse, a species of Pullman compartment, a cockpit, a radio station, and a den, it was an observatory, a TV studio, a music booth, a kitchen, a lab, a bedroom, bathroom, a gym and a clubhouse. It must also have seemed on occasion hardly larger than the inside of a theatrical trunk. It was undeniably a womb for triplets. On demand, three umbilical cords could plug into three space suits. Three men could breathe through three tubes and float in an enclosed volume. Yet if they had been equipped with laser beams they might have ventured forth as a celestial gunboat. With their banks of instruments, they could entertain the illusion they were playing an electronic organ. For its aura of austerity, Collins was later to call it a mini-cathedral. With a man and woman for passengers, the Command Module, red velvet laid all out, could have served as a boudoir or at the worst, a sultan's most private tent.

And this object had cost – who could even calculate the precise cost? Accountants would go to war over such a project. For how much would be apportioned to research, design, and simple error? how much to the invention of thousands of new tools without which the module could never have been built, and one had not yet begun to speak of the huge expense of creating the Mission Simulators, then improving them – did it cost a billion dollars to build the first Command Module? One would never know. Certainly, the Command Module of Apollo 11 had come a long way. Its pieces and parts assembled by North American Aviation had derived from sixty-two subcontractors and suppliers, its first models had been tested in seventy-five hundred hours of wind tunnel, it had gone into altitude chamber and airlock, been subject to tests for pressure and tests for vacuum. The prototype had been dropped free-fall with a crew into water tanks, then dropped on land to simulate the impact of unhappy parachute landings. It had undergone the laboratory equivalent of solar radiation and temperature extremes of heat and cold, its electrical power sources had been tested, it had been subjected to loads and stresses and bendings, its heat shield had passed through heat fluxes of as much as 25,000 British Thermal Units per pound in gas stream enthalpies. (That was a word not even to be found in the *Shorter Oxford English*, but enthalpies had to suggest the very entheasms of heat!) Then its radio and telemetry systems had been tortured in shock tests, exposed to dust, to sand, to rain, to salt spray and the corrosions of raw oxygen, its components were conducted through the rigors of acoustical shock, vibration, high acceleration; at Downey it had been originally assembled in the largest and cleanest clean room known to the world, a chamber longer than a football field and more than half as wide, an average of fifty feet in height, a room into which the Command and Service Module had been inserted through an air lock to keep all dust out; and therefore been put together in a volume whiter and more protected than the delivery room of a hospital, even examined by technicians wearing white smocks and sometimes white cloth masks over their nose and mouth. In theory it had been flown through simulations by astronauts, flown through volumes of imaginary space with a six-foot globe of the earth for reference, and computers to replay the movements through the trajectory. With the Service Module it had gone through 587,500 inspection points, been checked for its conformity to 8,000 drawings and 1700 sets of manufacturing and engineering specifications, it had even been roasted on one side and frozen on the other to anticipate the searing heat of the sun and the freezing cold of 320 degrees below zero in those shadows of space it would create for itself on its dark and sun-obstructed side. It had undergone every checkout American technology could devise and that was one full file of manned and unmanned simulations and flight readiness tests. Yes, it was some boy's dream of a habitat. It came complete with ten major subsystems, three for such special occasions as launch escape, thermal protection (which is to say the deployment of heat shields), and earth landing (or parachutes); it had seven subsystems for use at any moment over the eight days of the trip, subsystems under the respective titles of communications, electrical power, environmental control, guidance and navigation, reaction control, service propulsion, and stabilization and control, which probably could be abstracted to three major categories: communications, life maintenance, and flight.

Each system was designed with parallel and complementary functions, the equivalent of a network of roads between any two cities. A variety of routes existed whenever it was desired to shift some system in Apollo 11 from one condition to another, just as one always has a choice of superhighways, highways, country roads, feeder roads, and in emergency, simple dirt lanes in transit from one place to another. So the electrical system had three fuel cells for powerplants, which each drew on two separate supplies of hydrogen and oxygen and had for products electricity, water, and heat. There was actually need for only one fuel cell, but they were relatively delicate (as Apollo 13 would once again prove) so for safety not two, but three were installed, and since any one of them could be used at any moment, while the tanks of hydrogen and oxygen were doubled, there were in consequence six combinations for one source of power.

Every part of every system which might be switched into other systems was capable of being employed that way. The oxygen tanks for the fuel cells were also piped into environmental control to provide a breathable atmosphere. The waste product of the electrical process, we may remember, became drinking water, for the electricity had been generated by the action of hydrogen and oxygen on electrolytes.

It was the same with other systems. Guidance and Navigation was complete in itself; it had a gyroscope, a computer, and optical data, but Stabilization and Control, a system which enabled the crew to guide and navigate the ship, was also capable of taking over every function of the first and larger system, just as there were batteries charged by the fuel cells which could in emergency maintain the electrical load for a period. The Service Module had a complete set of little motors called thrusters to tip the ship, lift it, depress it, change its attitude of travel and its ability to roll – so did the Command Module have an independent set of thrusters which could do the same work, and in certain conditions where the main motor of the Service Module might fail to function, so the combined forces of the Service and Command Module thrusters plus the motor on the Lem could be employed to achieve an equivalent result. There were two methods of voice transmission, by Very High Frequency and by S-band, which was Ultra High Frequency. The transmission was controlled either by a push-to-talk switch in the astronauts' umbilical cable or by an automatic relay circuit triggered by the voice. For emergency transmission, the push-to-talk switch could be used like a telegraph key. If the VHF tended to be employed during the near-earth phases of the mission, and the S-band in deep space, still the VHF would serve for communications between the Command Module and the Lem when they were separated. It was composed of two complete and separate transmitter-receivers in one unit while the S-band equipment had primary and secondary transponders as well as an FM transmitter, and also had primary and secondary power amplifiers. Duplicates were everywhere. The VHF had two antennae, the S-band had five, consisting of one which was high-gain and four which were omnidirectional for backup. The more one pursued the details, the more they led into one another; it was apparent the aim of all design had been to allow the maximum of alternate routes around every conceivable malfunction. As if in imitation of the brain, each of whose lobes are capable under duress or injury of appropriating most of the functions of the other lobe, so redundancy was built into every aspect of every system and had become one of the basic words at NASA. Where English students might shudder at the comment 'Your style is redundant,' aerospace made redundancy a virtue. But we can listen again to Collins:

You can say here's an oxygen tank, and I want to have a pipe going out of that so you can sniff oxygen. So you have one tank and you have one line. Then somebody says well, maybe we need two tanks. So now you have two tanks and two lines. Then somebody says suppose you had a leak over here and you lost all this oxygen. Wouldn't it be nice if you could get it from that tank over to the alternative line? So now you've got two tanks, two lines and a cross-over tube with its own valve. Then somebody else says suppose you got a leak over there? You wouldn't want to leak oxygen from both tanks out through one place, so we'll put a check valve in each one of those lines. But suppose the check valve gets clogged up? Well, we'll put in two check valves. But what if you want to bypass the check valve? You have to put a line around it. So now you've got four check valves and a cross-over valve, and you've got a bypass line here and a bypass line there, and each one of those needs a valve. Now just to get oxygen out of a tube we've got seven valves; it grows like Topsy. And each one of those valves has a little switch and a lever, and you've got to remember which is which.

If a car had been made whose motor had two carburetors, a double set of tappets, two sets of dual spark plugs, two batteries, two generators, two coils, two distributors and condensers, a double bank of mufflers and twin exhausts, a gear shift and an automatic transmission, a double set of valves and tappets, a set of auxiliary axles and auxiliary wheels and tires to drop out of the chassis when a tire blew, a horn and a siren, a steering wheel and a control stick, a left foot brake and a right foot brake, a hand brake, and a transmission parking lock, horizontal windshield washers and vertical windshield squeegees, a door which could be opened and locked from inside and outside with two sets of locks functioning in connection or independently, and windows which were automatic or hand-operated, as well as a gas tank with two reserve gas tanks, an air-conditioning system for the front seat and another for the rear seat each capable of being used for the entire car, heaters which worked off the battery and heaters warmed by the temperature of the motor, double radiators, double air filters, a hood which lifted from the front, and motor panels which lifted from the side, seat belts and fenders built in eggcratelike compartments to absorb collision, an AM radio, an FM, a stereo, and a radiotelegraph system, all of which components were capable of sending back to the garage all available information on the rigidity of the chassis, the wear and heat of the tires, the temperature of the driver, his nearness to sleep, the octane level of the fuel, the conversations of the passengers, the compression of the engine, the fade of the brakes, this radio and telemetry in link with communications from garage back to car as well, if we think of what a wonder such a car would be, so too are we obliged to wince at its weight, its cumbersome solutions, its engorgement of available space, its difficulty to drive – we have not even given a word to the monitorings and alarums of its dashboard! But it would be safe, yessir! as safe as human ingenuity could make it, and its chances of getting through to its destination would have to be greater than the odds on the average car. If on the ground such safety was hardly necessary since one expected to take a few uncertainties into a long car trip, Apollo 11 was on the contrary moving through space, and it did not have three thousand parts like a car, but two million. And it was disrupting God knows what Valhalla of angels and demons, what eminences of benignity and eyries of the most refined spook essence – the future of the rate of acceleration of technology itself was at stake with what trillions of dollars and commitments of soul no historian of a later century could count, no, the value of the lives of the astronauts was not to be measured by ordinary lives, for the shock to the continuation of the technology of the world if the astronauts were killed on the mission would be greater than the shock to the Church if a maniac succeeded in murdering the Pope, or at least as great as the ripple of unrest through the nation if a dog were dismembered for an audience at peak time on television – and Cronkite for comment to follow. Yes, the trip was dangerous enough, the very conception of landing on the moon was blasphemous if the marriage of technology and the heavens was not part of God's design, decades of reverberating horror and indecision could result from a sudden mysterious and catastrophic end to the flight of Apollo 11. So the redundancies were obliged to go beyond that fine and impressive balance one sees between risk and safety in race cars and unmanned rocket ships; redundancy became a value in itself. Later, risks could be taken, men might be lost, but for the present, the fact itself was gamble enough: a ship was on its way to the pale graveyard of sleep, and in the awe engineers felt before men so free of primitive taboo as these pioneer astronauts, there was no precaution taken which could be great enough: upon the success of the flight now rested nothing less than the alleviation of dread in technicians whose libidos were enmeshed in all the winding nets of number.

viii

The Lem, plus the Command and Service Module, had completed its separation from the third stage of the launch vehicle at Ground Elapsed Time of five hours and six minutes into the mission. At that time the S-IVB, it may be recalled, was sent off in a slingshot maneuver to take it behind the trailing edge of the moon and on to solar orbit. The astronauts' last view of that empty lonely bucket was therefore about twenty thousand miles out from earth, the object seen across the glare and the haze and the rosy darkness of bright sun in black space. Given the light reflecting in all directions from the golden-bronze foil of the Lem and the polished steel of the Command Module, they could hardly be certain it was the S-IVB. It hardly mattered. They were settling in for the flight.

The hours would now pass in a fifty-wide variety of repetitive chores and duties. The thousand subdetails of routine rocket housekeeping were upon them, the plethora of tests, measurements, precautions, anticipations, routines, alerts, passing malfunctions and jokes which make up the schedule of their hours. Their radio communication was constant, which is to say that with the exception of their hours of sleep, not many periods of fifteen minutes passed in a day when they were not in conversation with the earth, and rare was the hour which did not have twelve or fifteen call signals to initiate conversation from ground to space or from Apollo 11 to Houston, then dialogue back and forth. So on an average, sixty or eighty separate transmissions took place each hour. If the first five hours of flight had been spent in just such conversation, ranging over such subjects as the check-out of their color TV and the alignment of their inertial table with the stars, communications occupied in turn with describing the sights below and copying down computer data delivered up from the ground, testing their thruster systems, repeating commands on the burn for Trans-Lunar Injection, then comments afterward on the splendid qualities of Saturn V, if the last of these first hours was spent dealing with the multitudinous details of the Lem extraction and the slingshot, a less eventful routine began on the long coast upward. They spent the rest of their day in battery charging and fuel-cell purging, in new star sighting and TV transmission, in descriptions of the workings and idiosyncrasies of the cryogenic fans and heaters. They talked about the functioning of the separator which removed the chlorine bubbles from the drinking water, and discussed the angles for deadband control. That was a complex matter which would have to be readjusted again and again. Much of the work in this phase of the journey consisted of setting the angle or attitude at which the spacecraft traveled so that the particular antenna employed would have an unobstructed line to earth. That was not so simple as it seemed. Since the sun would always remain on one side of the ship, a program had been set up to fire the rocket thrusters now and again in order to roll the ship around its own axis like a carcass on a barbecue spit. At a rate between two and three times an hour, the spacecraft rotated slowly about this axis, the better to keep itself toasted equally on all sides by solar rays rather than to suffer the heat to port and freezing to starboard which would ensue from voyaging with one side always in the sun, the other in darkened space, for such inequalities of heat and cold could affect the performance of the ship. These revolutions, called Passive Thermal Control, or PTC – this was the acronym used most perhaps on the flight – were usually conducted with the ship traveling Lem-first, and the S-band antenna (which was employed most often) projected out from the rear of the Service Module, and so had an unobstructed line to earth. But there were also periodic star sightings to take close corroborations of the settings of the gyroscopes and inertial tables – these sometimes demanded a fixed position where the ship did not revolve. Besides, the star sighting was sometimes obstructed by glare. So the ship had to be turned through a large angle and would proceed for a while tailfirst with the Service Module in front. Now the angle of the antenna would present a problem. Therefore, numerous conversations proceeded back and forth on the best angles for setting the ship and setting the antenna, then further conversations on the best way to resume PTC. So passed the first day with dialogues on star sightings, waste-water dumps, and altitude settings, gyroscope settings, Verb and Noun settings for the computer, and Passive Thermal Control modes. Teams of men came and went on eight-hour shifts in Mission Control. They had names: the White team, the Green team, the Black team. Radio readings were exchanged and comments sent back and forth on the quality of the TV.

Finally the day came to a close. It had been an extraordinarily long day for the astronauts. It was now 10:30 P.M., eastern daylight time, and they had been up since four in the morning and in space for thirteen hours. Doubtless they were ready for sleep. The center couch was dismantled and stowed under the left couch. The astronauts stretched out in sleeping bags under the left couch, under the right couch, on top of the right couch, and zipped themselves in. If the temperature of the cabin was an equable air-conditioned seventy to seventy-five degrees, they were still weightless and so the bags were obliged to be strapped down and the astronauts restrained from floating up and bumping the walls. Yet it was still their first night in space this trip. Not so quickly did they fall into sleep. By virtue of the bio-instrumentation belt, telemetric data sent to earth monitored their slumber. Given the rate of their heartbeat the onset of sleep could be gauged, and if Collins was quickly out, the others stayed awake for awhile, listening perhaps to the pumps and the quiet clickings, the tickings, the sound of the fans, the sense of silence vibrated by a hundred intent mechanisms all pursuing the routine of their much-studied functions, the life of the men as dependent on a few of those machines as the pushings and dilations of their own hearts, pumping now perhaps at the thought (first time this busy day) of the oncoming adventure of the moon. On the fourth day would they land. Only the most stolid of minds, insulated altogether from imagination, could be indifferent to the possibility that in four days they might be dead or by the moon fundamentally altered, by the moon in some as yet unknown fashion, yes, fundamentally altered. 'I hope I don't get a tender foot,' Aldrin had said at the press conference in Houston eleven days before.

An hour later, close to midnight, eastern daylight time, the Black team came on to replace the White team at Mission Control. They were the monitors of sleep and would not waken the astronauts or have conversations with them as the ship rotated through the hours in PTC, all window shades pulled down, all but instrument warning lights turned off. The sun kept shining in the darkness of space, the ship coasted upward, revolving slowly, revolving slowly, three times an hour revolving in Passive Thermal Control. The thousands of miles went by. Apollo 11 was sixty-four thousand nautical miles from earth when sleep began. By morning when the crew was awakened, they were more than ninety-four thousand nautical miles away. The day to come would be filled with duties and empty of events. So it might be ideal for study – there was a boredom to space, a boredom not unlike the lethargy of people on a long trip by car, but here a boredom incalculable to the layman who did not engage the details, the buzzings and the hustings of one hundred thousand miles of intercom and interoffice memo. Few of the words would be clear to all, few of the words would give pleasure. It was technology pure coming up.

3

A Day in Space and Another Day

On Thursday morning, July 17, the crew was up before they were called. Presumably they lifted the shades, let in the sun of full free space (what a shine in the eternal darkness!), and worked their muscles with a spring exerciser in obedience to the depressing discovery by previous flights that muscles go slack in weightlessness at a much accelerated pace.

There, all resemblance to America's good morning could end – the radio was working. They were on. There would be voices in their ear for the next sixteen hours.

CAPCOM: *Apollo 11, Apollo 11, this is Houston. Over.*

ALDRIN: *Good morning, Houston. Apollo 11.*

CAPCOM: *Roger, Apollo 11. Good morning. When you're ready to copy, 11, I've got a couple of small flight plan updates and your consumable updates, and the morning news, I guess. Over.*

COLLINS: *Standing by for your updates. Over.*

CAPCOM: *In your postsleep checklist and in all other postsleep checklists, we'd like you to delete the statement that says AUTO RCS jet select 16 to ON, and what we're doing here is picking this up in the procedure for exiting PTC that's in your CSM checklist and in the CSM checklist on page foxtrot 9-8 – if you want to turn to that – we'd like to change the order of the steps in that. Over.*

COLLINS: *Okay, page F 9-8. Go ahead.*

CAPCOM: *Okay, right now it reads to exit G and N PTC, then you've got a PAN 8 change that says AUTO RCS select 12 main A and B down to the second step, so the procedure would read Step 1 Manual Attitude 3, Accel command, Step 2 AUTO RCS select 12 Main A B, Step 3 would be verified deployed . . .*

A way to begin the morning! Now the time-line charge on Battery A is installed in their ear, then an updated set of angles in Program 52 for realigning the table on the Inertial Measurement Unit. Numbers come their way, more numbers, then instructions as to when to initiate a waste-water dump. It will be begun as soon as Program 23 concerning their midcourse navigation is completed.

The crew reads back the figures and their assent to the instructions, they delineate the attitude of angle of the craft, they offer numbers for the ground to check against Mission Control monitors. Now the crew presents crew status report. 'CDR 3, CMP 7, LMP 5.5.' If it is correct in transmission, it means Collins slept seven hours, Aldrin five and a half, and Armstrong only three. The figure passes without comment. What may be said about insomnia in a monitored capsule? One cannot even lie and pretend the sleep was good.

Now comes the consumables update, a report on the amount of oxygen, hydrogen and fuel consumed by the thrusters and the fuel cells. It is sent up from ground to space, and downline from Apollo 11 to ground comes a report from the spacecraft on their own measure of fuel left. They have not used that much the first day – they have 96 percent left in Alpha Quad thrusters, 87 in Bravo, 88 in Charlie and 90 in Delta.

Next a discussion takes place on the charging of the batteries. The batteries are exhibiting idiosyncrasies. A consultation with the electrical engineers at Houston assures the astronauts that the reasons are minor and concern no more than natural differences between individual batteries – all of this has taken no more than twenty minutes. The first twenty minutes of the first day of awakening in space has just passed. Think what a jolt those numbers would have been to a fine mind hung over with sweet recollections of the night before, acid stomach, whisperings of guilt, a disconnected head and dread remorse. Say, it is no light and easy load for the astronauts themselves – even virtue is incarcerated in the dungeon of numbers. Armstrong's wife would recollect that in early June after weeks and months of working ten, twelve and fourteen hours a day in simulators, 'Neil used to come home with his face drawn white, and I was worried about him. I was worried about all of them. Their morale was down. They were worried about whether there was time enough for them to learn the things they had to learn, to do the things they had to do, if this mission was to work.' It gives a hint of how killing is the work in the simulators. To use all of one's best energy hour after hour, working day after working day in order to keep up with a machine whose brain is more rapid if not more brilliant than one's own, plunking all of one's ambition, avarice, charity, pluck, discipline, and education into an electrical set of brains which will give back nothing but firm answers, can hardly improve the human brain at anything but the cruelest expense to the body. No wonder their morale revived when the flight date was made certain on June 17 – the work till then must have been privately unendurable, a feat next to taking IQ tests for twelve hours every day in front of a six-hundred-fold instrument panel of blinkers, switches, alarm lights and warning bells – food for psychosis in rats! yes, it must have been so unendurable that the thought one might yet pursue this huge effort to no purpose – for one's participation could always be washed out – was a way to lay the groundwork for the worst of diseases, the leaching out of one's soul. The thought that this huge effort of training, this courtship of numbers could end in nothing, must have fired up a torture at the core of their unconscious, even that stoic imperturbable core of the astronaut heart. They had begun as active men with quick reflexes and the impulses of athletes – now they were dial-bound and fingertip quick, naught but blinkers and tweekers in their working day – it was like making a Ping-Pong champion out of a great tennis player.

Besides, the massive regions of their technical ignorance had to weigh on them. They were by now in varying degree not only expert pilots but expert engineers, expert mechanics, expert technicians. It was still not enough. Just as a devoted English professor cannot pretend to be equally knowledgeable on every last poem, jotted whim, or letter wrapped in ribbon in the Descriptive Bibliographies of Wordsworth, Shelley, Byron, Milton, and Donne, no more than he can begin to cover the major sweep of English prosody from Beowulf to Allen Ginsberg, leave Shakespeare alone, Marlowe alone, Melville! think then of the astronaut who presides over a field of instruments whose engineering theory, while not necessarily so attractive to live with as English literature, nonetheless engages as many volumes of the mind as could any close-up survey of English, history, or medicine. Each astronaut must recognize that for all he knew, there were icebergs of mystery in all he did not know, all he could not keep up with. To a profound student of Shelley, the third beribboned letter might contain a phrase which could turn his notions of Shelley into new directions, but an expert on

Chaucer or Faulkner would pass it by with a quick reading. *There* was
the danger for the astronauts! To be insufficiently grounded in the subtlest hints of malfunction in any one of six hundred instruments which,
perish them all, they had not become sufficiently expert upon, was an
invitation not only to their own disaster but to the peril of the NASA
program, last chalice of Good Square Life. Every Square is aware of his
own devil – that is why he chooses to be Square. With failure, they
could never be certain the program had been immune to their secret
treachery. It is a fair anxiety, enough to turn men pale and eat their liver.
Of course, they had Mission Control behind them, and from Mission
Control quick links extended to all the experts in all the fields – there
were the engineering equivalents of experts on plainsong and experts
on Burroughs, experts on Gay, Cheever and Updike, Bellow and
Smollett, Swinburne and Walter Savage Landor, Ronald Firbank and
Baron Corvo, Ben Jonson and Babette Deutsch, name it, name it –
Mission Control was in touch with men who knew all about the subject
of toggle switches or television grids, knew them like an English honors
man knew Shelley's letters after forty years, Mission Control was in
touch with helium transfer men, honeycomb panel men, valve men and
monomethyl hydrazine men, data acquisition system engineers and
gigacycle deep space antenna fellows, solenoid experts and nuclear particle detection masters, single channel decommutator designers, and
pseudorandom noise ranging test set doctors, specialists in mass
flow-meters, astronomers, and air-conditioner men, flight direction system engineers, orbital physicists – the astronauts did not have to know
it all. Every malfunction could be reported, discussed on the ground,
considered, the solution of how to adjust to the difficulty radioed back.
But there were anxieties nonetheless. The response might take too long
to get. The crisis could accelerate. Any English professor entrusted with
the last safeguarding of the English language would be secretly appalled
at the dimensions of his own ignorance in such an apocalyptic mission
even if he had ten thousand experts and graduate students to support
him a few minutes of communication away.

Well, no need to assume the first twenty minutes on July 17 was even
remotely suggestive of the pits of this venture. Routine, if painfully
detailed instructions on exiting from PTC, on charging the batteries,
updating the programs, updating the record of consumables and

dumping the waste water had taken the time. Now the morning news came. Each morning Mission Control would give a five or six minute newscast –

WASHINGTON UPI: *Vice President Spiro T. Agnew has called for putting a man on Mars by the year 2000, but Democratic leaders replied that priority must go to needs on earth . . . Immigration officials in Nuevo Laredo announced Wednesday that hippies will be refused tourist cards to enter Mexico unless they take a bath and get haircuts . . . 'The greatest adventure in the history of humanity has started,' declared the French newspaper* Le Figaro, *which devoted four pages to reports from Cape Kennedy and diagrams of the mission . . . Hempstead, New York: Joe Namath officially reported to the New York Jets training camp at Hofstra University Wednesday following a closed door meeting with his teammates over his differences with Pro Football Commissioner Pete Rozelle . . . London UPI: The House of Lords was assured Wednesday that a major American submarine would not 'damage or assault' the Loch Ness monster.*

Dig it! What unheard comments on the Loch Ness monster, or what Rozelle must have said to Namath. What probable laughter at the hippies – long hair and funky unwashed caverns were the zone of demarcation in the physics of smell, stink was the Siegfried line between magic and technology. But the astronauts were plenipotentiaries of space so they chose to respond to Vice President Agnew. 'We hope we get a chance to see him when we return.'

'Roger,' says the Capcom, 'and I understand he was down there and really enjoyed watching the launch. We think it was pretty magnificent and you-all are doing a great job up there.'

Spacecraft: 'Thank you.'

Congratulations and good wishes having ranged back and forth, there was now a half hour with no conversation. Presumably the astronauts were eating breakfast. On DAY 2 the schedule called for Meal A: Fruit cocktail, sausage patties, cinnamon-toasted bread cubes, cocoa and grapefruit drink.

When silence was broken, a long set of instructions on how to avoid errors in angular readout on trunnion bias determination was given by the Capcom. This introduced a specific discussion on how to locate a reference termed the M-line in the ocular system. It seemed the M-line was required to be parallel to the earth horizon at the 'substellar point.' The spacecraft returned with the information that on the previous day they had not been able to bring the M-line into any kind of parallel.

> COLLINS: *Before we started marking for the first time it appeared that the computation of those three angles was somewhat off, and that I was wasting a lot of gas by going to those three angles and then having to make a large attitude change after that to get the M-line parallel, and in some cases it appeared to be just an accepted attitude required, and you all said it wasn't needed. So I was marking in some cases with the M-line not parallel. I thought perhaps you had some processor for computing that offset and making sense out of that data, but as far as I know we gotta have the M-line parallel.*
>
> CAPCOM: *Roger, 11. We don't have that capability. We do require that the M-line be parallel to the horizon in order to get a good mark. We feel that possibly the status vector information that you were using for your maneuver basis yesterday may have needed to be updated a little, and if you'll stand by a second we'll give you an evaluation of what we feel you'll get today by the auto maneuver.*

But then that discussion was interrupted in order to do an oxygen fuel cell purge – periodically a fuel cell would be shunted over to inactive status and nitrogen flushed through its oxygen or hydrogen pipes in order to clean out the detritus. Reports were then made on the flow rate during purge.

New discussion began of ways to attack the problem of the M-line, but it was again interrupted by a remark from ground that the level of carbon dioxide in the Command Module seemed to be increasing. Had they changed the canister? They had not. Would they then 'plan on accomplishing that after P23 is over?' Back to the M-line, where more maneuverings were commenced to offer positions for the telescope and sextant so that they might sight on stars which could fix their position. Let us follow that dialogue for the next quarter of an hour. It is impossible to comprehend altogether – one would need to work for a year in the Mission Control Room at Houston – nonetheless it has all the authority of discussion at vast distances about small measurements.

> COLLINS: *Everything looks beautiful except there is no star in sight. It is just not visible.*
> CAPCOM: *Roger. Is this for star Zero One?*

COLLINS: *That's correct.*

CAPCOM: *You are not getting any reflections or anything like that that would obscure your vision, are you?*

COLLINS: *Well, of course, the earth is pretty bright, and the black sky, instead of being black, has sort of a rosy glow to it and the star, unless it is a very bright one, is probably lost somewhere in that glow, but it is just not visible. I maneuvered the reticle considerably above the horizon to make sure that the star is not lost in the brightness below the horizon. However, even when I get the reticle considerably above the horizon so the star should be seen against the black background, it still is not visible.*

CAPCOM: *Roger, we copy. Stand by a minute, please.*

CAPCOM: *11, this is Houston. Can you read us the shaft and trunnion angle off the counters?*

COLLINS: *I will be glad to. Shaft, 331.2 and trunnion, 35.85.*

CAPCOM: *Roger, thank you.*

COLLINS: *It's really a fantastic sight through that sextant. A minute ago, during that automaneuver, the reticle swept across the Mediterranean. You could see all of North Africa absolutely clear, all of Portugal, Spain, southern France, all of Italy absolutely clear. Just a beautiful sight.*

CAPCOM: *Roger, we all envy you the view up there.*

COLLINS: *But still no star.*

CAPCOM: *11, this is Houston. Over.*

COLLINS: *Roger. Go ahead, Bruce.*

CAPCOM: *On our ground computer we confirm the shaft and trunnion angle that you have as being pointed at the star. However, it looks as if that shaft and trunnion angle is also pointing into the structure of the Lem so that while you will be getting the earth horizon, the star is obscured by the Lem. We recommend an automaneuver to the attitudes in the flight plan. Roll 177.2, pitch 298.2, and yaw 330.0. Over.*

COLLINS: *Okay, fine, let's try that . . .*

But this line of inquiry was interrupted.

CAPCOM: *11, this is Houston. While you're maneuvering, could we get an LM CM Delta P reading from you? Over.*

COLLINS: *Roger. Just a tad under 1, Bruce – .95.*

CAPCOM: *Roger. .95.*

PUBLIC AFFAIRS OFFICER: *This is Apollo Control. That reading was the difference in pressure between the Lunar Module cabin and the Command Module cabin . . .*

Now they resumed efforts to solve the position of the star, the instrument, and the horizon.

> COLLINS: *Okay, our maneuver is complete and at this attitude the M-line is exactly ninety degrees out of phase. It is exactly pointed along the vector toward the center of the earth instead of being parallel to the right.*
>
> CAPCOM: *Roger.*
>
> COLLINS: *I'm going to hold right there for your next suggestion.*
>
> CAPCOM: *Roger.*
>
> COLLINS: *Okay, Houston. It appears to be okay now. We've changed our attitude slightly and I have a star and I'm maneuvering to get the M-line parallel.*
>
> CAPCOM: *This is Houston. Roger, we copy . . .*
>
> COLLINS: *I gave it back to the computer for a second. I put the mode switch from manual back to CMC while I fooled with the DSKY, and the computer drove the star off out of sight, so the delay here has been in going back to manual and finding the star again which I've finally done, and just a second here, I'll go to enter and get a 51 and mark on it. As I say, for some reason the computer drove the star off out of sight.*
>
> CAPCOM: *Okay, Roger, out.*
>
> CAPCOM: *Apollo 11, this is Houston, over.*
>
> COLLINS: *Go ahead.*
>
> CAPCOM: *Roger, we show you as a little less than an hour to the midcourse Correction Number 2 burn, and we recommend that you terminate the P23 activities here, and press on with the waste-water dump which we need from you and getting ready for the burn, over.*
>
> COLLINS: *Okay.*
>
> CAPCOM: *And I have your midcourse Correction Number 2 pad when you're ready to copy.*
>
> COLLINS: *Stand by. Roger, Houston. Apollo 11, ready to copy MCC 2.*
>
> CAPCOM: *Apollo 11, this is Houston. Midcourse Correction Number 2, SPS G&N 63 zero 59'er plus 09'er 7, minus 020, GET ignition 026 44 57 9'er 2 plus 00 118 minus 00 003 plus 00 177 Roll 277.*

So they proceeded, the M-line never getting to parallel. Of course, there were other means of calculating their position. Emphasis now shifted to the midcourse correction. The Service Propulsion Motor, the main motor at the rear of the Service Module, was scheduled to be fired for three seconds. That would be time sufficient to slow their speed by

21.3 feet per second and thereby bring them nearer to the moon on their first orbit two days later. In preparation radios were readjusted. Then the burn took place. For three seconds, the ship now reversed, the astronauts looking toward earth, the motor pointing along the path toward the moon, a quiet churn of rocket fire came back to their ears. A little later, they commented on the readings of chamber pressure in the motor.

Work continued. It is not difficult to conceive of the three men pursuing their details in the small cramped volume of the Command Module, the sunlight glancing through the window, the view of space dark, neuter, numb, blank but glaring except when there is sight of earth. The sounds of the pumps and the ticking of a hundred instruments are in their ears mixed with the sound of static. It is time spent like the sense of time the inner mechanisms of a machine might possess if machines had a sense of time.

So they worked. With ground they discussed the entrance of Verb 66 into the computer, and chatted about the quality of the TV transmission the night before, then they mentioned the changing of the carbon dioxide filter. Having been obliged to quit Passive Thermal Control in order to set their attitude for the midcourse correction burn, they proceeded now to go back to it. Ground told them to use Quad thrusters Alpha and Bravo, 'from a propellant balancing standpoint.' S-band antenna angles next were given. A little later the spacecraft proceeded to report on the view of the Mediterranean and Europe through the monocular optic – as the earth turned, and the telescope swiveled, came descriptions of a cyclone over Brazil and panoramas of Central America, the Caribbean, of Greenland, then the east coast of the United States.

Discussion began again of simpler methods to find stars by which they could check their course. The fiasco with the M-line had not been altogether forgotten. Jokes followed. Since Collins would later remain with the Command Module while Armstrong and Aldrin descended to the moon in the Lem, so Collins had taken on a metaphor for himself: he was the storekeeper of the Command Module, the caretaker, the worrier, the passive sufferer, the little woman – it is the only clue to the deep bite in his competitive pride. 'I've been very busy so far. I'm looking forward to taking the afternoon off. I've been cooking, sweeping, and almost sewing, and you know, the usual little housekeeping things.'

But ground took him up on this to inquire on the condition of moisture condensation on the walls, which in turn suggested the functioning of the hydrogen filter on the hot-water squirt gun. Queries came up on the temperature of the water, reports went back and down on the character of the coffee 'not piping hot, but it beats stone-cold coffee.' The progress of Passive Thermal Control was reported up from the ground: 'PTC has started and looks good.' Adjustments followed on the oxygen flow transducer.

CAPCOM: *Okay, we want you to install the cabin vent quick disconnect which you'll find in compartment R6, that is Romeo 6 on the urine connector on panel 257. When this is completed verify that the waste stowage vent valve is closed and then open or position the waste management overboard drain to the pump position. Over.*

Well, it was no more difficult for a literary mind to follow these directives, injunctions, parleys and technological negotiations than to crack his way phrase by phrase through *Finnegans Wake*.

ii

Five hours had gone by since the crew had awakened. Now the difference in cabin pressure between the Command Module and the Lunar Module was measured, so too the oxygen flow on the on-board gauge. Further analysis of the after-burn chamber pressure in the Service Propulsion Motor came next. 'The SPS is definitely GO, over.'

ALDRIN: *Good to hear it.*
CAPCOM: *Roger. We thought you'd feel that way about it.*
ALDRIN: *We're right in the middle of salmon salad or something like that. That's probably why we're not answering you right away.*
CAPCOM: *Okay, well, we don't want . . .*
COLLINS: *My compliments to the chef. That salmon salad is outstanding.*

The food came in plastic bags. Some of it was hard like bread cubes or cereal cubes, cocoanut cubes, peanut cubes, or cheese cracker cubes.

Some of the food was wet pack and could be eaten out of the bag, some was freeze-dried, and water had to be inserted through a one-way valve, the bag then kneaded to make cream of chicken soup, or Canadian bacon and applesauce, a species of pot roast, or beef and vegetables, or ham and potatoes. The mash could then be squirted through another valve into the mouth. Obviously most of the chow had a consistency like baby food. Any attempt to eat in other fashion required much care, for in the weightless space of the cabin, food lifted to the mouth by a fork or spoon was in danger of taking off from any sudden move and floating through the air, later to enmesh itself in the smallest apertures of the instruments, or be inhaled to irritate the lungs. Eating in such valve-and-mash fashion, they were obliged to drink that way as well. Water ran from a dispenser through a seventy-two-inch coiled hose which abutted in a pistol placed between the lips, a button-actuated pistol – one pressed the button and water shot through the barrel into the mouth. A miserable mode by which to eat and drink, yet the food was a considerable improvement over early space flights when there had been nothing but bricks of processed nutrient and water: that original space food had been designed to produce the lowest fecal content. Now the food was better – as a corollary, the post-nutritive disposal substances (how *did* NASA spell shit?) was worse, for there was more of it. The astronauts had the straight embarrassment of squatting in front of one another on a Johnny Camper type of stool. If urine could be conveniently dispensed into a rubber roll-on cuff which fit over the penis and was attached to a one hundred inch flexible hose – that passing in turn to a urine transfer system bag and eventually out the waste-water dump – pure molecules of urea now floating in space – the fecal subsystem had its problems. It was built on bag assemblies, that is to say, inner and outer fecal bags with pouches holding germicide and skin cleaning towels. The rim of the inner bag was covered with cement and a thin plastic cover. The plastic peeled off, the bag was pasted smack on the buttocks. Baby mash for food, and technological diapers for a squat. Afterward, the germicide was put in the inner bag, and both were sealed in the outer bag, then kneaded to work the germicide and break up noxious gases. There were also odor removal vents and deodorant canisters in the cabin.

It was a self-contained universe, everything from cosmic ray detectors to split-membrane traps in the waste disposal compartment to keep those wrapped bags of feces from floating back into the cabin – what a nicety not to dump them in space!

CAPCOM: *Is that music I hear in the background?*
COLLINS: *Buzz is singing.*
ALDRIN: *Pass me the sausage, man.*
CAPCOM: *Okay.*
PUBLIC AFFAIRS OFFICER: *This is Apollo Control at twenty-nine hours into the mission. Apollo 11's distance from earth is now 115,837 nautical miles, velocity 4,788 feet per second. Spacecraft weight, 96,117 pounds.*

Three-quarters of an hour of silence followed. It was the first long silence since awakening. When they came out of it, Capcom had a long thoughtful anticipation prepared for them on the functioning of the Passive Thermal Control, which conversation was followed by queries from Apollo 11 about clouds which had moved over the Gulf of Mexico. Was it raining in Houston, they asked? Yes, it was raining in Houston. Armstrong replied, 'Well, it looks like it ought to clear up pretty soon from our viewpoint. The western edge of the weather isn't very far west of you.'

Now Capcom came back with a report that the flow-rate sensor on the oxygen tank was in fact malfunctioning, but that it was a minor malfunction.

Time went by. The charge on Battery A was terminated about seven hours after it had first commenced. There were updates in the data necessary for going around the moon if they were obliged for any reason to return immediately to earth. The new data was a reflection of the change in velocity acquired in the three-second midcourse correction burn that morning. Now, the astronauts focused their television camera on the Display and Keyboard of the computer, and the big antenna in Goldstone, California, reported its ability to read the numbers.

The White team came on duty at Mission Control. Of the White, the Green, and the Black teams, the White team was obviously the most important on this mission, for its shift would be on duty Sunday, the fifth day of the trip, when the Lem would separate from the

Command Module and descend to the moon. So their responsibility would be the greatest, and their tension the most. Today the spacecraft greeted them cheerfully, 'How's the old White team today?'

'Oh the old White team's bright-eyed and bushy-tailed. We're ever alert down here.'

'Ever alert and ready,' came back the spacecraft.

It was the self-conscious kind of joking one could expect when old drinking buddies were carrying on a conversation in the full hearing of churchgoers.

COLLINS: *Hey, you got any medics down there . . . I'm trying to do some running in place here. I wondered just out of curiosity whether it makes my heart rate act up.*

CAPCOM: *Well, they will spring into action here momentarily. Stand by.*

CAPCOM: *Hello 11, we see your heart beating.*

COLLINS: *Okay we're all running in place up here. You wouldn't believe it.*

But these fine spirits were replaced by a long sober discussion of flaws in the TV transmission. For the next half hour a dialogue about horizontal bonding and strayed lines, or bending of vertical lines followed by waviness in horizontal bands progressed back and forth – word eventually arrived that the TV people consulted on earth had said such aberrant lines were inherent in the design of the camera. Then the oxygen purge was terminated. Omnidirectional antenna was checked out for television transmission and found to be unusable. So only the high-gain antenna would be employed. That meant Passive Thermal Control must be halted. A complex discussion of the shift from PTC to the best TV attitude continued for a time, and was eventually decided. The PTC was terminated, the move was made, the earth was where they wanted it. The high-gain had its unobstructed line. Here came the TV.

COLLINS: *Okay, world, hold on to your hat. I'm going to turn you upside down.*

CAPCOM: (Charles Duke): *11, that's a pretty good roll there.*

COLLINS: *Oh, I'd say sloppy, Charlie. Let me try that one again.*

CAPCOM: *You'll never beat out the Thunderbirds.*

CAPCOM: *Apollo 11, Houston. That practice did you some good. It's looking – real smooth roll, there.*

COLLINS: *Oops.*

CAPCOM: *Spoke too soon.*

COLLINS: *I'm making myself seasick, Charlie, I'll just put you back right side up where you belong.*

CAPCOM: *Roger.*

COLLINS: *You don't get to do that every day . . .*

CAPCOM: *We can still see the earth through the left window and it appears that we can see a floodlight off to the left, either that or some sun shafting through the hatch window.*

COLLINS: *It's sunlight.*

CAPCOM: *Rog.*

CAPCOM: *Now we're coming in. Can't quite make out who that . . .*

ALDRIN: *That's big Mike Collins, there. You got a little bit of –*

COLLINS: *Yeah, hello there, sport fans. You got a little bit of me, plus Neil in the center couch, and Buzz is doing the camera work just now.*

CAPCOM: *Roger; it's a little dark, 11. Maybe a bigger F-stop might help.*

COLLINS: *Yeah, that should work.*

CAPCOM: *It's getting a lot better now, 11. Mike, you're coming in 5 by. I got a good –*

COLLINS: *I would have put on a coat and tie if I'd known about this ahead of time.*

CAPCOM: *Is Buzz holding your cue cards for you? Over.*

COLLINS: *Cue cards have a no. We have no intention of competing with the professionals. Believe me. We are very comfortable up here, though. We do have a happy home. There's plenty of room for the three of us and I think we're all willing to find our favorite little corner to sit in. Zero g's very comfortable but after a while you get to the point where you sort of get tired of rattling around and banging off the ceiling and the floor and the side, so you tend to find a little corner somewhere and put your knees up, or something like that to wedge yourself in, and that seems more at home.*

CAPCOM: *Roger, looks like Neil is coming in 5 by, there, 11. Mike, see you in the background. The definition is really outstanding. The colors are good. Armstrong . . .*

COLLINS: *And Neil's standing on his head again. He's trying to make me nervous.*

CAPCOM: *Roger.*

COLLINS: *He's disappearing up into the tunnel . . . going into the Lunar Module, only backwards.*

The show went on further, it was on in America and the world for thirty-five minutes, and was watched by the astronauts' families in

Houston. Aldrin spoke for a while on stellar navigation and showed the instruments and Armstrong said very little. During the telecast, reporters in the homes of Mrs Armstrong, Mrs Aldrin, and Mrs Collins took notes. The comments of the wives and children were put down in all the fullness of History recording itself, caught in the act of laying out a groaning board of gold plate and Melba toast, Irish linen and Hawaiian punch. 'They look great, don't they,' said Mrs Collins. 'Look, Mike is growing a mustache!'

iii

The day went on to its close as it had begun. PTC was recommenced after TV, and Guidance and Navigation data came up the miles from earth to be duly repeated by the astronauts before entrance into the computers. Verbs and Nouns were sent to them for further installations in the same computer. The programs with their Verbs and Nouns were, of course, demands upon the computer to be in readiness to give answers to certain kinds of problems for which the data had already been inserted. It was as if in a program designed to measure the size of rations in an Army kitchen, data consisted of the multiplication table and the inventories of food, the Verb was an order to divide, and the Noun directed the Verb toward what it should divide, which in this case happened to be the portions for supper. Apollo 11's programs were naturally more complex. There was room for ninety-nine Verbs and ninety-nine Nouns in the computer, such Verbs for instance as those which called for Display or for Monitor, for Load, Request, or Recycle, for Mark or Calibrate, Update, Initialize or Start, Reject, Enable, Perform. The Nouns specified just what would be the subject of the operation, be it the Address in the computer, or the Alarm Data or Alarm Codes, the Time of an Event or the Time *from* an Event, the Apogee or Perigee of an orbit, the Angles of the Vehicle, the Change in Velocity, the Pitch of the S-Band Antenna. Naturally, Apollo's programs concerned whole hierarchies of phase like Pre-launch, Earth Orbit Insertion, Coasting, Thrusting, Alignment, Moon Entry, and Abort – there would be, for example, seven distinct programs on Moon Entry covering each particular condition of separation from the Command Module through entrance into detectable moon gravity including several programs to choose for the moon landing itself, programs to offer much or little control by the pilot as he descended in collaboration with the computer.

So a host of numbers were passed back and forth again. Then the astronauts proceeded to report the amount of radiation they had been subjected to while passing through space that day. It was within proper bounds – no more than the radiation offered by the dentist who takes an X ray of your teeth – no medication was suggested. As usual, Armstrong had received the most radiation and Aldrin the least, but this was characteristic – Armstrong's pulse rate and heartbeat were always the highest of the three.

Now came final instructions for the night on the cryo heaters and fans. The liquid oxygen and liquid hydrogen were required to be heated before reaching the electrodes, but any excess heat was obliged to be dissipated. The process of warming and cooling, adjusting the flow, and purging the lines, went on constantly, and commands to turn specific heaters and fans on or off had come up to the astronauts all day. Collins had a comment later:

Those fuel cells . . . are funny things. It's not that they either work or don't work. They are like human beings; they have their little ups and downs. Some of them have bad days and then they sort of cure themselves. Others are hypochondriacs, they put out lots of electricity, but they do it only bitterly with much complaining and groaning, and you have to worry about them and sort of pat them and talk to them sweetly.

In fact the fuel cells were very funny things – they were an element as untried as any piece of equipment in rocket engineering for they derived their power from the mysterious ability of hydrogen and oxygen atoms to travel as hydrogen and hydroxyl ions through melted potassium hydroxide at close to 450 degrees Fahrenheit and a pressure of four atmospheres, a careful balance to be all the while maintained between the formation of water vapor and the building of electrical potential – no easy matter to strip electrons from oxygen and pass them over as a voltage loading to the hydrogen electrode – the fuel cells were still another step into the caverns of not completely charted electrical phenomena.*

Now their day was finally done and the crew had supper and went to sleep about 9:30 at night, eastern daylight time, with the spacecraft 134,000 nautical miles from earth. They had traveled a little over forty thousand miles in their working day, and would travel another 25,000 nautical miles during the twelve hours of sleep. There was at this stage so little to do that no call was put in to them until 8:40 in the morning at which time they had approached within 64,000 miles of the moon.

The morning of the third day proceeded like the morning of the second day. Let us live with that routine once again. Batteries were charged and waste water was dumped. Midcourse Correction Number 3 was canceled. Capcom reported consumables of fuel, hydrogen and oxygen; spacecraft answered with their own percentages, which had been independently measured. Then a report on the hours they slept: Armstrong 8, Collins 9, Aldrin 8. Good, but not a record. The Apollo 10 crew had reported ten hours of sleep one night. Next a description of what could be seen on earth. Then a discussion of PTC mode.

CAPCOM: . . . *As a result of your waste-water dump, it looks like the PTC mode has been disturbed somewhat. We're showing you about twenty degrees out in pitch right now on about six degrees in yaw which is significantly greater – about twice as much – a little more than twice as much as the deviation you had prior to the waste-water dump. We're watching it down here, though, and we'll let you know if we think any corrective action is required. Over.*

ALDRIN: *Okay, maybe next time we ought to split that in half. We could put half on one side and half on the other or something like that.*

CAPCOM: *Yes, we could do that. We were actually pretty interested in seeing what the effects on PTC would be in a waste-water dump. We don't recall ever having performed a waste-water dump during PTC on previous missions. Over.*

ALDRIN: *Well, now we know.*

CAPCOM: *Roger.*

* No need to draw on the adventures of Apollo 13, nor the postponements of 14.

A little later:

CAPCOM (Bruce McCandless): *Apollo 11, Houston.*

COLLINS: *Go ahead, Houston.*

CAPCOM: *Roger. We have been working under the assumption that we would take about an hour for the interference from a waste-water dump to dissipate to the point where you can reasonably take star sightings for platform alignment navigation or something of this sort. If you have a spare minute or two, could you comment on the observation conditions now. Over?*

COLLINS: *Yes. Stand by one, Bruce.*

CAPCOM: *Okay.*

COLLINS: *My guess would be that a telescope is rather useless, but you can differentiate in the sextant between water droplets and stars by the difference in their motion.*

CAPCOM: *Okay, Mike. I guess that we've still got – what you are saying is that we've still got a lot of water droplets visible, but you can pick them out and distinguish them in the sextant then.*

COLLINS: *Right. I think so, but Buzz is looking through it now. Just a second.*

CAPCOM: *Okay.*

ALDRIN: *Houston, Apollo 11. It looks like at this time the sextant would be quite usable for any alignment. There's actually very few verticals that need to be aligned.*

CAPCOM: *Roger, Buzz. How about the telescope? Is it useful now?*

ALDRIN: *Well, it's not quite as useful. It doesn't seem to be. Depending on the position of the sun it's got that band that seems to go across the center. I don't think it's because of the waste-water particles that it would lack its effectiveness. Over.*

CAPCOM: *Roger. What – is this band something that's deposited on the outside of the optics? Over.*

ALDRIN: *No, it's a reflection from the sun.*

CAPCOM: *Roger.*

COLLINS: *The sun bounces off the Lem structure. With the Lem attached, the telescope is just about useless. Those star charts that Ed has provided I think would be most useful if we had to use the – if for some reason we had to burn through the telescope we could use those as a guide for what we're looking at and say, well, that bright blob over there has got to be that star because that's the position we're in, but so far we've not been able to pick out any decent star patterns while docked with the Lem using the telescope.*

CAPCOM: *This is Houston. We copy.*

The waste-water dump set up a cloud of spherical icicles bright as diamonds in the void, and the particles traveled like a cloud of insects around the ship and dissipated only as slowly as the speed of the dribbling velocity with which they had trickled through the vent. It would of course have been simple to have propelled the waste water out at speed, but the greater force employed would have affected the trajectory of the ship, even as it had affected the roll this morning in PTC. In space, since there was a relative void of relations, so every little event was related to everything else. Droplets of urine could look like stars, droplets of urine could serve as projectiles to correct a rocket in trajectory.

But it was a routine day. The charge on Battery B was terminated, the parameters of Luna 15 were reported. The morning news was also reported – at 2:30 in the afternoon. An item about the world's porridge eating championship in Corby, England, inspired some happiness among the astronauts. Hemingway was crossing Andy Hardy.

COLLINS: *I'd like to enter Aldrin in the oatmeal eating contest next time.*

CAPCOM: *Is he pretty good at that?*

COLLINS: *He's doing his share up here.*

CAPCOM: *You all just finished your meal not long ago, didn't you?*

ALDRIN: *I'm still eating.*

CAPCOM: *Okay, is that –*

COLLINS: *He's on his nineteenth bowl.*

A little later:

CAPCOM: *Roger. Are you having any difficulties with gas in the food bags like the janitor reported?*

COLLINS: *Well, that's intermittently affirmative, Bruce. We have these two hydrogen filters which work fine as long as you don't hook them up to a food bag. But the entryway into the food bag has enough back pressure to cause the filters to start losing their efficiency. A couple of times I've been tempted to go through that dry-out procedure, but we found that simply by leaving the filters alone for a couple of hours, their efficiency seems to be restored.*

CAPCOM: *Roger, we copy.*

COLLINS: *Their efficiency ranges anywhere from darn near perfect to terrible just depending on the individual characteristics of the food bags we're putting through it. Some of the food bags are so crumpled near the entryway that there's no way we can work them loose to prevent back pressure.*

CAPCOM: *Roger.*

This endless preciosity of specification was necessary. In relation to their equipment, the trip was not unique, but merely another store of information in the continuing line of missions from the past which would lead toward expeditions in the future. So everything was important – the malfunctions in the oxygen transducer, and the glare from the foil wrappings of the Lem, the time it took the waste-water cloud to disperse, and the hours they slept, the unexpected reactions of the computer. Everything was important. After a while everything began to seem equally important, even the crumpling of the food bags. Like narcissists, like children, like old people, the astronauts all exhibited a single-minded emphasis on each detail which arrived before them, large or small. Just as prostitutes or male trade give equal attention to each unwinding detail, 'and then I bought ham and eggs and the ham was too rubbery and I got disgusted and crossed the street and bought a paper and a guy came along and pulled out a gun and held up this old lady right in front of me and then the cops came and there was a hassle and I skipped off and went window shopping and saw this leather suede jacket' – just as old people measuring the length of their days in the constriction of each breath cannot therefore imagine anything more important than their diet, the fluctuations of their disease, and the regularity of family visits; just as children are absorbed in the sensuous exploitation of each instant, for any new experience may turn to magic or the commencement of a new habit (and thus nothing can be ignored); just as narcissists can be indifferent to no shift in the mood, for each ripple of mood is a wave which will illumine or erode their beauty, just as a prostitute or male hustler takes each detail in full and equal measures of readiness and indifference, for each new encounter is either the law, sex, money, violence, boredom, outright rejection, or new adornment, so everything in the astronauts' finicky man-supported and prone-to-malfunctioning, unhealthy, plastic, odor-sealed, and odor-filtered environment was of equal news and interest because not only did their lives depend upon such constant and promiscuous concentration, but the style of their endeavor was implicit with it. Embarked on a heroic vault and subjected to a monotonous round of monitoring and mechanical housekeeping in relation to objects they could never comprehend sufficiently well, they existed in capsule like the real embodiments they were of technological man, forever engaged in activities whose controls he wields until he controls them no more, powerful, expert, philosophically naïve, jargon-ridden, and resolutely divorced from any language with grandeur to match the proportions of his endeavor. Constantly watched, every last function they performed monitored in one way or recorded in another, it was a little like working with God: every last detail of one's moves was observed. So they had burst clear through paranoia – they could tell themselves they were collaborating with the intents and detections of the Almighty. No wonder they spoke in the mood-smashed, random-item sequences of the modern world, where the depth of the thought was not nearly so important as the ability to brook interruption, and then interruption upon interruption, to live in an environment so form-less and externally directed that weightlessness was the next and logical step, to suffer shortwave radio and the life of static like deaf people communicating in a factory which produces nothing but noisemakers and celebration horns, yes, let us quit this accounting of their third day, let us dispense with their next television show, their LM Delta P, their O_2 flow alarm, their platform realignments and further water dumps, their Delta H updates, their cryo checks, their reports on the docking latches, the disassembly and reinstallation of the probe, drogue and hatch to the Lunar Module, their entrance into the Lem, their examin-ation, their television humor – now near to the humor at a drunken party, 'Hello, there, earthlings,' let us pass over the new data for the Alternate and Contingency checklist, the switch positions on the high-gain antenna, the changes in the glycol flow for the radiators, the small shifts in the Lunar Module mission rules, the supper and the music played, the check on the placements of the landing site obliques, the correction of typos on the APS DPS fuel card, the listing of three adjustments in the Mission Rule's GO–NO GO card file, the check on the gyroscope drift, the discussion of abort programs above or below low gate with consequent loss of thrust axis, the status report, the Delta

P again, it all comes in with lists and pieces and interrupted parts, comments on the tone in the alarm signals and the ubiquitous use of the word 'great' for anything that works as well as it was designed to work, great, great, great, the television show was great, and the last waste-water dump, the Saturn performance and the Service Propulsion Motor performance, the burps and the bursts of the thrusters. It was the small-town reaction to the grim miracles of the modern world, everything was great, a bite of steak, a chocolate bar, a movie which made you laugh, a high focus on a television screen – great, great, great, great. The famine of American life was in the sound of the word.

Still up they climbed, up that narrow pass between their diminishing acceleration and the decreasing force of the earth's gravity, and as they climbed, up from the great mass of the earth, so they came nearer to the influence of the moon. That night in sleep, third night out, they would pass over from the earth's domain to the pull of the lunar sphere, and as if to tip a lance in recognition of the oncoming moment, the Passive Thermal Control began to act in curious fashion.

CAPCOM: . . . *Mike, could you give us some help? This PTC is strange, it's not like anything we've seen before. We were wondering if you all have had any vents or any odd data that could help us out, over.*

COLLINS: *I didn't understand that. Say again.*

CAPCOM: *Roger, we're looking at a, sort of a funny-looking PTC. We've already drifted out to seventy degrees in pitch and we're wondering if you all had any vents or any such thing as that, that could have caused us to pick up these rates to drive us off, over.*

COLLINS: *Negative, Charlie. We don't know of anything.*

CAPCOM: *Roger.*

COLLINS: *Unless it's got something to do with that entry from the position that we want to be in. I don't know.*

CAPCOM: *Roger, when we started off it looked real fine to us, now it's drifting off with a funny pattern that we haven't seen previously on a flight, and we're just trying to figure out, I think we'll probably start it over again. We'll be with you momentarily, over.*

COLLINS: *Okay.*

CAPCOM: *Apollo 11, Houston. We hate to say it, but we'd like to terminate this PTC and start over again. We have no assurance that we're going to get it through the* sleep period. With this funny configuration, or funny pattern. We'd like you to stop it now and go back to pitch 090 yaw O and roll, whatever you stop on, over.

COLLINS: *Roger.*

An hour later, still bothered by the memory of the inexplicable wobble, the following conversation took place:

ARMSTRONG: *Do you have any idea where the S-IVB is with respect to us?*

CAPCOM: *Stand by.*

CAPCOM: *Apollo 11, Houston, the S-IVB is about six thousand nautical miles from you now, over.*

ARMSTRONG: *Okay, thank you.*

COLLINS: *Houston, Apollo 11, how is the PTC?*

CAPCOM: *Stand by.*

CAPCOM: *11, Houston. The PTC looks great to us, over.*

COLLINS: *Hey, do you have any idea what happened to the previous one?*

CAPCOM: *We have absolutely no idea, over.*

COLLINS: *Okay. Did it look like it was all right and just all of a sudden start diverting?*

CAPCOM: *Negative, if you look at the plot, which we'll save for you and let you see it postflight. It started off immediately on the first rev and just spiraled out to about oh, twenty degrees in pitch, and then it seemed to be setting up a spiral around an offset pitch point of about twenty degrees off from ninety degrees, but we didn't want to take a chance that it would become stable at that point. We thought it might diverge so we told you and started over again, over.*

COLLINS: *Okay, no complaints. I was just curious as to what had happened.*

Beyond the psychology of machines was the unformed psychology of space – was there some equivalent of haunted houses and creaking doors which now laid odd changes of attitude into Apollo 11? Consider your psychological state when you are weightless and therefore as disposable to floating as a particle of dust, conceive of an environment where your shoes are covered with a hooked-fiber plastic in order to keep them stuck to the floor, an environment where the easiest way to look through a telescope is to float toward the eyepiece like a swimmer, head down, eye raised to peer through at contact, yes the astronauts' idea of space was obliged to be more filled than the layman's with a sense of the domains of space.

A space which consists of domains – it is like thinking of a house with no walls and no ceiling. How can we call it a house? Yet if we conceive of heat so constantly generated that it keeps a particular volume of air at the same temperature, if we think of a field of force about the roofless house which keeps off rain and snow and repels all wind, if we conceive of this house as a sanctuary in open air protected by invisible zones of demarcation quite the equal of any wall or roof, then we can begin to think of the real nature of space. For there are redoubts and radiation-free harbors in the seas of space and the earth resides in one of them. The astronauts knew there were belts of radiation around the earth, and a field of magnetic force generated out of the depths of molten iron in the core of the earth itself. Beyond the atmosphere, out, forty thousand miles out, was the magnetosphere, a kingdom of sub-atomic particles thought to have originated from the sun but now charged by the earth – the earth had thus formed an electrical cavity about itself, an enclave of space wherein to protect itself against the gales of cosmic rays. But then all of the solar system could be said to inhabit still a larger vessel in greater space, for there was the indisput-able phenomenon of solar flares which swept out on special occasions from the sun in storms of electrical activity which collapsed the mag-netosphere, brightened the aurora, extinguished all long-distance radio communications, and acted like a great cloth or broom to sweep out the solar system, yes the sun swept the great spaces between the planets from sudden onslaughts of galactic cosmic rays which had originated in the explosion of stars far out beyond the sun, stars conceivably malig-nant, as if the sun were not only the source of life, but the lion at the gates of life defending the bastions of its planets with a fierce tongue to lick the infected space between. And in all of this, three men, and a spacecraft which proceeds for reasons they cannot yet discover, to have tipped that once in PTC, then wobbled just a little with the intimations of a drunken top – it had been nothing, but now, sleep period upon them, it would have been legitimate if the faintest sense of terror had come to visit the long curving aisle of their trajectory.

At any rate, less than an hour later, the crew still not able to sleep, the spacecraft passed over into the sphere of influence of the moon, and all imperceptibly, light as the breath of a bird, an acceleration began. The gravity of the moon pulled upon the ship, and the force of lunar gravity, whatever was that force, drew upon the ship, pulled gently, perhaps as gently as that first almost inaudible ripple which pulls along a shore when the tide shifts and begins to go out. Baleful or benign, palpable or utterly without influence, the moon beckoned with the weight of its matter, the computers on earth changed over to the parameters of the moon, passed over into the divide of lunar sleep and began to live in her domain.

Enough of engineering! Enough of reading dials and setting knobs. Let us rather take a good look at the moon. She is now a presence outside the drawn shade of the capsule's window, she is . . . she is not thirty-four thousand miles away and every instant nearer, and her pull upon the craft comes greater. Yes, let us look at the moon.

4

The Near Side and the Far Side

On the fourth morning, a half hour after the usual routine of awakening on consumable updates, sleep reports, drift checks, REFSMMATs, fuel cell purges, PTC maneuvers, and radiator-flow checks, there was finally a conversation about their port of call. They had traveled through the night, their speed accelerating as they came nearer to the pull of their destination, and at 71 hours, 31 minutes out, a half hour less than three days from the time of lift-off, on seven-thirty of this fourth morning, they were now no more than 11,000 nautical miles away and traveling at a velocity of 4,141 feet per second.

ARMSTRONG: *Houston, you read Apollo 11?*
CAPCOM: *Roger, 11. We're reading you loud and clear now . . .*
COLLINS: *Roger. What sort of F-stop could you recommend for the solar corona? We've got the sun right behind the edge of the moon now.*

When there was no answer, the spacecraft spoke again.

ALDRIN: *It's quite an eerie sight. There is a very marked three-dimensional aspect of the corona coming from behind the moon glares.*
CAPCOM: *Roger.*
ALDRIN: *And it looks as though – I guess what gives it that three-dimensional effect is the earthshine. I can see Tycho fairly clearly – at least if I'm right-side up – I believe it's Tycho in moonshine, I mean in earthshine. And of course I can see the sky is lit all the way around the moon . . .*
CAPCOM: *Roger. If you'd like to take some pictures, we recommend you using magazine uniform which is loaded with high-speed black and white film. Interior lights off. We're recommending an F-stop of 2.8 and we'd like to get a sequence of time exposures.*

It's quite an eerie sight. What an absence of technology in the remark! We need not even guess at what a panorama they had. Armstrong was later to report: 'Of all the spectacular views . . . the most impressive to me was on the way toward the moon when we flew through its shadow.' The moon was three times nearer than it had been at the hour of sleep nine hours ago when the shades were drawn. Three times nearer, it was three times larger, and filled their circular window – the sun was behind and so throwing a halo several times the size of the satellite.

They had had glimpses of the moon before, of course, there had been occasions to study it all the way up, but the occasions were imperfect and full of glare. In the flaring lights of sun and black space, every reflection from the spacecraft dazzling in their eyes, as hard on visibility as driving into the sun, there were no stars to see (except with every difficulty we have noted) and the moon was often no more than an area of darkness in the brilliant haze.

But now the sun was back of the moon and the halo, like a nineteenth-century painting of heaven, was three-dimensional to their eyes, a borealis of golden light with shafts and vales and mansion of light and gardens of light back of the moon. And in the center of the celestial corona was the land of their visit, visible at last, the moon now as clear to the eye as earth on the night of the fullest moon, no, far more bright than that, brighter far, for the moon was now in earthshine. The light of the earth reflected on the blue-gray face of moon highlands and deserts and craters, the earth reflected back a light eighty times more intense than the brightest light of the full moon. It was like the light at early evening . . .

ARMSTRONG: *Houston, it's been a real change for us. Now we are able to see stars again and recognize constellations for the first time on the trip. The sky is full of stars, just like the nights out on earth . . .*

Yes, there was the moon before them, as visible finally as lands of the horizon in the endless twilight nights of a northern summer, the satellite of the earth, a body mysterious beyond measure, unique in the solar system, a moon whose properties and dimensions resisted all categories of classification between planet and satellite, that moon whose origins remained a mystery, whose lunar features were shaped – no one could prove quite how they had been shaped – the moon lay revealed beneath them in its multiplicity of design. Whether dead record of the forces at work in the heavens, or something else not altogether so dead, there beneath them turned some darkened world of blue and silver-gray with color of a subtlety in its corners, and craters luminous to the eye. It was an eerie sight, eerie as a presence, eerie as a strange and desert shore emerged across a dream of sky and glassed-up surface of waters. How to row? How to breathe? The blue and desert shore approached across the impalpable space, cathedrals of light bent around the rim of its edge.

What a land was now there for study! If dead, the death was with dimension. It was a heavenly body which gave every evidence of having perished in some anguish of the cosmos, some agony of apocalypse – a face so cruelly pitted with an acne would have showed a man whose skin had died to keep his heart alive. What a burble of lavas and crusts, of boils on the pop and buds in frozen blight; what a scale of extinguishments; what a mystery of lines and rays and rills which ran from the coil of one burned-out crater to another; the moon was like a crazy old-fashioned computing machine with a tangle of wires all burned, a mute battleground of blows and hits and concussions and impacts from every flying or voyaging body or particle or radiation of the solar system and beyond. The moon spoke of holes and torture pots and scars and weals and welds of molten magma.

Punched-out, eviscerated, quartered, twisted, shucked, a land of deserts shaped in circles fifty and eighty miles across, a land of mountain rings higher some than the Himalayas, a land of empty windings and endless craters, craters within craters which resided within other craters which lived on the mountainous rim of very large craters, craters the size of an inch and craters to the depth of a mile, craters so vast Grand Canyon could have resided as a crater within the crater: There is a crater known as Newton and it is eighty-five miles wide and almost thirty thousand feet deep – the rim lifts up thirteen thousand feet above all surrounding mountains, and there are chains of mountains so high and vast they are called the Alps and the Apenines or the Caucasus and the Carpathians. There were also clefts, flattened rounds, ghost craters on the plain whose existence was distinguished only by a ring of lighter colorings as if the moon, every other death already available to her, was also a photographic plate of explosions, impacts and holocausts from other places. Scoops out of the lunar soil were to be seen, and pocks and cracks and scums of wrinklings on the plains, domes and bowls and hollow cones, blackheads and whiteheads, walled terraces and cataracts of random rock, hundred-mile spews of boulder, eggcups, table mountains and rims, mudholes, clamholes, spouts, gashes, splinterings of formation faults and extrusions, chains of craters, long mysterious slashes, long as endless roads from one vast crater to another, dark craters and bright craters, craters bright as phosphorescence in a moon-lit sea, and long mysterious inexplicable networks of rays – there was no better word nor way to comprehend why lines flew out across the surface, thousands of lines from certain craters, lines straight, and lines which wobbled, lines which stopped short and lines which seemed to skim from peak to peak like a pencil drawn across the grain of a rough plank, lines which continued as a hundred separate little flutterings, and thick lines, thick as brush strokes scumbled across the ridges of an old oil canvas, then lines which wove in and out of valleys – these lines, these rays, hundreds of miles long, even thousands of miles long, were without vertical dimension, they were not ridges or grooves, it was merely that they possessed some special property on the moon soil – they reflected light in a different way, as if they were a different kind of moon dirt and dust, an overlay or powder of some species of mind or order which had visited the moon after the early mind of the moon was gone, some species of hieroglyphic to record the history of relation between the moon and the earth, yes, studying the moon was enough to encourage curious thought, for the moon was a phenomenon, the moon was a voice which did not speak, a history whose record all revealed could still reveal no answers: Every property of the moon proved to confuse a previous assumption about its property. Yes, the moon was a centrifuge of the dream, accelerating every new idea to incandescent states. One takes a breath when one looks at the moon.

ii

It was large for a moon; relatively, it was the largest moon of any planet – its mass was one-hundredth of the earth. Ganymede, the major moon of Jupiter, had a mass only one-twelve-thousandth of its planet, and Titan, the greatest of Saturn's satellites, weighed in at the ratio of 1 to 4700. Even Triton, heaviest of all moons, was only one-two-hundred-and-ninetieth of the mass of Neptune. For further comparison our moon had a diameter larger than a quarter of the earth's whereas other moons in the solar system varied from one-ninth to one-thirtieth. So it was easy to think of moon and earth as a double planet. Indeed, there was no clear evidence the moon had been torn from the womb of the earth – it could as easily have begun as a separate body and wandered through space until that apocalyptic hour when it was captured.

The astronauts had lived with the moon: for years in increasing tempo they had studied lunar atlases, worked through geology, read the theories of the vulcanists and the impact men – in their privacy and their sleep they brooded upon it, we may assume, more perhaps than they knew, for the meanings of the moon were arrayed in all the caverns of sleep – was the moon a dead body or the dwarfed equal of earth? The nearer one came to a full contemplation of such mysteries, the greater was the temptation to think not at all. Perhaps in consequence, the astronauts were back immediately into talk of high-gain antennas and secondary loop checks, pericynthion burns and the morning news. *Pravda* had called Armstrong the 'Czar of the Ship.' Capcom had added, 'I think maybe they got the wrong mission,' and there would be jokes about this later in the day – Collins would remark that the Czar was brushing his teeth.

No, they would not dwell overlong on descriptions of the moon. It lived outside the hatch window, it filled almost all the view, its roughed-up hide paraded below – that skin of craters set upon craters which would inspire every simile from the popholes in pancakes cooking to barnacles upon a rock. One could say it looked like molten metal or blistered paint, one could speak of the erosions of bacteria culture in a Petri dish, or leukemia cells in an electron micrograph – one could also be looking at the heavens on a hazy night for the rays from the crater Tycho and the crater Copernicus spread out in such profusion that the moon also looked like a photograph taken across space of the streams of the Milky Way, even its dark *maria* – those dark seas of moon plain – Imbrium, and Tranquillitatis, Nectaris, Fecunditatus, Nubium, Humorum, Serenitatis, especially the vast dark plain of the Oceanus Procellarum looked like the dark and empty spaces of the sky where one saw no stars. And the multitude of craters were like a multitude of dots and rings of light, were like the overlapping luminescence of stars, as if the moon, properly read, could betray as much of the real character of the heavens as the lines on a man's hand could enrich an eye which understood a world where histories might be written in the hieroglyphics of some universal form neatly concealed in the crack of the palm.

The moon traveled around the earth and both traveled around the sun; the moon moved therefore in a path, which if drawn, would have been not unreminiscent of the outline of an old gear with rounded teeth. The moon had a period of twenty-seven days seven hours forty-three minutes, and eleven-plus seconds – one could be tempted to predict that the interval of normal period of all the women in the world if taken for average would come to the same eleven-plus seconds, forty-three minutes, seven hours and twenty-seven days – there was a hieroglyphic from the deep!

There were others hidden no doubt on the far side of the moon. As she turned about the earth, the moon kept herself like a subject before the king – her face was always presented, her back always hidden – so from earth one saw only her face. The far side had remained a mystery until the first unmanned Soviet spaceships passed around the satellite and sent back photographs by television. Since then Apollo 8 and Apollo 10 had taken scores of pictures. Now, the far side of the moon was no longer a complete mystery. But it was different from the face seen on earth, it was powerfully different, and the astronauts would soon see the far side – they were indeed approaching toward rendezvous with the leading edge of the moon. Before too long, their trajectory, caught by the accelerating pull of the moon's gravity, would begin to bend about the moon – at the appropriate moment they would be drawn into orbit around the back, they would see the far side, they would be the seventh, eighth and ninth men ever to see the far side of the moon, and then flying motors first, they would fire their main engine to brake their speed. When they came around the moon again they would not go whipping back to earth – rather they would be in lunar orbit, there to circle the moon ten times and more while preparing for the descent. So they would be in orbit about a moon which was in orbit about an earth, and if moon and earth were both in orbit about the sun, well the moon would soon have its satellite as well, a tiny satellite with three men.

Different, however, were the orbits. If the moon took twenty-seven days and some hours to go around the earth, that period was its lunar year just as it took the earth a year to go around the sun. Yet this lunar year of twenty-seven days plus seven hours was also the moon's lunar day. Since the moon kept its face always to the earth, then as it went around the earth, the sun would shine on our visible side of the moon when the earth was between the sun and the moon, but then the sun would shine on the far side of the moon when the moon was between the earth and the sun. So daylight on any area of the moon lasted for fourteen days, a long fourteen days of sun beating on the desert and the craters and the plains of the dead seas, shining on the pinnacles and turrets of the mountains, and the temperature went up as the fourteen days went by, those fourteen earth days which were but one day of sunlight to the cloudless moon. Each terrestrial day here on earth was then by lunar measure of time no more than the rough equivalent of an hour, and the heat increased each lunar hour, each part of the fourteen continuous days of sunlight. At any given point on the moon, the temperature went up to as high as 243 degrees Fahrenheit when the sun was at zenith. Then the moon, always wheeling with her face to the earth, would pass that face out of the sun and into the dark of earthshine and the long fourteen-day night of the lunar night would begin and the temperature would drop. Down to 279 degrees Fahrenheit below zero it would drop in the depth of the long night, cold as the liquid oxygen in the tanks of Apollo-Saturn before the lift-off, and the cold of that one night would last for the equivalent of two weeks of days and nights on earth. And there was no air and no wind.

Perhaps the moon had been once in rotation, and had had a day and a night not unlike the earth. But there were signs to indicate that the moon on approaching the gravitational grasp of the earth had had its cooling skin seized and twisted by the ferocious tides of the earth's pull. Lines of mountain had been pulled up and the moon's rotation had slowed. The moon staggered away into space, went out nearer to Mars, then caromed back and was captured again. One history of the moon conceived of it entering relations with the earth three times in much such a way before its final capture; what spirit of earth and lunar forces must have been released, exchanged, and conceivably not lost forever? It was a theory among others, but it was not without relation to the difference between the near side of the moon and the far, and it offered an explanation for the moon's craters, since the molten lavas and boiling waters of the young moon's interior would have been profoundly disturbed in such a courtship of the spheres, and volcanic eruptions would have been the predominant order of events. That was one of the major theories, but in fact there were two fundamental and antagonistic hypotheses to account for the terrain of the moon: each was in difficulties before the critics of the other. There were men who believed the craters of the moon had come in their entirety, or almost entirely, from the impact of meteors in those remote ages when meteors abounded as planets formed. Such theorists were termed lunar impact men, and their thesis had to live with objections so thoroughgoing as the fact that craters sometimes presented themselves in a clean row like strings of artillery shells; worse! impact men were obliged to explain how craters could be spaced like pearls around the rim of a larger crater: how indeed could the meteors have landed in such order? But the largest of the objections was to be found in the size of the dead seas. The Oceanus Procellarum was better than a thousand miles across – a meteor comparable in size to a moon for our moon might have had to collide for such a scar to be left.

The masters of the meteoric hypothesis had other difficult answers to give. Most craters were not a simple bowl, rather a bowl with an umbilical at its core – a separate mountain peak at the center of the bowl surrounded by a ring of plain. Even the smaller craters were rarely without a pit, a core, a vertical elevation at the center, or a root like the pintle on an old bottle. Such centers suggested a now extinct fountain of lava. So theories of impact were not comfortable, no more than the scientists who believed the craters of the moon were the product of volcanoes could explain with ease why even the largest volcanoes on earth were small in comparison to moon craters, and why the floor of a terrestrial volcano was elevated clear above the surrounding countryside, was often indeed a steep cone with no more than a modest vent for crater, and so most unlike the terrain of the moon. On the other hand the largest meteoric crater seen on earth was only two miles across. Models drawn from the earth simply did not suffice for either vulcanists or impact men. How to explain that the moon was either the subject of vast meteoric bombardments from ages past which the earth had somehow escaped, bombardments with an occasional perfection of aim equal to shooting pearls onto the circular points of a crown; or equally how to convince others that the moon was the product of its wanderings between the earth and Mars, child of regimens of boiling and cooling, geysers of water, mud and lava, a world of subterranean seas and trapped gases, a cauldron of upheavals and subsidences in some romantic, even catalytic relation to the earth, a theory therefore in furious conflict with the ideas of all meteoric theories, for the impact men were attached to the idea that the moon was ancient and unchanging, and the major bombardments had occurred eons ago, billions of years ago. It was the best explanation why such bombardments were not to be found on earth – the earth, equally pitted then, had had its face subsequently altered by geological upheavals went the hypothesis. It allowed one to induce that the impact men were classicists, positivists, traditionalists, upholders of the public common sense (since the moon by all common sense *looked* as if it had been bombarded). The vulcanists were romantics, Dionysiacs, existentialists, animists – at their most adventurous they would even search for some faint hope of life on the moon.

Neither could secure an explanation for the contradictory facts, and before the phenomenon of the rays, no meteorite man nor apostle of the geyser-volcano could offer an embattled theory. The rays had no explanation which could explain all the facts attached to them. The rays emanated from the craters, but before one could even speak of a stream of white powder flying out, it was necessary to note that the rays were sometimes tangential to the ring of the crater and sometimes radiated out from the center. Sometimes a crater had but a single ray which led to a particular place, often to another crater. Yet if the rays were some physical embodiment of an idea or a communication, they were yet singularly sensitive to obstacles – even a low ridge across their path could stop them short – yet in other places they would extend across a thousand-mile sea. They were even described like 'snow thinly drifted by a strong wind across a black frozen lake.'

Imperfect, unhappy, unsatisfactory theories abounded within the two schools of theory. And other theories abounded, even the theory that the moon was a dead civilization, its endless rings the record of a last atomic war – it was said the earth would look like the moon after a nuclear holocaust had destroyed us.

Yet it was with the partial hope that their projected experiments on the moon and the rocks they brought back would begin to apply answers to these questions that the astronauts doubtless now received for the thousandth time the order of their assignments on the moon ground and the probabilities of vulcanism, formation by meteor, or some other theory altogether. Revolving below them, coming each minute nearer was the dead beast of the moon ground, the mute mysteries locked in the formation of form itself. Would the moon yet answer the fundamental question of form – that all forms which looked alike were in some yet undiscovered logic thereby alike? – which is to say that if the skin of the moon was reminiscent of boiled milk and cancer cells and acne, so then – would a theory yet emerge which could revolve at some ease through the metaphors of the moon and find the link of metaphysical reason between cancer, acne, blisterings of paint and the wrinkled ridges of a boiled and skin-thick milk?

iii

One can dismiss the enigma of form too quickly. Let us propose an artist who draws a face, and assume that the man in the portrait is bald, and the artist on a whim draws a wen on the back of the head, then the scar of an old boil. Finally he puts in a mean wrinkle high on the neck. Abruptly, a second face emerges on the back of the bald head. There with the wen and the old boil for eyes, the wrinkle has become a mouth. But the artist has lived with form for so long that he is far from surprised the back of this bald dome has revealed a face: he is already concerned with what that embryonic expression has to say – he would indeed go so far as to assume that the wen, the boil and the wrinkle are exactly where they are because the flesh has desired to become a draftsman, because the back of the head secretly wished to draw a particular face on its own skin, as though to say, 'If I must suffer a boil, let it at least do some work for me, let it establish that I have not one head, but two; one biological, and one somewhat less tangible – perhaps a second spirit to be discerned only on the back of my skull.'

Painters are not invariably articulate. They can live quietly with such thoughts rather than look to express them. An argument ready to claim that the vulcanism of the skin is designed not only to draw attention to the imbalance of boils, eczema, and acne, but to the secret urges of the man to display a few items he cannot otherwise display – a hint of buried horrors if he is timid, or of humor if he is sufficiently pompous – is an argument beyond the average artist's desire. For the discussion would have to confront the apostle of common sense, and the apostle, an impact man, would be quick to point out that the boil is where it is because of where the dirt was in the collar. Not every painter would have the wit to answer, 'Dirty collars are many, boils are few,' or be ready to suggest that the site of the boil, while important, is not necessarily as significant as the visual statement offered by the form of the boil as the crater dried to a scar. There might lie the right to assume the flesh drew its own design.

The painter is probably suggesting that form is a language which seeks to express itself by every means. If man has his voice, nature has wind, it has thunder, the sound of running waters, it has its variety of cataracts and Krakatoas. If we agree that the urge to create a language is basic to man, so basic that one can begin to define his nature by saying he is an animal with the irrepressible desire to develop formal speech, why must it be altogether uncomfortable to assume other categories of nature would not attempt to shape forms that could delineate their inner meaning, why indeed is it not as natural for nature to shape itself as for man to speak? On that assumption could commence a metaphysics of form. If common sense would ask immediately why nature would not also wish to conceal itself, in fact elaborate a labyrinth of false and misleading form to protect itself, it can be agreed that not all form reveals, form may also be designed to betray meaning. Still there is an economy to nature one might as well assume, for that is easier to comprehend than lack of economy – a prevalence of wasteful and misleading communication might prove a luxury the cosmos could not necessarily afford.

So Aquarius, happy with this supervulcanism, would have the moon not only responsible for writing much of the record of its own history, but in fact could go so far as to search for evidence that the face of the moon might be a self-portrait which looked to delineate the meanings of its experience in that long marriage with the earth and its long un-insulated exposure to the solar system and the stars.

Well, Aquarius was in no Command Module preparing to go around the limb of the moon, burn his rocket motors and brake into orbit, no, Aquarius was installed in the act of writing about the efforts of other men, his attempts to decipher some first clues to the unvoiced messages of the moon obtained from no more than photographs in color of craters, chains of craters, fields of craters and the moon soil given him through the courtesy of the Manned Spacecraft Center, photographic division of public relations, NASA, yet in the months he worked, the pictures were pored over by him as if he were a medieval alchemist rubbing at a magic stone whose unfelt vibration might yet speak a sweet song to his nerve. But he had all the failures of the occult in all the ages for model, and so he knew as he wrote that if the riddle of the ages was at the root of every form – a pure medievalist is Aquarius! – so, too, was every temptation of insanity. The profligacies of thought – total irresponsibility of connection, and complete loss of the ability to convince – sat in every widening ripple of contemplation over form. If he would travel into the inner space of his brain to uncover the myster-ies of the moon, he could dignify that expedition only if he obeyed the irritatingly modest data of the given, the words, the humor, the reso-lute lack of poetic immortality in the astronauts' communications with the earth, say, even more than that, would have to certify his respect for the particular endeavor of Apollo 11 by returning to the room where the Capcom speaks.

iv

If we would talk of mysteries, forms, projects, riddles and all of their roots, it is as well to recognize that the root of the Command Module as she approaches the moon is a room on the southeastern plains of Houston in a building called the Mission Control Center of the Manned Spacecraft Center, and this room, the Missions Operations Control Room, MOCR, a room perhaps thirty feet deep and sixty feet wide, is divided into a floor of desks and consoles and television monitors with a gallery of raised seats for dignitaries to observe at one end, and a set of projection screens and maps at the other, an air-conditioned windowless room lit by fluorescent light, with nothing but illumined graphs, quotations of number, and the intent locustlike feeding of flickering electronic dots on empty screens, the very gems of static, to come back to the visitor. Men in the uniform of NASA, white sleeveless shirts, dark ties, dark pants, sit before the consoles and confer, and on the illumined wall maps, a little figure of green light shaped like the silhouette of the Lem with the Command and Service Module attached, that flying-bug-with-a-bullet-up-its-bung goes crawling across the screen. The map up for display is the equator of the moon. In sixty minutes the bug will travel some ten feet across a lunar map. It is the most interesting sight in that intent space.

The spacecraft now is almost a quarter of a million miles away, but the MOCR is its neighbor. Closer than that, Apollo 11 and Mission Control in MOCR are as close as dialogues between conscious and unconscious mind. Here, like every other broad expanse of brain, ready to work on each problem the conscious mind can provide, are the collected resources, habits and themes for the flight. Outside the MOCR are Staff Support rooms, and on the first floor is the Data Processing Area – a large and open room which could be a diorama of a housing development of the future, its ceiling a mammoth artificial sky of square panels of fluorescent light, its boxlike machines equal to windowless buildings. It is the Real-Time Computer Complex, containing five IBM 360/75 computers plus facilities for flight dynamic analysis, telemetry, processing, acquisition predictions, flight controller display generation with call-up capability, etc., etc., a particularly interlocked nest of jargon to describe the total management of the mathematical and visual displays of the flight on the floor above.

There, back in the MOCR, is the Capcom, his formal title the Spacecraft Communicator, and he sits before a console in the second of four rows of desks and consoles which seat representatives in the Mission Control Team – the officers, directors, and engineers of each relevant department and function of the flight. Just behind the Capcom in the third row is the Flight Director, situated as near to the center of these sixteen responsibilities as the four rows permit – the critical decisions of the flight are to be made by the Flight Director. Almost within reach of his arms are the desks of the Mission Director, the Director of Flight

Operations, the Assistant Flight Director, the Capcom, the Vehicle Systems Engineers, and the Experiments and Flight Planning Officer. Near the Capcom are the desks of the Surgeon, the Booster Systems and EVA Mobility Unit, the Retrofire Officer and the Flight Dynamics Officer. Further away from both Capcom and Flight Director are the Public Affairs Officer, the Department of Defense Officer, the Network Controller, and the Guidance Officer. These sixteen men with their links and taps and lead-ins and consoles and commo loops, their commands of data received and data processed by the computer complex below, their reference slides, their formats, combinations, pictorials, their alpha-numerics, analog plots, plotting data, their nerves and their experience, their knowledge of flight parameters, their red-lines and qualifications for GO or NO GO in each and every separate department are the base for one constant run of comments and queries and warnings and advertisements of future warnings about the functioning of the equipment, the condition of the flight and the oncoming tests and procedures of the flight plan, all funneled into the receiver and ear of the headset of the Flight Director who holds this web of sixteen rays of information in the bowl and dome of his skull and relays to the Capcom (not much more than the tap of an arm away) what piece of information might next be advanced up the quarter of a million miles to the minds and controls of the astronauts. And the Capcom in turn relays the requests, local data, appreciations and complaints of the astronauts. The Capcom, astronaut himself, is thus the advocate, mediator, lifeline, counselor, coxswain, gripebox, court wit, and expediter for the styles and habits of the astronauts, their modes of procedure, the intimate who recognizes their separate voices and attempts in these static-filled limits to cater to the rhythm of their duties and their literal positions in the cramped quarters of the Command Module with which, as an astronaut himself, the Capcom is familiar.

On this hour, as the spacecraft came nearer to the advancing moon, now nine hundred miles from the outer edge, now six hundred miles from the limb, as the ship of space came closer and closer to disappearing around the back, so a tension began in Mission Control; the spacecraft would be out of radio contact on the far side. Radio waves were rarely modest in their properties, but they did not bend like waves of water around obstacles nor could they penetrate a sphere. On the far side, for a period of forty-seven minutes, there would be silence – it was called Loss of Signal. In that hour no ordinary anxiety would be felt. Once again the nightmare of technology was aroused. For if something happened to the astronauts on the far side – if the spaceship disappeared without a sound, then there would be no report to give on the source of the malfunction, no logic to the destruction, no pattern to the failure – only misery, the misery of exploring a thousand equally mysterious possibilities. Worse by far than failure was failure for undetermined reasons. So even the head of NASA himself, Dr Thomas Paine, had asked the astronauts not to fire the Service Propulsion Motor and thereby brake their speed into the lower velocity of moon orbit if they thought anything was wrong; in such a case it would be better to coast out on the other side of the moon and swing back to earth. So concerned was he that the astronauts not take chances behind the moon that he promised them another flight soon if they were forced to return immediately.

PUBLIC AFFAIRS OFFICER: *We are three minutes away from loss of signal. Apollo 11 is 425 nautical miles from the moon, velocity 7,368 feet per second, weight 96,012 pounds.*

CAPCOM: *Two minutes to LOS.*

CAPCOM: *Apollo 11, this is Houston. All your systems are looking good going around the corner and we'll see you on the other side. Over.*

ARMSTRONG: *Roger. Everything looks okay up here.*

CAPCOM: *Roger, out.*

PUBLIC AFFAIRS OFFICER: *And we've had loss of signal as Apollo 11 goes behind the moon. We were showing a distance to the moon of 309 nautical miles at LOS, velocity 7,664 feet per second. Weight was 96,012 pounds. We're 7 minutes 45 seconds away from the Lunar Orbit Insertion number 1 burn, which will take place behind the moon out of communications. Here in the Control Center two members of the backup crew, Bill Anders and Jim Lovell, have joined Bruce McCandless at the Capcom console. Fred Haise, the third member of the backup crew, has just come in, too, and Deke Slayton, Director of Flight Crew Operations, is at that console. The viewing room is filling up. Among those we noticed on the front row in the viewing room are astronauts Tom Stafford, John Glenn, Gene Cernan, Dave Scott, Al Worden, and Jack Swigert. With a good Lunar Orbit Insertion burn the Madrid station should acquire Apollo 11 at 76 hours 15 minutes 29 seconds . . . This is Apollo Control at 75 hours, 49 minutes. Apollo 11 should have started this long burn, duration 6 minutes, 2 seconds, DELTA V 2917 feet per second. Given that burn we expect an orbit of 61 by 169.2 nautical miles. We're 24 and one-half minutes away from acquisition of signal with a good burn. The clock has not yet started counting for the other acquisition time. We'll take this lying down now and come back just prior to the acquisition in time for no burn. This is Mission Control, Houston . . . We are past the burn acquisition now and we have received no signal . . . It's very quiet here in the Control Room. Most of the controllers seated at their consoles, a few standing up, but very quiet . . . We are 4 minutes away now . . . There are a few conversations taking place here in the Control Room, but not very many. Most of the people are waiting quietly, watching and listening. Not talking . . . That noise is just bringing up the system. We have not acquired a signal. We're a minute and one-half away from acquisition time . . . 30 seconds . . . Madrid AOS, Madrid AOS . . . Telemetry indicates that the crew is working on the antenna angles to bring the high-gain antenna to bear . . .*

SPACECRAFT: *(Spacecraft signal very weak – inaudible)*

CAPCOM: *Apollo 11, this is Houston. Are you in the process of acquiring high-gain antenna? Over.*

CAPCOM: *Apollo 11, Apollo 11, this is Houston. How do you read?*

COLLINS: *Read you loud and clear, Houston.*

CAPCOM: *Roger. Reading you the same now. Could you repeat your burn status report? We copied the residuals burn time and that was about it. Send the whole thing again, please.*

ARMSTRONG: *It was like – like perfect. DELTA T O, burn time 557, ten values on the angles, BGX minus .1, BGY minus .1, BGZ plus .1, no trim, minus 6.8 on DELTA VC, fuel was 38.8, OX 39.0, plus 50 on balance, we ran an increase on the PUGS, NOUN 44, show us in a 60.9 by 169.9.*

CAPCOM: *Roger, we copy your burn status report, and the spacecraft is looking good to us on telemetry.*

PUBLIC AFFAIRS OFFICER: *Burn report was by Neil Armstrong.*

V

If tension had been palpable in the Public Affairs Officer's voice, one can easily imagine the forty-seven minutes of dread experienced in the MOCR when the flight of Apollo 8 took Borman, Lovell and Anders over the hill, and men saw the far side for the first time. There had of course been photographs received from unmanned spacecraft, but those photos had been transmitted back to earth by television data and were blurred. They had hardly been a full preparation for the sight turned up to the eyes of the crew of Apollo 8, and Apollo 10, and now repeated for Apollo 11.

The side of the moon which faced the earth had features, it had oceans and seas and mountain chains and straits. If there was no water in the seas, and they were in fact dark and desert plains, still they possessed features. One could speak of the man in the moon for his face could be found in the contrasts between the highlands and the seas. But the far side, on superficial view, was nearly without distinguishing marks, an endless waste of craters laid upon craters. If the moon had kept her face looking toward earth, the back side of the moon was as undistinguished as a head of hair. Later, out of better photographs would come maps, and subtle features might begin to emerge, but for the present only a few huge craters stood out in all that near hemisphere of hitherto unplotted terrain – a great crater with a great peak in the center and a very dark moon floor had first been seen by Lunik 2, first Russian unmanned spacecraft to circle the satellite, and had been as quickly named the Crater Tsiolkovsky after the father of Russian rocketry. Other craters of real dimension appeared here and there, Mare Moscoviense for one, another was even named Jules Verne, but the mass of terrain appeared to be little but a mire of endless holes, a barnyard trod by countless hooves, a beach with hollows and mounds from thousands of feet. Borman was to describe it as like a battlefield and since this hidden moon land was obliterated of variety, an apparent dump and blasting ground of all the angers of the heavens, the meteoric hypothesis regained force here where the ground of the moon seemed subjected to every size and variety of meteor. The ground humped and holed, it writhed and twisted like a spill of sand thrown over nests of snakes, it seemed to boil, it was as trackless as the rough bark of a tree, as filled with the holes of craters of every size as a molten slag boiling in a pot. Yet the longer one looked, the less was the impression of meteors, the stronger a sense of volcanic forces which had once boiled beneath and emerged in poppings and blowholes of crust. Search the lip of every large crater and there in the center of the circumference of each round ridge was a little crater so perfectly placed it must have boiled up out of the lip, and indeed nothing for hundreds of miles before the eye but swellings and distensions of the terrain like a skin beneath which furies must have wrung themselves, a bewildering end-lessly worked-over expanse almost without rays, a stretch of bumpy knobby pockmarked upthrown churnings equal to the view from a low boat – without horizon one could never sight a level, and direction was hopeless, a windtwisted choppy sea had been frozen on the instant to stone. So one had no sense of scale. Staring down on a photograph of the far side it was not possible to tell without text whether the picture was of a square mile or of a square five hundred. Craters the size of New York were indistinguishable from craters the size of a house. All orders of magnitude were gone. Giving oneself to these studies of the moon, there followed that hypnotic sense of falling out of human mag-nitude into other magnitudes. It came upon the senses that in the hour of death, consciousness might separate into other dimensions, dissipate into other orders of the immense and the minuscule, consciousness might at last be off on terminal voyages to microbes, molecules, or the stars. Aquarius had been devoted to painting for close to thirty years; an amateur of the mysteries of form, it took him close to thirty years to comprehend why Cézanne was the father of modern art and godfather to photographs of the far side of the moon.

But it had come to him at last in one fair burst of appreciation, and he had a glimpse of why Cézanne's work had been obsessive to so many painters. For hundreds of years, they had worked to capture the sheen and texture, the hairs, the dust, the flickering motes of light on the sur-face of a drape. Miseries and glories of apprenticeship had gone into painting a velvet sleeve, a pearl, a drop of dew on a grape. Western civ-ilization arrived at a materiality of forms where every surface was recognizable in its own right. Did a painter work on canvas properly? Then one could cut out a square inch of canvas, show it to an unfamil-iar eye, and the response would be that it was a piece of lace, or a square of velvet, for the canvas had been painted to look exactly like lace or velvet.

Cézanne, however, had looked to destroy the surface. A tablecloth in any one of his still lifes, taken inch by square inch, resembled the snow-fields of mountains; his apples could be the paint-stained walls of a barn, or the clay roundings of a rock; the trunks of his trees were stems, or pillars, or hairs beneath a microscope. His skies, patch by patch, could be taken for a sea as easily as a light-blue throw cloth; the skin which ran from a man's eye to the corner of his mouth was like the sun-beaten terrain of his hills. In one lifetime of work, Cézanne disqualified the virtuosity of the craft and brought painting away from the capture of light on material. He showed instead a panorama of rises and depressions on every surface, the similarities between surfaces now more profound than the differences. As he succeeded, so the orders of magnitude vanished in his painting, and one could not know, looking at a detail, whether he was representing the inside of a flower or the inside of a tent. Something in that vision spoke like the voice of the century to come, something in his work turned other painters out of their own directions and into a search for the logic of the abstract. Art had embarked on an entrance into the long tunnel where aesthetics met technology. Picasso and Cubism would pour through that hole in the old love of surface until one could not tell which wall was near and what floor had begun to recede. It was as if the century to come was already anticipated in the veins of its artists, as if the century to come would go out to explore the dissolution of all orders of magnitude and so begin a search into the secrets and unwindings of death. Little surprise that before the century had finished its seventh decade, the artist had crossed from the brush to the wind machine and blew up walls of plastic through which the patron all blindfolded would creep. Art and theater were ready to view the dimensionless dimensions of the moon on its far side.

vi

If we have been about to assume that objects are shaped in a way which offers meaning, not only scientific meaning, but existential meaning; if we even press on to the notion that a firm sense of magnitudes is characteristic of human life as much as the loss of such a sense will take us through the arcade from sleep to death, we can at least be certain that whatever the astronauts are thinking about, they are not close at this instant to Cézanne. Fair enough. They have more than a little work to do. As they come about and acquire signal again, they are having the pleasure of all soldiers, athletes, young surgeons and novitiates – they are at last working upon the material they have studied for so long, and using the tools for which they were trained. So animation is in their comments. As they go around the moon in this first orbit, and then again in a second orbit, Armstrong particularly is lively. He has not been volatile the first three days, usually restricting himself to remarks about the look of weather on earth, and dutifully taking down each series of numbers in the data pads. Aldrin has talked a little more – his specialty is celestial navigation and his comments have been frequent on the availability of star checks – listen to him long enough and you might hear him name Fomalhaut and Arcturus and Vega, Polaris, Canopis and Capella, Rasalhague, Regor, and Diphda, Dnoles, Nunki, Sirius, and Peacock. Still it is Collins who has done three-quarters of the communicating with earth. The Command Module has been his. If he were not to land on the moon, he was still the Command Pilot. While Armstrong was Commander and ranked him technically, still Armstrong had exercised a most quiet authority. It was Collins who had been in contact with the Capcom, Collins who made the jokes, traded quip for quip, proved concerned about the equipment, inquired about minor deviations from any expected result, offered metaphors for Apollo 11's housekeeping, commented on the quality of the food, and worried each detail until time to worry the next detail. He was like an actor who has every early scene in the play and yet is doomed in his role, for he is not in the big scenes to come and they will be so big that his own scenes must be drowned in the wake – an ambitious actor in such circumstances works harder, as if the cumulative emoting of his presence may intrude itself into the results.

But Armstrong had been near to silent. Well described by the Press for his silences, his air of loneliness, his desire for privacy, one had the impression of Armstrong drawing silence about him even in the Command Module, staring through his window for hours, or for the minutes which would make up the hours in the pauses between chores on the long trip up. He was a pilot, he was a pilot first and last, and so he studied the weather all the way – that we must assume – for the love of flying demands the attention of a lover to the moods of weather. A rich lover, connoisseur of love, might give an hour to the choice of flowers for a new love and another hour on falling asleep to the nuances of her reaction at hue of mother-of-pearl in the orchid; so, aviators would be drawn to that deep sense of balance they could feel between themselves, their machines, and intimations of the universe come upon in the winding and unwinding of the clouds, even as the lover found his universe in the rose gardens and hailstorms of the new lady's emotion. That was one of the reasons they were aviators first – something incommunicable, therefore to be talked about lightly, was to be found in the experience of searching the weather. For at such moments they could as well have been oracles envisioning futures in the entrails of a goat, or palmists reading a life in the lines of the hand, then they were interpreting the mood, the whim, and the shifting temper of a presence which must have seemed at times on the edge of revealing itself to them. Consider a moment in the life of a flyer when he is high in the air alone and on that peak of focus to his reverie where emotion and logic come together – how could it fail to affect him if the weather were to shift mysteriously at the same instant he had a bold thought.

Armstrong, who had spent a life studying weather from five miles up, from ten miles up, would then hardly have ignored the earth from 100,000 miles up, nor a quarter of a million miles away, not that view of the earth as a planet! He had been quiet wherever he could, a silence within the silence, a man composing his mind to give quick recall to a hundred plans, ten thousand names, a multitude of alphabets and numbers he might yet have to consult and so must order one more time in his brain. But now it was the moon before him, a globe without atmosphere, a cloudless sphere with a ground so humped and wrinkled that its horizons were never round but forever uneven, a compressed presence of a planet no more sibling to the earth than a head shrunken by aborigines is near a normal head.

Yet they had come up over the hill, they had come around the far side of the moon, over the last of the trackless craters, and now like pioneers coming out of the hills they had the equivalent of daylight again, their radio was working, and the terrain was there to be recognized. There on the moon ground, one hundred twenty miles below, were plains as well as craters, chains of mountains and ravines beginning to emerge from the endless pitting of promiscuous craters. Their maps were coming together, their training, the films they had studied, the charts, and the advice of the astronauts from previous flights. What a pleasure to recognize the full measure of a piece of topography with a name, a full landmark twelve miles across. A happy voice spoke '. . . going over the Taruntius crater and the pictures and maps brought back by Apollos 8 and 10 give us a very good preview of what to look at here. It looks very much like the pictures, but like the difference between watching a real football game and watching it on TV – no substitute for actually being here.'

CAPCOM: *Roger. We concur and we surely wish we could see it firsthand also.*
PAO: *That was Neil Armstrong.*

They passed over the Messier series of craters and a crater called Secchi. Next was Mt. Marilyn, named for the wife of the astronaut Jim Lovell from the flight of Apollo 8, Marilyn Lovell, tall, big, dimpled, cuddly, happy in social manner as a Texas cheerleader, the proud possessor of a mountain on the moon. From Mission Control came word that her husband was smiling – it is clever to name a part of the moon after one's wife. Much will be easier later.*

They came to the Sea of Tranquility, near whose western edge they would land, passed over some flat plains with occasional craters now predominant, a flat dull expanse glaring like desert in the sun beneath, an area as indistinguishable mark by mark as the quietly rusting plates on an old freighter. Here a blister of paint, there a pock – nothing like the barnacles on the back side of the moon. They named landmarks, Boot Hill, Duke Island, Sidewinder – a long twisting ravine. It was an area honored by no outstanding feature but the fact that they would approach along this way on their landing tomorrow. They were coming into the terminator now, the line of evening between day and night on the moon and the glare of the desert diminished – at the terminator the moon had a color of ash and gray, the landing site was on ahead to the west and well into the dark. From that height, looking down on the plain, little could be discerned – they would need to be nearer than sixty miles up to be certain of the intimate character of the land they would approach.

They passed Triesnecker Crater and Rima Hyginus, a doglegged ravine one hundred miles long. They passed Eratosthenes and Copernicus

---

* With the wife at least.

to the north, passed Copernicus without comment, Copernicus! the most outstanding crater of all *maria*, mighty crater in the Mare Imbrium! From Copernicus extended mountain chains, Carpathians to the left, Apennines and the Caucasus to the right, the crater Archimedes was among them and the rays of craters Kepler and Encke. More visible, the view would have been remarkable, some of the most individual features of the moon were in the Mare Imbrium, the Mare Cognitum and the vast Oceanus Procellarum. There, great craters emerged from the plain of the sea like the ruins of ancient cities, with runs of highland thick with combs and cores of small craters, then sea again. It was the part of the moon most like earth. Toward the horizon, the lines of craters looked like gunboats and battleships steaming in formation, clouds of stippled smoke hanging over the armada, or so at least you could try by a trick of vision to transform the churn of the highlands which ran from Kepler to Encke to Kunowsky.

They were now on the search for Aristarchus. Far in the distance, four hundred miles north, and next to invisible in the earthshine, still they thought they could see it. Excitement showed. Aristarchus was the most mysterious of the observed craters. Its reflected light seemed brighter than any other crater, and it had given evidence over the years of possessing some kind of local atmosphere, some hint of clouds or emission of gases. Areas of the crater would glow bright or grow dim in unaccountable fashion, as studied through the telescope, and its most luminous period would tend to come at the beginning of the lunar day when the change from cold to heat was greatest, rather than at midday when the sun was overhead. The astronomer V. A. Firsoff, who had studied Aristarchus most closely, would state: 'A suspicion is aroused that at least some of these white features are atmospheric veils or jets of finely divided white matter emitted by the crater or some parts of it.'

Armstrong: 'Hey Houston I'm looking north up toward Aristarchus now, and I can't really tell at that distance whether I'm really looking at Aristarchus but there's an area that is considerably more illuminated than the surrounding area. It seems to have a slight amount of fluorescence to it . . . one wall of the crater seems to be more illuminated than the others.'

Some minutes later the spacecraft went over the hill and loss of signal occurred while Apollo 11 traveled back of the moon on the beginning of its second orbit. When the astronauts, forty-seven minutes later, came around the corner again, they began to offer details of what they viewed.

The comments now came fast. They were beginning another TV show. The audience would be offered a view of the moon no later than one orbit after themselves.

SPACECRAFT: *We're about 95 degrees east, coming up on Smyth's Sea . . . Sort of a hilly-looking area . . . looking back at Marginus . . . Crater Schubert and Gilbert in the center right now . . . a triple crater with a small crater between the first and second, and the one at the bottom of the screen is Schubert Y . . . Zooming in now on a crater called Schubert N . . . very conical inside wall . . . coming up on the Bombing Sea . . . Alpha 1 . . . a great bright crater. It is not a large one but an extremely bright one. It looks like a very recent and I would guess impact crater with rays streaming out in all directions . . . The crater in the center of the screen now is Webb . . . coming back toward the bottom of the screen into the left, you can see a series of depressions. It is this type of connective craters that give us most interest . . .*

CAPCOM: *We are getting a beautiful picture of Langrenus now with its really conspicuous central peak.*

COLLINS: *The Sea of Fertility doesn't look very fertile to me. I don't know who named it.*

ARMSTRONG: *Well, it may have been named by a gentleman whom this crater was named after, Langrenus. Langrenus was a cartographer to the king of Spain and made one of the early reasonably accurate maps of the moon.*

CAPCOM: *Roger, that is very interesting.*

ARMSTRONG: *At least it sounds better for our purposes than the Sea of Crises.*

CAPCOM: *Amen to that.*

vii

With the end of the television broadcast, the mood altered abruptly. In a few minutes less than twenty-four hours ahead, their attempt to land on the moon would be successful, or aborted, or a disaster – they would be successful, or aborted, or a disaster – they would be alive and heroes, or alive and failures, or dead. Just as if the clock of some inner schedule had rung a change, as if the act of finishing their television show was the end of some last view of a parlor and the sounds of summer on the lawn, so they stepped back into their worries again. Not everything was proceeding automatically. They had had trouble earlier with the nitrogen-driven ball valve in Bank B of the Service Propulsion System. It had developed a small leak and was a few hundred pounds under normal pressure. Since it remained at 1960 pounds per square inch and the red-line was at 400 pounds, Bank B would have more than enough room to develop its leak further without being out of commission. And, of course, the system was redundant. Bank A in the absence of a working Bank B would open and close all the ball valves for the hypergolic fuel. Still, it was a malfunction, and not accounted for. It had to produce the same species of mild anxiety as a strange noise in one's car engine.

Besides, something more mysterious was going on. It had begun before the television broadcast when Collins reported: 'The Lem wants to wander up and down. I can get it completely stabilized in data and let it alone and in another couple of minutes it will have developed its own rate . . . several minutes ago I was exactly steady on data and since then I have been moving forward, the Lem pointed straight down toward the radius vector and that's been despite a number of down minimum pitch impulses.'

Ten minutes later Collins was reporting again, 'The tendency seems to be to pull the Lem down toward the center of the moon . . . it may have something to do with mascons or it may be just the peculiarity of the DSKY display.' In another fifteen minutes, he remarked, 'Interesting data on thruster firing versus pitch angle. It looks like that Lem just wants to head down toward the surface is all.'

'I have a comment here,' replied the Capcom, 'that says that's what the Lem was built for, I believe.'

Collins was grinding through the anxiety that the Lem was behaving most peculiarly, not unlike a dog on a leash who keeps leaning in the direction of a new and fascinating scent. There were unsatisfying explanations for the inexplicable. The moon had mascons, mass concentrations of dense material buried irregularly in its sphere. It was a little as if it had not one core but several. So its gravitational effects were a hint displaced. It would pull on the orbits of the spacecraft in irregular fashion, even distorting the orbit over a period of time. So the mascons might pull on the spaceship itself. But why pull on the Lem in preference to the Command and Service Module? Of course, these movements were minuscule. They appeared only on the computer rather than in the senses of the astronauts. So perhaps the malfunction was in the computer. How disagreeable either way – a mysterious fluctuation in the computer, or a mascon which attracted one of two connected spacecraft and ignored the other. Well, 'that's what the Lem was built for,' the Capcom had said. It was a joke, but the psychology of machines was in the wit. What if the moon would yet prove a womb to the chromosome of psychology in the bowels of each machine? Why did the Lem tip toward the moon? The answer did not come back from the Capcom and it lived as one more anxiety to take through the afternoon and into the evening and sleep.

The second burn for Lunar Orbit Insertion was fired on the far side, and the valves worked exclusively on Bank A. When they came out, their orbit had an apolune of 65.5 nautical miles, a perilune of 53.7. That orbit had been calculated to respect the eccentricities of the mascons so that the Command Module, two days and twenty-four revolutions later, would be traveling in something close to a perfect circle at the hour the Lem would ascend from the moon and seek to rendezvous.

Work continued. There was more than a little work to do. The burn report on the second Lunar Orbit Insertion was relayed downlink, and the Lunar Module was pressurized. The radio was switched to high-gain after a temporary loss of voice communication. Then a Program 22 for Lunar Surface Navigation was sent from the ground. The probe and drogue were removed from the Lem, and a small puddle of water was reported on the floor of the Command Module for the first time. The EECOM at MOCR was consulted on the best way to dispose of it, and Capcom told them to stand by for advice. The spacecraft went over the hill again and Loss of Signal occurred for the third time as they went into their third revolution. When they came around the corner again, the air-to-ground communications were noisy. An awkward spacecraft attitude had been required in order for them to be able to sight out the window on the landmark tracking in Program 22. So the omni-antenna was being used once more instead of the high-gain. Conceivably, the psychology of machines gave a wink, for DAP, the Digital Auto Pilot, the pilot in the Lem computer, now showed signs of independence. '. . . thing that was a little odd is that there was some DAP thruster activity . . . roll and yaw got excited and the DAP went into a flurry of thruster firing . . . We've noticed the same thing in the Command Module System and just written it off as a CSM peculiarity.' There was undeniably something eerie about a rocket thruster system which would fire from time to time out of unforeseen reactions from the computer.

But they were in the Lem, and proceeding to check it out. The various systems in the Lem were being turned on, and data was being sent by telemetry. Now from two spacecraft, still connected, the LM and the CSM, were messages being sent independently to the ground.

Capcom: 'We got some beautiful data here, Eagle. All those guys are looking at it – systems guys.'

Beautiful data was clear and thorough data. An engineer's idea of beauty was system perfection. Beauty was obviously the absence of magic.

The data flowed, the check-outs continued. There were as many instruments, controls, switches, handles, circuit breakers, dials, panels, and displays on the Lem as on the Command Module, indeed by actual count there were more, for the Lem, being more delicate than the Command Module, had more redundancy built into it. Besides the Lem was two ships in one – even as the Command Module and Service Module were attached to each other before ever being connected to the Lem, so the Lem in turn had a descent stage and an ascent stage. Both ascent and descent stage would go down together to the moon – only the ascent stage would rise up off the moon and into rendezvous.

Now the data being sent downlink by Columbia was lost for a time and communication proceeded between Eagle and ground to restore it. Further communication checks went on between Eagle and ground via S-band OMNI. 'You're beautiful in this mode, Buzz. We're reading you 5 by.' 'You're gorgeous also,' answered Aldrin.

The cameras were checking out and the eighty-five-foot dish antenna on earth and the two-hundred-and-ten-foot. The voice on S-band was found to be 'truly beautiful.' What a quiver in the fundaments that a voice would vibrate loud and clear over a quarter of a million miles. (But then light could travel for millions of years at one hundred eighty-six thousand miles per second.)

The rapid tempo of work continued. They had much to do before dinner and sleep. The Eagle was powered down to twenty-seven volts, and Trans-Earth II was sent uplink to Columbia. Each time they went around the moon a new set of Trans-Earth Injection data was delivered to the computer. If an emergency occurred, they could press a button and the rocket would be fired into a trajectory which would bring them back from the far side of the moon with velocity sufficient to return to earth. Next was a request sent up to do the waste-water dump in thirty minutes. Instructions were also sent for the water spilled on the floor. It could be sopped up with sponges and deposited in the waste stowage area. 'If it's too much, then we recommend using the procedure in the checklist on page F10-14.' A new S-band configuration was suggested for Columbia in order that reacquisition of signal would be automatic when they came next around the moon. Then ground sent instructions to Eagle to give the Oxygen Pressure System reading. Loss of Signal soon followed. The fourth revolution was begun. A discussion of Lem systems and warning lights followed, then fuel cell purges and time for battery charges, a balancing of the cryo tanks once more. The degree of deadband or limits on minor deviation in PTC were established and the radiation report on the astronauts' dosimeters was given. High-gain lock was lost and reacquired manually. Tests were made for automatic reacquiring. Optics were zeroed for the night, and the crew had dinner and prepared for sleep. Music was heard in the background. The crew played taped music while eating dinner. Last chores were worked upon. Signal was lost for the fifth time on the trip around the far side, then acquired again. During that interval the fuel-cell purges were finished, and the problem of automatic reacquisition of signal was finally solved to the satisfaction of Mission Control. The bunks and zip-in hammocks were set up. Around midnight the crew went to sleep. They would be up in six hours. It was their last night before the moon landing. They were obliged to wonder whether two of them would sleep tomorrow on the moon or sleep forever. Yet calmly they went to sleep. So recorded the monitors on the ground. There are souls whose health is to sleep upon the edge of profound uncertainty, men whose greatest calm resides in the edge of danger. Who indeed can understand the psychology of astronauts? Let us try to comprehend how men can be so bold yet inhabit such insulations of cliché. As they sleep, we are forced to think again about the mysteries of makeup in these men who are technicians and heroes, robots and saints, adventurers and cogs of the machine. Let us try to think of astronauts.

5

The Iron of Astronauts

Aquarius turning over scientific ideas he had not considered in years, was startled by the sudden disclosure that 'any piece of iron you pick up on earth is likely to be older than four and a half billion years because it's made of the interior of some star.' (The remark was dropped in passing by Dr Edward Anders at the Lunar Science Press Conference in Houston in January 1970.) Sieving the transcript for lunar gold, Aquarius was struck and struck again. If iron was the interior of the stars, iron was also – its molecules aligned in one direction – nothing less than the seat of magnetism. That suggested some intimate relation between the stars of the farthest galaxies and the turn of a magnet in one's hand. But, then there was always in the force of the smallest magnet as intimate a sense of some stirring in far-off dominions as there is a capture of the sea in the roaring of a snail shell held next to the ear.

Yet if every magnet had an invisible field of force surrounding it like a mood, that exact field we experience in our fingertips as we keep another piece of iron from touching the magnet, so a metal wire cutting across that field (any metal – it can be one of the ways to define a metal!) will have a sudden electric current passing through it. No physicist has ever explained the phenomenon satisfactorily, yet on that phenomenon is built the generator: all the electric power of the world can be seen as the translation from physics to electrical engineering of the controlled interruption of a magnetic field. So if magnetism derives ultimately from some communion in the interior of the stars, electricity may be nothing less than the interruption of that communion. Yet run a current through a wire wrapped around a bar of iron, and the iron bar will become a magnet. Electricity, passed through a coil, becomes the agent by which such magnetism is restored. It leaves one to think of iron as a spine on which electricity breaks and restores the resonance of the stars.

Now we speak of the iron of astronauts. What is one to do with the metaphor? Iron in its finished state, ground and polished, is a material of much strength, near to impenetrable surface, shining appearance, limited flexibility, and must be kept insulated from the corrosion of the atmosphere, usually by a thin layer of oil: astronauts are men of much personal strength, moral and physical, ground and polished to a turn by years of training, the depths of their character are kept hidden by the impenetrable qualities of their personal surface, and they shine in appearance. (Indeed, eleven of the sixteen astronauts who had been up in space and were still active had blue eyes. To better the item, fifteen of the sixteen had blue, gray, green, or hazel-colored eyes. Only Collins had dark-brown eyes.) That their flexibility is limited can be demonstrated by their powers of description before superlative sights and well-functioning equipment. That they were insulated from the caustic atmosphere of America could be taken for granted. A fine layer of public relations put oil over all unnatural exposure.

But the measure was not to be found in formal properties. It was rather that the astronauts were the core of some magnetic human force called Americanism, patriotism, or Waspitude, and if they were finally the men of all the men on earth to take the first step on the way to the stars, who was to say it was not the first step back to the stars, first step back to joining that mysterious interior material of the stars, that iron of communion with cosmic origins? Indeed, who could be absolutely certain that the astronauts were not men finally forged out of some powerful equivalent in their blood of the stars' iron – but in fact there is iron in the compounds of the blood – men created out of some impulse so deep that the metaphor of iron is re-engaged; we may yet have to lean on the notion that the astronauts, strange, plasticized, half-communicating Americans, might still be the spine on which electricity breaks and restores the resonance of the stars. It was at the least a thought that the mysteries of America, the urgencies of the national itch to dominate the world, the contrasts and harsh comedies of race and war and pollution, the schizophrenia of the land growing more Faustian and more Oriental each season with ABM and million-footed folk-rock festivals at the poles, yes, all the incomprehensible contradictions of America might yet come to focus on the possibility that races were at war in America like forces from the cosmos, and it was no accident we were on our way first to the moon, no, that had begun to shape itself in the cauldrons of the stars, and so the astronauts could even be men with a sense of mission so deep it was incommunicable even to themselves, as if they had signed on as the core, no, rather as the most finished product of a human ore whose purpose – despite all thoughts it had found that purpose – was yet undiscovered. Which is to say that one could also hold the thought that the real function of the Wasp had not been to create Protestantism, capitalism, the corporation, or a bastion against Communism, but that the Wasp had emerged from human history in order to take us to the stars. How else to account for that strong, severe, Christian, missionary, hell-raising, hypocritical, ideologically simple, patriotic, stingy, greedy, God-fearing, nature-despoiling, sense-destroying, logic-making, technology-deploying, brave human machine of a Wasp? It was a thought with which to begin to look at astronauts.

ii

Aquarius, being a firm believer in intellectual husbandry, had gone to work of course on his own statistics. If the astronauts had been accepted by NASA in six groups from 1959 to 1967, to a total of sixty-six men of whom eight had resigned or transferred, and eight been killed, Aquarius chose out of the fifty who remained a group of sixteen for most of his comparisons. They came from the first three groups and consisted of all the men who had flown Mercury, Gemini, and Apollo flights and were still active astronauts. They were Aldrin, Armstrong, Bean, Cernan, Collins, Conrad, Cooper, Cunningham, Eisele, Gordon, Lovell, Schweickart, Scott, Shepard, Stafford and Young. He could have included Grissom, Glenn, Carpenter and Schirra from the first group, McDivitt and Borman from the second, and Anders from the third, but Grissom was dead, and Glenn, Carpenter, Schirra, Borman, Anders and McDivitt had either resigned or were doing other work at NASA. He suspected that it did not matter, for the personal characteristics of the astronauts did not vary much. All sixteen were for example married, and had a total of forty-seven children, almost three to a couple. Only one astronaut had ever been divorced and he was married again. The youngest of the astronauts was thirty-five, the oldest, Alan Shepard, was forty-seven, but eleven of the sixteen had been born between 1930 and 1932. They varied in height from Pete Conrad, who was 5'6" and weighed 138 pounds to Scott and Stafford, who were 6 feet and weighed 175, but most were between 5'9", 5'10" or 5'11" and exactly half had a weight between 165 and 170 pounds. They had all, but for Schweickart, been in combat, been test pilots, or done both. Ten of them had a bachelor or masters degree in Aeronautical Engineering or Astronautical Engineering, the rest had a college degree. Four had gone to Annapolis, three to West Point, and three had done graduate work at MIT. Most of them had commissions in the Air Force, the Navy or the Marines.

If one studied their preference in sports it was directly significant that they neither listed spectator sports nor played them now. There was a line in their biographical records for Recreations and Hobbies, and among the sports in which they participated, baseball was not listed, nor football, nor wrestling, nor hockey, nor boxing, and only one astronaut mentioned basketball, and that in the course of listing a number of other sports and pastimes.

Their choices marked and tabulated, the order of preference in sports for astronauts was handball, swimming, fishing, hunting, golf, water skiing, sailing, and skiing. A few mentioned squash or gymnastics or jogging or tennis. A few did skin diving, or scuba, or surfing. One liked shooting, one did weight lifting, two bicycling. Of them all, Armstrong was the only one to list soaring, and it was the only sport he listed.

Aquarius added the choices of another eighteen men. His selection was now derived from all the remaining astronauts but for the scientist-astronauts, a special group he decided, since some of them could not even fly a light plane at the time they were accepted. It did not seem to affect the findings. The younger eighteen astronauts liked handball and fishing and hunting and swimming and golf with about the same frequency as the older men. He thought if there was a perception to extract from these modest facts, it was that the astronauts seemed to like water and were drawn to the principle of striking a ball or hitting a target. In fact, twenty-one of the thirty-four astronauts liked all three. Since five of the astronauts did not list their sports, and one said he liked all sports, the adjusted count gave twenty-two out of twenty-eight or three out of four men who were happy to be in the water or aiming at a target. It is not so hard to comprehend their being attracted to handball, golf, and hunting. Astronauts were not going to put on boxing gloves or play tackle football any more rapidly than surgeons or musicians – they had no desire for personal injury. It would delay their careers. But coordination between hand and eye was crucial, the cool nerve of the hunter was crucial, and competition among all those keyed-up egos competing with their quick reflexes for the same few limited Apollo seats – well, handball was no surprise. But the water. Why should men who fly in air be drawn to water? Well, Aquarius could hardly adopt his name without passing one grudged respect to the astrologer. So he noted that nine of his astronauts were Pisces and five Scorpio. Even with one Cancer, signs of water prevailed. Fifteen astronauts out of thirty-four was almost twice the expected average.

It was as if water was a clue to the difference between astronauts and other men, for it hinted at a readiness for displacement. We tend to forget that not many primitives love water. As recently as a hundred years ago, few men could swim. It is as if we turned to the water when technology began to capture the land, as if we explored our bodies in the sea about the time we learned to fly in the air. Should there be an oncoming self-destruction for the species, it is not unreasonable that unconscious warnings would exist – if a partial destruction, all the more urgent might seem the alarms. In a landscape of pollution and disease when the roots themselves are poisoned, the healthy man might begin to prosper on the very condition which once had been next to death itself – he could learn to live on the stimulations of the uprooted and the displaced. In morphological terms, it is as if the cancer becomes the continuation, as if life leaps a step and the anomaly acquires the art of reproduction. So why not turn to water? To live with style and pleasure in such displacement is to get ready for space stations, hydroponic food, and air which derives from a canister. An afternoon on the water may be a bridge to their ultimate mission.

Too farfetched? Is it too extortionate finally to deprive an astronaut and his family of a Sunday on a powerboat or an afternoon of fishing? Is it finally necessary to explore into these recreations? There are hours when Aquarius, thinking of the astronauts he had met, would wince at his own ideas, for he knew the astronauts were likely to say they believed in death-to-party-poopers. To read the transcript of Apollo 10 near the moon was to hook up with room service on a three-day wowser.

'Hello, Houston, this is Snoopy.'

'We is going. We is down among them, Charlie.'

'We are right there. We're right over it.'

'I just wish we could stay.'

'I'm telling you, we are low, we're close, babe!'

'Oh, Charlie, we just saw an earthrise and it's just got to be magnificent.'

'We have been down among them, babe.'

He had only to think of Jim McDivitt of Apollo 9, who had the genial and side-calculating air of an urbane and successful Irish executive who had risen in a predominantly Protestant corporation. What had his personality to suggest which had any relation to this calculus of the soul Aquarius would inflict on astronauts? And there was Gordon Cooper, who had taken the last Mercury flight through twenty-two orbits of the earth, a record at the time, and he had been sufficiently relaxed to sleep in the Mercury capsule while waiting for the launch. Cooper was blue-eyed with a sun-burned nose and a red neck. He looked like a man who played on a semipro football team because he wasn't big enough for the major leagues, and worked in a gas station the middle of the week. He had the look of a player who would saunter into the dressing room half an hour before the game with his helmet cocked on the back of his head. While playing, he would roll up his sleeves. After the game he might take a shower with an unlit cigar. What would he think of Aquarius' theories of displacement?

And Schirra would laugh. Schirra had the keen cut of nose and cut of upturned lip which had him always looking about a room for the next fly he could catch. He looked more like a ruddy young movie executive than an astronaut. He never failed to smile. He smiled the way other people shake hands, or a cop directed traffic.

Or Jim Lovell, who would command Apollo 13. He was big for an astronaut. Stolid. Imperturbable. He was reminiscent of the kind of clubfighter who usually came out of Canada – durable would be his middle name. You could hit such a fighter all night, but he would win the fight. Not by knockout. Points. Lovell would listen solemnly to such theories.

And Pete Conrad, quick as a little tiger, but sensitive about his looks for he had protruding eyes, a small hooked handle of a nose, and a big friendly space between his two front teeth. He had a reedy voice guaranteed to pierce all static. 'That's an interesting theory, Norm,' he would doubtless say, 'now, let me tell you mine.'

And Frank Borman, hard as hand-forged nails, Borman of Apollo 9, first to go behind the moon. He had read Genesis to America at Christmas while in lunar orbit. 'In the beginning God created the heavens and the earth,' Borman had read with all the sober respect and uneasy pronunciation you bring to a stock market report. Borman was the man who had come back from his trip and said, 'We had a horseshoe up our ass.' Now he was adviser to Richard Nixon on NASA affairs. Borman had been indignant when a misconstrued medical remark had prevented the President from having dinner with the crew of Apollo 11 the night before the launch. 'He can't possibly see them now before the launch,' said Borman, rapping his hand against his drinking cup while talking in the sun. 'If one of the astronauts was even to sneeze, they'd blame the President forever.' Borman had the look of a man derived from five generations of German farmers who had never failed to do a hard day's work. Aquarius did not have to wonder how Borman would receive his theories.

Face to face with these onslaughts of the matter-of-fact, the practical and the philosophically immune, still Aquarius would not relinquish that other intuition which insisted it was more than a life of high risk, great technique, and enormous demands of commonplace phlegm, more than a natural graduation from fighter pilot to test pilot to space, no, the logic of these matters did not add up, the logic crossed a gulch. The life made no sense unless they were men of enormous even unnatural ambition or men attuned to a mission they could barely confront in their dreams.

iii

A man who becomes a good fighter pilot does not think anything can ground him. Fighters are like downhill racers – their sanity is that they do not look to pick up a sane option. Their balance is at the edge of balance.

Fighter pilots growing older sometimes become test pilots. The chances they used to take casually now collect around the act of flying the most dangerous aircraft to be found. But they have developed into engineers.

Once they become astronauts, however, they hardly test anything. Where once they might have been testing planes every two or three days, now they do not have more than one rocket flight every two or three years. They still fly, they fly T-38 jet trainers for hundreds of hours a year, but they are not testing the T-38, they are merely flying in order not to lose their reflexes. And astronaut training is in simulators or classrooms. They are on call for public appearances. As test pilots they flew rocket planes at four thousand miles an hour and lived as they pleased, free to carry a drinking party to the dawn or to search for solitude; now that they were astronauts they were obliged to live in homes of a certain price in suburbs of an impeccable predictability in a world of public relations where they were rendered subservient to propriety by a force of mysterious propriety within NASA itself, a force which might just as well have seeped from every door of every office until anything spontaneous in a man was stuffed into the cellar of his brain. They worked long hours, perhaps an average of fifty hours a week, and hardly knew when they would have their first space flight, or once up, whether they would ever have a space flight again. They were the best pilots in their profession, but now they flew only for practice, and they could not know for certain whether they would ever be able to practice their new profession. (It is as if Truman Capote gave up literature because he wished to write opera and suddenly could not find out whether any of his music would ever be sung.) All the while the astronauts were obliged to live in intense competition with one another, yet had to exhibit every face of good spirit and teamwork to the world. Stories were common at the Manned Spacecraft Center of astronauts who had shared the same flight yet hardly spoke to each other in the months before, so intense was their mutual dislike. Still they kept their animosity private for fear they could lose their seat on the flight. Then all of them had had to swallow the wrath they might have felt for the contractors connected to the fire which killed Grissom, Chaffee and White. Say it worse. Eight of them altogether had already been killed. Besides the fire, four had gone in fatal crashes of the T-38, and one in an auto accident. (Astronauts were quietly famous for driving their cars a foot apart at a hundred miles an hour.) One could say that the demand for order, hard work and propriety in such competitive near-violent men had produced their deaths, as if the very tension of their existence as in a game of musical chairs had pushed the escape from death of one man over into a higher potentiality for accident of another. Being an astronaut was perhaps the most honored profession in the nation, but of a total of sixty-six astronauts accepted since the program began, eight were dead and eight had resigned. Since seventeen of the original sixty-six were scientist-astronauts, and only three of them had separated from NASA, it meant that thirteen of the forty-nine flying astronauts were no longer present. The resignation and mortality rates are not so close in other honored professions. Of course, most of the astronauts work for only thirteen thousand dollars a year in base pay. Not much for an honored profession. There are, of course, increments and insurance policies and collective benefits from the *Life* Magazine contract, but few earn more than twenty thousand dollars a year.

So we are obliged to consider why a man would divorce himself from his true talent – which is to test a new jet or rocket plane – and live instead in propriety, order, competition, and tension for twenty thousand dollars a year, knowing he could make three to five times that much in private life, and not be afraid to utter a resounding opinion, get drunk in public, yell at his children before strangers, or be paralyzed by scandal or divorce. Can it be that any man who takes up such a life for thirteen thousand dollars base pay a year is either running for President, patriotic to the point of mania, or off on a mission whose root is the field of the magnet in the iron of the stars?

Some may have been running for President. John Glenn had been campaigning for senator before his accident in the tub, Borman was now close to Nixon, Schirra was a television commentator (a holding position) and Collins was yet to enter the State Department. And there were bound to be others. If an astronaut had political ambitions he did not necessarily announce them.

Then there were men for whom a celebrity as astronaut was preferable to the professional anonymity of the test pilot. And some were patriots. There is no need to diminish the power of this motive. Once, in a meeting of astronauts, NASA executives and scientist-astronauts, the NASA administrator, then James Webb, had told them there would be a hiatus in the Space Program during the early 1970's due to budget-cutting. The scientist-astronauts were gloomy. Last to arrive in the program, unscheduled for flights, they saw a delay of a decade or more before they could even go up. Scientific examination of the moon and space by experts such as themselves would be again and again delayed. One of the scientist-astronauts said, 'Mr Webb, this hiatus you've been referring to – how would you say that the scientific community –'

'To *hell* with the scientific community,' Frank Borman cut in. The astronauts laughed. The attitude was clear. They were not in astronautics to solve the mysteries of the moon, they were astronauts to save America.

Nonetheless, if two-thirds of the astronauts were politicians and patriots, the remainder might still be priests of a religion not yet defined nor even discovered. One met future space men whose manner was friendly and whose talk was small, but it was possible they had a mission. Like Armstrong or Aldrin they were far from the talk at hand. If they followed the line of a conversation, they still seemed more in communion with some silence in the unheard echoes of space.

The Director of Flight Crew Operations, the chief in effect of the astronauts, Donald K. 'Deke' Slayton, was a man with powerful big-knuckled hands and the rugged grin of a good mechanic. He had his own burden – he had never been up. One of the original seven astronauts, he had been scheduled for a Mercury flight, but a faint heart murmur had been detected. So he was grounded. Because he had not gone into space, he had not had the dubious opportunity of the others to come back down and tell about the experience to family and friends, go through debriefings and talk to the Press until half the life of the event was gone. He had had instead to listen to the accounts of others. So of all the astronauts his personal intensity was perhaps the greatest – indeed he was certainly the astronaut Aquarius would have supposed had been longest and farthest in space. Instead, he was the man who directed the Astronaut Office and was therefore responsible for picking the crews.

Slayton always pointed out that no one could be certain when schedules might be changed for reasons beyond the control of the Astronaut Office. If he had selected Armstrong to be first on the moon, a small shift in events could have made it Stafford of Apollo 10 or Conrad of

Apollo 12. Nonetheless, it turned out to be Armstrong, and so it is hardly unnatural to assume that he was the particular pilot Deke Slayton wished to see there.

iv

Naturally, there was more to selecting a crew than divining who might be the men most suited to fill the aesthetic outline of a mission. Seniority had to be taken into account and a matching of abilities. Even chance dictated how certain astronauts were teamed with others. Collins, for example, had been joined with Armstrong and Aldrin out of a process which was close to automatic. Originally with Borman and Lovell on the crew of Apollo 8, Collins had been obliged to give up the flight because of a bone spur in his spine which required a serious operation. By the time he was fit again, Anders had replaced him. Collins, however, was entitled by the logic of these affairs to get on the first crew available. That happened to be Apollo 11. Whether he was particularly suited to work with Armstrong and Aldrin was probably not considered too long; indeed Slayton may have estimated that Collins was likely to get along with any crew. Besides, it is equally possible a man in Slayton's position would look to eschew fine psychological match-ups. The logic of the missions demanded that the men, like the machines, be to a degree interchangeable.

Still, if you have the power to choose astronauts for a mission, you are most likely to exercise such a function when the mission is as important as an initial landing on the moon. Let us make the assumption in the face of every published statement to the contrary, that Armstrong was Slayton's carefully considered choice and then go in for the further presumption of trying to enter Slayton's mind.

What would one look for? Obviously any astronaut selected would have to be a superb pilot, better if possible than his peers. On the other hand, it would not hurt if he were a man who would appeal to large numbers of Americans. The size of NASA activities in the future was going to be determined in some degree by the response of Americans to the moon shot. While the flight of Apollo 11 would certainly excite America in the summer of '69, how profoundly could the public be reached? Would it be deep enough to arouse desire for more space travel in the cold months of budget cuts in the years ahead? Or would it merely flip a thrill for one hot summer, after which the murderous laws of fashion would take over? To withstand that time when the moon shot could pass into discard, an astronaut able to capture the imagination of Americans was required. Slayton was in trouble already. He needed a man like Lindbergh.

Then came other categories of selection. They were less definable. All astronauts were brave men and all astronauts lived with death: landing on the moon, however, might require a special sensibility. While a man could get killed undergoing a routine check-out on the pad, while a man was certainly in danger going out into space for a walk, where the crew of Apollo 8 had stepped right over to the unknown when they had gone behind the moon, still none of that was equal to landing on the moon! A man ready to do that would need not nerves of steel but some sense of intimacy with death, conceivably some sense of death as a pale ancestor one had met before and known for years, or, failing that, a sense of tradition so profound, a faith so great, that the moon could reside in some outlying yard of that faith. A more ordinary astronaut, no matter how brave, might be dislodged from normal command of his nerve by the psychical dimensions of the event.

Finally, there had to be motivation beyond measure, some need to succeed which would keep a man pointed to his target, locked into his target, ready to dare the very explosion of his flesh before he would give up his destination. It would have to be a motivation powerful enough to take him through training, public exposure, inhuman tension, and the dread weight of being responsible for NASA's effort over ten years.

Slayton picked Armstrong. He had been one of the best test pilots in America, yet unlike many of the other astronauts, his personality did not suggest that there was a hive of possible occupations he had left behind. He would not have been a politician, a professional athlete, an Air Force general, a top corporation executive – no, something in his personality and in his history would insist that he seemed born to be a pilot. He had learned to fly before he could drive a car; he had paid for his lessons, nine dollars each, out of the forty cents an hour he earned from deliveries around a pharmacy. He had built model airplanes, read every issue of *Air Trails*, and had excelled at not too much else. He had not been an athlete in high school, nor the center of any high-riding social life – if he played baritone horn in a jazz band, he was still not winning dance contests, and if he went to college it was by the thrifty route of Navy scholarship to study aeronautical engineering, then on to Pensacola for flight training. Yet at twenty-one, he was flying Panther jets in the Korean War, had seventy-eight combat missions and three Air Medals, and had almost been lost twice. Once he flew a crippled plane back to the carrier *Essex*, another time he brought a plane with one wing half-lost back far enough from enemy lines to parachute to safety. As there are bullfighters who will go back and fight bulls so soon as they recuperate from their last wounds, so Armstrong was always back in a plane, no matter how many experiences he had had which would have encouraged a man to quit. It is worth repeating that he was the only astronaut who did soaring in his spare time. It was apparent that his life was founded upon the act of being aloft.

Of course Armstrong's qualifications as a matinee idol were not monumental. If he had a face as American as any astronaut in Houston, and a small-town background to match, if he would appeal to all of the silent majority in every town in the Midwest and South, he was still too perfect, too polite, too reserved, finally too pinched in manner to interest that part of America not partisan to space and inclined to give priority to the accelerating needs of the cities. Of course, none of the available astronauts were likely to satisfy. NASA had picked them in the first place for more functional qualities than the ability to arouse a continuing fascination in the public. Armstrong at least would not offend too many sensibilities for he would obviously respect the indefinable chastity of the moon ground; there were all too many astronauts who thought chastity was something which came with a belt. On the other hand, Armstrong had too many qualifications in the next regard. Over the years, he had had not only an intimate acquaintance with the nearness of death, but been in the literal presence of it. List not only the crashes and near-crashes in planes, there was also the death of his young daughter from a brain tumor, the near loss of his family when their house caught fire one night and burned to the ground. This was Jan Armstrong's account:

Neil told me to call the fire department. I couldn't get the operator on the telephone at 3 A.M. . . . I tried dialing 116, because I had had a first aid course in

California. Then I realized that number was local only for the Los Angeles area. So I put the phone down, and Neil had gone in for Marky. I ran to the back of the house, and I was banging on the fence calling for Pat and Ed White . . . It was a six-foot fence. The Whites' air conditioning wasn't working . . . and they heard me calling. Ed came bolting over the fence. I don't know how he did it, but he took one leap and he was over. He got the hoses out immediately, and by this time I had run around to the front of the house for Neil to hand Marky out the window. But no: Neil didn't do that. They were little windows, and Neil would have had to break one of them. He brought Mark back down the hall, back to our bedroom and out. He was standing there calling for somebody to come and get Mark because he was – what, ten months old? – and he couldn't put him down because he was afraid Mark would crawl into the swimming pool and drown. By this time I could hear the fire engines on the way – Pat White had turned in the alarm. This whole wall was red, and the glass was cracking in the windows. I can remember Ed White calling me. He was saying: 'Here you hold the hose; I'll get Mark.' Neil had gone in for Ricky, who was just awakening at the time. And I was standing with the hose, the concrete was burning my feet, and we had to keep watering the concrete so we could stand there.

Guess at the subsequent shock when Ed White was burned to death in the fire with Grissom and Chaffee.

Armstrong must have been a man for whom dread was as near as breath. Search for evidence in the very lack of emotion with which Armstrong relates what was probably the most unnerving malfunction of a spaceship up to that date. In his Gemini 8 flight with David Scott, the mission had called for rendezvous, then docking with an unmanned Agena capsule previously fired into orbit. The rendezvous proved successful, so did the docking at first. But in a short period both vehicles began to spin, first slowly, then rapidly. Attempts to reduce the movement failed to work. Here is Armstrong's account:

We felt something that Dave was to describe later as 'constructive alarm.' We were aware of a serious emergency. A test pilot's job is identifying problems and getting the answers. We never once doubted we would find an answer – but we had to find it fast.

Although we had no way of knowing for sure, we were concerned that the stresses might be getting dangerously high – that the two spacecraft might break apart. We discussed undocking, but we had to be sure that the tumbling rate at the instant of separation would be low enough to keep us from colliding moments later.

As we unlatched, we still hoped to rejoin the Agena. At this point we figured that the trouble was in the Agena, but it wasn't. After separation, the Gemini spacecraft stopped responding to the controls and rotated more rapidly than ever – the sun flashed through the window about once a second. The sensations were much like those you would feel during an aircraft spin. Neither Dave nor I felt the approach of loss of consciousness, but if the rates continued to increase we knew that an intolerable level would be reached. The only way to stabilize the spacecraft would be to shut down the regular control system and turn on the thrusters in the reentry control system.

I made that decision reluctantly – reluctantly, because once the decision was made, the mission had to be terminated. That excluded Dave Scott's EVA, the two-hour walk in space scheduled for later in the flight – and that hurt.

After a check of all the electrical circuits, we finally pinpointed the problem: The Number 8 thruster had been firing on its own.

What a horror; what a paltry air of description! Two men in a cramped capsule revolving around once each second, the sun flashing with the delirium of a shield in combat, the dials turning with the revolving eye. How difficult to avoid the conviction that one's existence was finally spinning into a vortex. Unspoken by any astronaut had been a covert wonder about the benign receptions of space. Was there no curse in space, no buried storms in the oceans of space? At that moment, Armstrong and Scott must have felt as if they had blundered past the last taboo into the whirlpool of all fury.

By pressing the abort button, Armstrong saved them. Reentry control began a set of operations programmed for landing. That stabilized the craft. Afterward, Armstrong was considered to have made a brilliant move, for a thruster had been stuck in firing position and there had been no other way to turn it off.

We cannot speculate on the dreams which followed, but the wipeout of all feeling from his description can hardly drive us away from the point that a man who has been through such an experience will not soon cease living with it in his sleep. Worse. Engineers were never able to explain the malfunction of the thruster. The psychology of machines is trial enough, but nothing next to the suspicion that one might have an unnatural influence on insensate instruments. What a suspicion for a man to hold of himself. Afterward, it would not be so automatic to remain an astronaut: one might need some of the monomania of Captain Ahab. There is never a hint Armstrong ever thought of trying another profession. If Slayton was looking for a commander with motivation, Armstrong had been offering his credentials. There is either something close to schizophrenia in his lack of reaction to the dangers about him, or we must go right past the competence of any psychologist to use the word for a man like Armstrong and recognize that some of the brave have a lover's sense of death as a partner whose nearness is comfortable, as if on those occasions a quiet voice declares, 'You, sir, have nothing to fear if your ship passes over the bar,' no, Armstrong must have been brave about flying and wise about the nearness of dying ever since he had learned to solo on his sixteenth birthday, for he was not much older when he saw one of his friends crash a plane into a power line and be killed. No wonder Armstrong's nerves were occasionally thus taut that a Houston journalist who inadvertently broke the news to an astronaut's wife about her husband's death was never to have any luck with Armstrong. 'Keep that man away,' was the comment. 'He's a ghoul.'

One wonders at last at the selection. The Director of Flight Crew Operations is obviously a man with his own nerve. For what avalanche of criticism would come on Slayton if the end of Apollo 11 was tragic. What a hint of the curse of the ages would feature writers find in details about Armstrong. Why, he had even crashed the Lunar Landing Research Vehicle – a species of steam calliope and helicopter on stilts which gave a simulation of what it might be like to land with one-sixth of gravity on the moon – yes, the LLRV had gone out of control on Armstrong, and he had been forced to use his ejection seat and be landed by parachute. He was the only astronaut who had crashed in the LLRV. Still, Slayton picked him. Living in the privation of sending other men on a job he wished to do himself, Slayton's sense of the abilities of particular astronauts to perform particular expeditions into the extraordinary must have taken on levels of acuteness one does not easily conceive. Almost any man other than Slayton was bound to classify Armstrong as accident-prone and a bad risk. Slayton, however, may have been working on the thesis that only a man who had been in and out of death as many times as Armstrong could be entrusted to pass through the unpredictable minutes of a descent to the moon, down to that untenanted and eventless ground, that dead body whose light, when full, inhabited the switches of human sanity on the living earth, yes, perhaps only a man familiar with that twilight of the sleeping sky which may be known in the soul when death lays near and the psyche begins to hover, only a man who has become familiar with how soon the psyche will lift and begin to depart, yes, only a man who has lived with death since a boy with a recurrent dream ('I could, by holding my breath, hover over the ground. Nothing much happened; I neither flew nor fell in those dreams. I just hovered.'), only such a man, hovering for all one knew next to dying without a struggle in his sleep on many a long night of his childhood, could be the man to enter the sanctums and veils of the moon. For he had inhabited them since he was a child. He was familiar with the awe of his task. He could be capable of calm where other brave men, feeling the terror of emotions never felt before, could spin into a vertigo of the will at that dread which comes not from dying but from stepping into death and feeling no immediate urge to retreat.

The inquiry has obviously pushed too far. It is not likely Armstrong went about talking of an old recurring dream to Slayton, but then it was hardly necessary. Slayton was a man who took off on hunting trips when he could – his decision would come from the presence of qualities he would not even bother to name. So Aquarius would report, and he had no more for evidence than a night at dinner with the Director of Flight Crew Operations and his wife, and it was a night in which Slayton with long practice had not made a single remark which could be quoted in the morning. Yet, in the resonance of his silences, the weight of his smile, in the sense of his gravity, yes, even in his maintenance of a pause at the depth of man's adventure to the moon, was all that fiction of unspoken evidence upon which novelists throw themselves and journalists snarl. After this dinner, Aquarius was always certain Slayton had picked Armstrong with care, picked him out of some equivalent of the reasons Aquarius would give, yet they never mentioned his name once, nor the name of Aldrin either, and Aquarius was equally certain Slayton had picked Aldrin with much the same measure of care.

V

'Neil and I are both fairly reticent people and we don't go in for free exchanges of sentiment,' Aldrin was to say. 'Even during long training we didn't have many free exchanges.'

One may be certain of that. They were a curious mating of opposites, a team picked most probably for their complementary abilities and aptitudes. Temperamentally they suggested the equivalent of one of those dour Vermont marriages where the bride dies after sixty years and the husband sits rocking on the farmhouse porch. 'Guess you feel pretty bad, Zeb, that Abigail is gone.' The chair keeps rocking, there is a long puff on the pipe. 'Nope,' says Zeb, 'never did get to like her much.'

It was not a case of Armstrong and Aldrin disliking each other. How would one ever know? It is almost as if the question never occurred to them. They had each, after all, been married for years and their wives did not pretend to know them easily. They were men who liked to be alone in their thoughts so much that once teamed up with each other, there may have been no need to speak for more than functional purposes. It is possible they never had to decide whether they liked one another or not. Those who wish to live within their own minds ask for nothing more perfect than a companion whose presence is not felt.

Yet how remarkable if their personalities did not impinge upon one another, for they were profoundly different men. Armstrong, as we have seen, was a virtuoso of a flyer; Aldrin was a powerful personification of organized human intelligence. Of course, he could fly a plane well, just as he could give a superior performance at any number of activities from pole vaulting to celestial mechanics, but in relation to other astronauts, he was not among the most accomplished of aviators, he had not in fact even been a test pilot. Yet of all these men he was the one whose command of mathematics and complex statistical operations was so tuned to the logic of information systems that he was doubtless the nearest human equivalent to a computer at NASA, and had a Doctor of Science degree in Astronautics from MIT, the only one in the first three groups of astronauts to have earned his doctorate. Back in April 1966, the Director of Flight Operations, Chris Kraft, had spoken of Aldrin in these encomiums:

In the early stages of the development of the Gemini rendezvous mission plan, Major Aldrin almost single-handedly conceived and pressed through certain basic concepts ... without which the probability of mission success would unquestionably have been considerably reduced.

A well-known geologist spoke of Aldrin as 'the best scientific mind we have sent into space.' Ted Guillory, an engineer engaged in designing the detailed trajectories and orbits of a flight plan, said, 'He carried a slide rule for his Gemini flight on the rendezvous, and I sometimes think he could correct a computer. I can remember him saying things like, "If the computer says I'm twenty feet out of plane, I'll believe ten of that, but not all twenty." He's one of the few people who can figure out all those rendezvous things in his head.' The accolade from his wife was one straight statement, 'If Buzz were a trash man and collected trash, he would be the best *trash* collector in the United States.' And Aldrin speaking of himself remarked, 'At West Point the name of the game is, "Do what people tell you to do, keep your nose clean, and work out your academic progress." I fitted into that pretty well. I'm a sort of mechanical man,' and then added, 'or I was.' We can return to his exception later. It is enough for now to say that if his personality suggests the loneliness of the computer in a man of enormous will, he is yet a complex figure. It is not easy to recognize, however, for he is so prominently a man who has become the instrument of his own will. All his biographical details emphasize loneliness, self-sufficiency, eccentricity. He was the son of a strong father, a man who had been an intimate of Lindbergh and Robert Goddard, a species of minor-league Rickenbacker; for his aid in handling Italo Balbo's flight from Italy to the Chicago World's Fair in 1933, Aldrin Sr. was made Commendatore in Mussolini's Air Force. We can picture a father full of force and full of guidance, but usually away on business trips to tend his interests in commercial aviation, so that his only son grew up among maids and sisters, a boy prodigy in his powers of digestion, for the son could consume cans of tuna fish, packages of Jell-O powder dry, and sandwiches of peanut butter and sliced banana sprinkled with powdered chocolate, a mystery until one assumes he was already processing carloads of hitherto undigested information on the parameters of elemental nutrients in abdominal rendezvous. His activities in school offer similar sidelights on his will. Barely able to read one year, he could dominate a class the next. Yet once interested in football, his marks deteriorated. Warned he would not be accepted at West Point or Annapolis if they did not improve, he gave up sports for a year and moved from C's and D's to A's and B's. Halfback and quarterback originally, he also moved to center. At West Point he was first in his class at the end of plebe year but on advice from his father, 'in my experience number-one graduates become more or less freaks,' he determined to slow down. He succeeded. He graduated third. The irony is unavoidably present – change for Aldrin was never grace but challenge. If challenge is usually the imperative to do more in an unfamiliar situation, it is also – given the momentum of much human mass applied to a problem – a challenge to reduce such momentum once it is begun. It is apparent in everything he comments upon. For example:

I became a counselor at a summer camp – Trout Lake Camp, about sixty miles north of Portland, Maine. I had been going to this camp since I was nine years old, and I continued to go there until the summer before I had to take the examinations for West Point. I look back on my experiences there as being quite instrumental in leading me toward what I call competitive appreciation for associating with other people: having standards set for you, set by other people, or standards you would set for yourself.

When I first went up there, before I became a counselor, I think it was the first exposure that I had had to small groups of boys my own age. We lived together, and were given challenges in swimming, track and baseball.

The only challenge which ever seemed to bring him close to malfunction were the first years of his marriage. His wife was also in part a stranger at their wedding – they had corresponded, but they had seen each other only five times, and the early years were not happy. 'In my whole life,' Joan Aldrin said, 'I had never been alone, and all of a sudden I found myself in Montgomery, Alabama, with a husband who was off flying all the time and leaving me home. I was very naïve, an only child, and spoiled. He was so terribly inarticulate during the first several years of our marriage, and I was just the opposite. He complained because I was so direct, and I did have this tendency to bull my way through everything, even after we were sent to Bitburg, Germany. I look back at those years as the period in which I was always pregnant and always mad at Buzz for being away.'

They finally were struck down with yellow jaundice in 1959. It came 'from drinking dirty wine in Italy,' diagnosed Joan Aldrin, but jaundice is that infectious disease beyond all other which comes to strong people when they live too long in an environment alien to their will, work with all their power to solve the complexities of that environment, and fail. Nothing in the organizational and logical powers of Aldrin's brain was equipped to comprehend the subtler abdications from reason of a woman's brain. No unfulfilled powerhouse of an actress with a half-career frustrated behind her would be able to sympathize with the emotional indestructability of a man able to perform any task he set himself. They went down into jaundice together. It was conceivably a sign of love – a mating of devils is more likely to result in but one wound.

vi

They are both religious men, yet Armstrong suggests the mystic, whereas Aldrin, with his prodigious will and senses oriented to techno-logical rather than sensuous perception, is nonetheless an elder and trustee of the Webster Presbyterian Church, a formerly 'mechanical man' who will go far as to smuggle consecrated bread and wine aboard the Lem in order to celebrate Communion on the moon, and so is not only a technologue but a high priest, indeed is the pure spiritual ances-tor of that line which runs from Calvin, Luther, Knox and Wesley to Edison, Ford, and IBM's own Watson. So he illumines something in the mystery of the Wasp, gives us purchase on that dichotomy between technology and dogma which inhabits their lives. It is as if ceremony, formal repetitive somber ceremony, has become the communicant between technology and the dream, as if the road back from the machine to the primeval can be accomplished only by transporting the machine-oriented senses into the machine of ceremony. So Aldrin was a traditionalist with a faith that never seemed to alter. Perhaps he did not often use his faith to explore any inner space, but rather to restore emotional depletions. Serving a powerful heart, faith was as predictable as a flow-chart he had designed himself, faith was a perfect proposition.

'I wouldn't,' stated Aldrin, 'classify myself as a fatalist or anything like that. I just think when I'm engaged in one of these things, I'm in no danger at all. It may be a question of faith, a belief that I wasn't brought here to meet with some untimely occurrence ... I would feel worse about not doing the right thing than I would about any danger that is involved ... But why do you do anything? Maybe because you were selected to do it.' And his mother's maiden name, when all was said, had been Marion Moon.

Still, his faith was not merely a support. If Aldrin's sense of ceremony could excite his wife to remark – hearing his voice across space – 'Doesn't he say numbers beautifully?' he had still gone out to march on Palm Sunday of 1968 in downtown Houston in order to pay his respects to Martin Luther King just assassinated. In the political straits of NASA it was a brave act. It could have affected his career. People in his office even called the march a peace parade; little could damage a man more than to be considered sympathetic to peace in Vietnam while working in Houston.

It was apparently part of his philosophy that you must not only use your religion but must make it work as well. With his minister, Reverend Dean Woodruff, the man with whom he had marched in Houston, Aldrin looked to prepare some explanation of the moon trip 'which would have universal appeal ... The symbolism of the flight – of what we were looking for, of what I was interested in – seemed to transcend modern times. I searched for some words, or some symbol to be representative of man's expanding search.' Woodruff took up a thesis of Ernst Cassirer and wrote a paper: *The Myth of Apollo 11: The Effects of the Lunar Landing on the Mythic Dimension of Man.*

Man's capacity to symbolize and to respond to symbols is the central fact of human existence ... The Apollo event will be the kind of occurrence that will reach down to this level ... Science has created a worldwide technical civilization and, as yet, has not given birth to any cultural symbols by which man can live ... We need now a paradigm of the 'experience of the whole.'

The hopes Aldrin contained for this thesis were not necessarily so small. It is natural for Americans to believe that an idea will shake the world if every newspaper prints it the same day on the front page. The

Woodruff-Aldrin manifesto was however relatively unremarked and Aldrin was later to say, 'I was a little disappointed in this.' Still, Aldrin was one astronaut who would not have to wonder why he was on his way to the moon.

vii

But the night will soon be over. Already Apollo has lost signal for the sixth time, reacquired it, lost it, regained it, gone around again and again. It is into the eighth and the ninth revolution about the moon and still the astronauts sleep. It is four, and now five in the morning back in Houston.

Before they wake, let us give a last pass at their psychology by inquiring into the objects they bring to the moon in their Personal Preference Kit (which is a white diddie bag two inches by four inches by eight inches closed by a string). Aldrin has brought first-day philatelic covers, and gold olive-branch pins of peace he has ordered specially, one as a gift for his wife. There are also a few rings and medallions, and a gold bracelet which belonged to his mother Marion Moon. The names of the children and the grandchildren in the family have been attached to the bracelet on little gold discs. Carried along as well has been the small bottle of wine, the bread, and a little Eucharist cup. One salutes the lack of fear in Aldrin that the moon might ever be malign or the powers of ceremony insufficient before it.

And Armstrong? He has taken packages for friends, and for members of NASA with whom he had friendly relations. 'I don't even know what was in them,' he would say later. 'They were wrapped and I stuck them into the bag, then gave them back when we got home.' It is a perfect expression of I Ching. The priest is not to perform the ceremony nor to intervene between the subject and his relation to nature – the priest is there to make it possible, no more. But then Armstrong is not without Oriental bent – a large Buddha stood on a table in his living room and he told a visitor, 'I believe you would find in a Buddhist house that the central beam . . . would be painted red to ward off evil spirits.' For his wife and mother he would bring back no packages from the moon as if objects to which he might here attach personal desire could be affected by strange forces.

And Collins, whom we have ignored out of the cruel reality of his inferior position (snobs may seek to rise not because they love elegance so much, but for fearing inattention more!), what did Collins take? Well, his position was certainly cruel. The others had two white bags, one for the Command Module, one for the Lem, Collins had only a single Personal Preference Kit. He took 'mostly small flags and little medallions of our shoulder patch – a whole bunch of small things for people in the program, people we work with.' There were personal items as well, a college ring, a locket, a cross, but it was the items he would dispense later to others which remain in the mind, all those cuff links, tiepins and crosses. He had the politician's knowledge that politics is planting and a handshake is better than a vote, for a handshake breeds votes in the man you greeted. So a little gift from a large mission is the equal of a hundred handshakes. Collins was already on his way to a new profession – he would join the State Department when this mission was done and seek to rally the young to the merits of the war in Vietnam. It is difficult to conceive of anyone performing such work as a labor of love – it seems more like the beginning of a new career. Still, Collins is the man who said, 'It's been one of the failings of the Space Program . . . that we have been unable to delineate clearly all the reasons why we should go to the moon. I think the key to it is that man loses something if he has the option to go and does not take it.' That was the same Collins who could confess with charm that in the days before World War II when his father, Major General Collins, was commanding a post in Puerto Rico, he used to sit on the stone walls of an old Spanish fort outside his home, and look down on a brothel on the other side. 'They did a thriving business. I used to . . . talk to the girls, and I can remember they used to – oh, I was a rat fink – they used to toss me money to get me to come down and visit them, but I never would. I was scared to death to.' He was also ten years old. But he had learned already that a 'man loses something if he has the option to go and does not take it.' What might be his subtle intellectual consternation when he learned the young rebels he would talk to shared his belief. They thought they should go to Canada to avoid the war in Vietnam. But then all of America except the Indians and the Blacks could share such a belief, for they had once taken the option to come here. Now, not a few hours away, was the event which might yet begin the forging of a new migration.

6

The Ride Down

It is time to look at the Lunar Module. The astronauts have only to remove the overhead hatch cover in their compartment, unscrew the probe and drogue, then go to work on opening the overhead cover of the Lem. When they are done, they can float through a tunnel connecting the two ships. It is about thirty-three inches wide, and probably a little more than three feet long, and when it is open, there is nothing to keep them from taking a small jump off the floor of the Command Module (providing the center couch had been removed and so made a piece of floor available) and in such a weightless environment, if aim is good, they will pass on a trajectory up the six or seven feet to the beginning of the tunnel, then will stream through, fingers steering lightly along the handholds on the walls of the cylinder which form this short passage, before they debouch into the open volume of the Lem. Now, they can either finish the trip on the engine cover of the Lunar Module ascent stage, or with a half gainer proceed another few feet to the floor. So it is a leap from one floor to the other; the two spacecraft are attached head to head like the most unhappy version of Siamese twins.

Once inside the Lem, few would claim they had improved their condition. The habitable volume of the Command Module was two hundred and ten cubic feet, something equivalent to the old army sump hole six feet by six by six enlisted men would dig. If in that arena they were obliged to sleep, sit, stand, eat, even take a step or two, their volume in the Lem was reduced to one hundred and sixty cubic feet. It consisted of a slice of cylindrical barrel not eight feet in diameter and only three and a half feet wide; it had not as much room as one of the cages on a ferris wheel, and if there was a view out of two triangular windows less than two feet high, the walls were covered with an array of circuit breakers, knobs, switches, controls, displays and alarm panels quite the equal of the Command Module. In addition they could not sit and could hardly sleep. It was a spacecraft designed to be flown while standing up, only armrests, cables and restraints to keep them in position at the controls. Since the line of sight of their radio antenna and radar would twist them into some most unusual attitudes, there would be times when though they might believe they were standing up (for their bodies were straight and their feet in Velcro boots were attached to the Velcro nap of the floor), they were in fact moving along in positions which relative to the surface of the moon had them standing on their heads or riding on their backs.

This had all been calculated to save weight. It cost an expenditure of sixty pounds of fuel on the Launch Pad for every pound of spacecraft to go into moon orbit, and five hundred pounds of ground fuel for each pound the Lem would bring to the moon. It suggests that a saving of one hundred pounds (perhaps the weight of two couches) is, by the most conservative measure, a net gain of six thousand pounds of fuel. It used to be that for every dollar spent on a Hollywood movie, eight dollars had to come back at the box office to pay for prints, distribution, publicity, overhead and theater exhibition before a profit was shown. These fuel figures were more impressive by far!

Yet this concept of a Lunar Module had once been considered revolutionary. It had been forced to fight its way against earlier NASA plans which called for a brute rocket able to leave the earth, reach the moon, land and come back again. Such a rocket would have been as large as a battleship. Probably it could not have lifted off the ground with anything less than four times the fuel Apollo-Saturn was to carry. Therefore another solution was proposed. Two Saturns would go up. One would carry a spacecraft, the other a cargo of extra fuel to attach to the spacecraft. The bolstered ship could then proceed to the moon, land, and come back. Some argued it would take three Saturns, not two, for the job. Then, from a group at Langley Research Center, headed by John C. Houbolt, a suggestion was offered which said in effect: 'Why land a yacht on an unprotected beach? – send in a rowboat with an outboard.' But feelings in NASA were delicate in 1960 and 1961, NASA officials were still in the bureaucratic nightmare of embarking on a multibillion dollar adventure which still might produce nothing but flameouts and failure. In such untried situations, the bureaucratic instinct is to look for precedent: there was evidence the big boosters could at least function – there was still no certainty rendezvous could be managed. In the mathematics of celestial orbits, unknowns were always more numerous than equations, and the computers were still too heavy to be installed on board. Aldrin in fact was working these years at MIT on precisely this problem of making rendezvous feasible. No surprise if some NASA men were devoted to the idea of a battleship which could go all the way and never have to rendezvous at all. Yet the real argument proceeded deeper. Men like Von Braun were willing to take the gamble that rendezvous could be accomplished, but he was in favor of Earth Orbit Rendezvous with the two Saturns. When Houbolt's concept was put forward, Von Braun was quick to say, 'That's no good.' The hostility of others was intense. There was no question that in terms of economy of performance, the idea of Lunar Orbit Rendezvous was clearly superior to Earth Orbit Rendezvous, but more than bureaucratic rigidity was at stake. One can assume that a number of the most practical men at NASA did not really believe in 1961 that we would get to the moon by 1969. Still, they were rocket men who were attached to the idea of a continuing rocket program. A complete commitment to Lunar Orbit Rendezvous would show a complete loss if they failed to get to the moon. But Earth Orbit Rendezvous, under the guise of going to the moon, was the perfect program for developing a technology which would certainly result in elaborate space stations orbiting the earth. That offered a powerful appeal to the scientists in the program and to the military minds on detachment from the Pentagon. Going to the moon would be the most monumentally expensive gesture ever made, it would change the dreams of man himself, but it would not put an eye on the activities of the Russians, nor from the point of view of many scientists would a manned moon landing mean nearly as much as a space station for astronomical observations free of the earth's atmosphere. Or a space factory for electronic components best manufactured in a vacuum.

So it may be that Houbolt's struggle to get Lunar Orbit Rendezvous approved was in fact one side of an undeclared war in the minds of NASA officials to decide whether they really wished to commit themselves to the moon. It could be said that when Von Braun changed his mind, and Houbolt's scheme was soon after approved, that the lunar program had then truly begun. For the first time in history, a massive bureaucracy had committed itself to a surrealist adventure, which is to say that the meaning of the proposed act was palpable to everyone, yet nobody could explain its logic.

ii

It was also the first time in history that a manned vehicle had been designed which would never function in the atmosphere. The Lem was the pioneer man-carrying machine of the deep vacuums of space – it was designed to work in nothing but a vacuum, it would have collapsed in an instant if it had ever been obliged to travel unshielded through the air; rather, it had been brought up from earth like an embryo in the womb of the SLA, and was only later connected up to the Command Module and then transported in moon trajectory like a babe at the breast. The image is not likely to please, yet it is hard to conceive that the astronauts would not personify the Lunar Module, not see it as a creature just born, for in a few hours they were going to direct this craft never flown before right down to the moon. Yet they had never operated such a vehicle, only crude simulations of it.

If there was a psychology of machines, and a machine could actually be (even in one part of ten million) a creature, then the Lem was a creature. But whether it was attractive was another matter. It had been designed from the inside, and so was about as ugly as a human body which had shaped itself around the excessive development of a few special organs. Conceive of a man whose only function in life was to win pie-eating contests – what a stomach would he develop, large as a steamer trunk! So the Lem was a craft to carry two men in a minimum of space with the maximum of fuel. It would have guiding systems and communication systems and engines, and since it would come to land first by hovering, then settling, it had four long spiderlike legs rather than wheels. It even came in two connected parts. There was a lower portion containing the motor and fuel for its descent. The upper half or ascent stage, which the astronauts inhabited, would later blast off from the lower or descent stage, and go back to the Command Module, where its drogue would meet the other's probe in rendezvous. The descent stage, which would be left behind on the moon, was more or less symmetrical in appearance since its function was to nest tanks of gas about an engine, allow space for storage and serve as a base for its four extended legs. If one ignored the extended legs its height was ten and a half feet, its diameter fourteen feet. The ascent stage was also fourteen feet wide, and twelve feet high, but it was irregular from every angle, as knotted and bumped and pouched and protruded as a comic-strip drawing of Popeye or Palooka after a fight. There were protrusions everywhere. Four banks of rocket thrusters stuck out like pin cushions on outriggers at four corners except they were not so much corners as prisms or pouches. A cluster of spires and swiveling dishes and elaborate stiff-wired prongs and probes were there as well – VHF, EVA and S-band In-Flight and Steerable Antennae. They projected at all angles like a rooftop view of leaning television aerials, and there was a docking target and a docking window, and a hatch, and headlights, and two triangular windows set on a slant with the top further forward than the bottom. That gave the Lem the look of a demented cat, a spider on LSD, or some new species of king crab.

Indeed the Lunar Module was designed so completely from within that the outside could not really be seen; it was covered with plastic foil tied loosely around its legs and part of its body; it could have passed for some derelict of a machine with a cape drooped over the body and wrappings attached to the feet in lieu of shoes. Conceive that the Lem not only had no streamlining, but its wrappings did not have to be smooth, for it had no atmosphere to pass through, so it only required a thin skin strong enough to keep pressure in and occasional micrometeoroids out. Since the heat of the sun would sizzle it during the moon landing and other periods when Passive Thermal Control would be impossible, the Lem was wrapped with insulation, with black, orange, silver, aluminum, yellow, red, and gold coverings which allowed it to gleam like a scarab in red reflected light, gleam like an old harridan in a Halloween of rags and scraps and every tinsel.

That was its appearance. In function it was as complex as the Command Module with virtually as many systems and subsystems, a plumbers' warehouse of miniaturized equipment for environment control, a communications system with redundancy in S-band and VHF, a Reaction Control System, propulsion systems for the descent engine and ascent engine. There were caution and warning systems, tracking and docking lights, there was radar. But that brings us to Guidance, Navigation, and Control – it brings us to the computer, and the computer was to give a turn or two on the descent to the moon.

iii

The astronauts had not had a full night. In bed by midnight after the long chores of the day before, they were awakened at six in the morning before Loss of Signal on the ninth revolution. When they came around on the tenth, they were busy with breakfast. It was less than six hours to the separation of the Lem from the Command Module, less than eight hours from the beginning of powered descent, but the day had begun as all the others. The Biomeds were read, and data was given on the contingency burn of Trans-Earth Insertion 30. The consumables in the quads were read, the sleep report was given as CDR 5.5, CMP 6.0, LMP 5.0. The morning news was read. It was Sunday on earth and there would be church services around the world.

President Nixon's worship service at the White House is also dedicated to the mission, and our fellow astronaut, Frank Borman, is still in there pitching and will read the passage from Genesis which was read on Apollo 8 last Christmas.

Loss of Signal came again. On the eleventh revolution, Aldrin opened the tunnel and passed into the Lem for the final check-out. When they came back over the hill, the pace of their Lem activities was quick. In the next hour, they adjusted the radio, began to use the call name Eagle, took landmark updates and Digital Auto Pilot data, tested the steerable antenna and the glycol pumps, cycled the Optics off and on, received a time signal or Mark on when the switch to Lem power would occur, tested communications with Columbia, and more. Reference material, instructions, and new data were being dispatched up to the computers on Columbia and Eagle, and performance data was being telemetered down to the ground – the computers were now comparably as busy as the flow of transactions in a bank on the busiest day of the year. Up to Eagle came the gyro torquing angles for copy, up came Verb 42 – the fine alignment for the Inertial Measurement Unit – down went the battery report. The capture latch was checked, and minimum deadband attitudes given, antenna position instructions were forwarded. Now Columbia and Eagle, still connected, but called separately by radio, went over the hill. Loss of Signal. Beginning of the twelfth revolution.

Armstrong was in the Lunar Module now, his pressure garment on. Aldrin was back in Columbia putting on his own space suit. It was only in these stiff and massive white pressure garment assemblies that they would fly down to the moon, equipped for the vacuums of space with the exception of helmet and gloves. The suits were necessary even in the breathable atmosphere of Eagle's cabin, for the thin skin of the Lem was vulnerable to micrometeoroids, and air might blow out any such hole as fast as a leak in a balloon. The helmet and gloves would require but a minute to put on, but the suit took a quarter of an hour at least. In that time they could be dead, dead from one vacuum caused by one random bullet of a micrometeoroid.

Close to four hours of the busiest day they might ever spend in space had already gone by. Pushing every item on their checklist, ticking off each detail of the hundred and then the thousand details they must pass between their fingers or through their minds in the constant mounting adrenalin of these hours, the pages of the transcript become dense with communications back and forth. Humor is at a minimum, impatience growing. Functions are colliding. An overlong exchange between Columbia and Mission Control about a conflict between flight plan and cue card is followed by:

CAPCOM: *Eagle, Houston, could you give us an idea where you are in the activation?*
ALDRIN: *Roger. We're just sitting around waiting for something to do.*

The Reaction Control System was pressurized in the Eagle; the landing gear, stowed all the way in a semiretracted position, now fired its detonator cartridges. The landing gear was deployed. Columbia completed taking marks. Noun 49 passed downlink. Abort Guidance Section figures came up. High-gain antenna was reinstalled. A flood of data pads and computer addresses, state vector loads, minimum deadbands, holds and fire checks. The Lem was ready to undock. Its thrusters had been blooded, its electrical systems tested, its environmental controls, its communications, its data. As it went behind the moon on the thirteenth revolution, it was GO for undocking. The drogue was reinstalled in the tunnel of the Lem, the probe in the Command Module. Eagle closed her hatch. Umbilical connections between Lunar Module and Command Module were disconnected. The twelve ring latches were manually cocked. Hatch on Columbia was closed.

Now by remote electrical release, Collins throws open an extend latch on the probe assembly. The vehicles undock. The tiniest toot on the thrusters pushes them apart. Slowly they drift out from each other, at about the speed of a leaf moving downstream. The Eagle has wings. They will announce it as they come around the moon.

Yet there has been an error in all this work. They have performed the checklist to perfection, they have made no mistakes. Still, a mistake is now buried in the bowels of the computer. It is waiting to be discovered that the checklist, finest product of some of the finest engineers in the land, has an undetectable error which like a disease will come roaring up out of the computer in the next couple of hours as they fire their motor to descend to the moon. The Eagle has a computer pregnant with alarm.

There is no way to avoid the difficulty. We must dip into a disquisition on the property of computers, or we will comprehend nothing of the melodrama which follows.

iv

Consider a list which includes the discovery of the wheel, the extraction of iron from earth, the invention of gunpowder and the printing press, the steam engine, the powder loom, the electric generator and the internal combustion engine. The names of Christ, Mohammed and Buddha; of Augustine and Luther; Darwin, Marx and Freud.

We may as well recognize what is common; it is the power of the man or of the invention to create other inventions, other styles of life, not even conceived in the origin. So a history of cultural revolutions could also include paper money, tobacco, and the manufacture of an inexpensive mirror since each of these was also a vehicle to enable man to become the instrument of his own will.

Now, the digital computer had become one more revolution in the history of culture. Unlike the automobile or the airplane, however, the digital computer was not a machine which would force men to think in new ways about the environment, it was rather an electronic mode of calculating which might yet change the nature of thought itself: the digital computer had the power to run man's mind through an accelerator which could catapult him out to the universe or explode the remains of his mind on earth; perhaps for this reason, Aquarius could never pass through a room containing a bank of computers without a moment of woe, as if he had just walked through an amphitheater where some species of higher tapeworm was quietly ingesting the vitals of God.

The digital computer was a diabolical machine, or the greatest instrument ever handed to man, but it could hardly be both for it was constructed on the implicit promise that all phenomena might yet be capable of capture by statistics. The digital computer was based upon millions of switches, all primed individually to say one or nothing, yes or no. If the switch allowed a current to pass, the answer was one, or yes; absence of current was zero. 1 and 0 became the simple building blocks upon which numbers could be recreated. The entire decimal system was replaced by two numbers, 1 and 0, rather than 0, 1, 2, 3, 4, 5, 6, 7, 8, 9. Thus the figure of 111 in this new system stood for 7 because it was a way of writing $2^2+2^1+2^0$ or $4+2+1$. And 8 in turn was 1000 or $2^3+0+0+0$. The binary system was then obviously more unwieldy than the decimal system, much more unwieldy – a number like 437 could only be written as 110110101 – but since every number, no matter how large, became some combination of 1 and 0 and since one or zero could always be indicated by a pulse of current or its absence, numbers could therefore be transmitted in a stream of irregular electrical impulses like Morse Code. Since electricity traveled at the speed of light, so an extraordinary number of arithmetical computations could be manipulated at just that speed of light through millions of switches. And a world of wonders opened to that speed. Those long combinations of 1 and o were extraordinarily adaptable to operations of addition, subtraction, multiplication, and division – huge amounts of such work could be performed in the giant computers.

Where once these switches had been vacuum tubes, much like the tubes in radios, then had become solid-state transistors, and would eventually be molecules in a crystal lattice, the computers in which they were installed had reduced in size from buildings the size of barns to units the size of a refrigerator and then a cosmetic box, and would yet be the size of a cigarette lighter. If the functions of the digital computer were limited to solving only those problems which could be set in numerical form or which could be laid out in steps of logical analysis built upon Yes, No, And or Or, it was nonetheless amazing how many problems of the most extraordinary variety could be converted to the point where one was no longer certain whether the extremities of the problem were or were not amenable to solution, for as the employment of the digital computer became pervasive in business, industry, government, the Army, the Navy, the Air Force, the universities and in the mass media, and as profits and efficiency came to some industries by using computers and irritating losses to others who thereby sought more computers to solve the problems the first computers had created, as hundreds of thousands of computer operators formed a new profession, as computers themselves proliferated into a race for more and more exceptional properties of operation and powers of dominion over activities formerly virgin to the computer, so a breakthrough occurred. The computer became the new frontier, a frontier of airconditioned windowless rooms with fluorescent panels in the ceiling and electronic whirrings and gurglings. The men who rode the limits of this range were computer programmers who wore horn-rimmed glasses and shirts of synthetic fabric; the horses were computers with a natural tendency to convert the nature of all human problems. If a problem could be programmed, that is, if it could be broken down into a form which could be inserted into the computer, then the problem could be solved. Sometimes it could only be solved to a degree. Translations from foreign languages were lamentable, computers which played chess were only average players and the poem which follows – well, it was written by a computer.

CHILDREN

Sob suddenly, the bongos are moving.
Or could we find that tall child?
And dividing honestly was like praying badly,
And while the boy is obese, all blast could climb.
First you become oblong.
To weep is unctuous, to move is poor.

Perfect plastic poem, for it left the same aesthetic satisfaction in the heart as the smell of vinyl.

The subtle imbalance in the computer was that it had all the power to push man out of a reasonable relation to his environment and into an accelerated relation. It had created plastic brainpower ready to flow in the direction of every problem even as plastic sometimes gives signs of pushing toward every manifestation of flesh and form. The digital computer was crude in comparison to the brain. The mind had ten billion neurons each equal to a switch in a computer. The mind could store two hundred billion billion bits of information, the largest computers could hold but a ten-thousandth of that. The mind worked on a tenth of a volt of electricity – its economy of power was ten million times less wasteful than the computer.

But the computer worked in nanoseconds and picoseconds, billionths and trillionths of a second, the computer was one hundred thousand times faster switch for switch than the brain. So the computer could perform calculations in a minute which would occupy the brain for ten weeks, or indeed for almost a year if one assumed the brain was to work no more than forty hours a week. A minute to a year. There was the strength of the computer, even greater than the horsepower of Saturn V compared to a single man, yes, the digital computer was one hundred thousand times more powerful than the mind, even if it achieved that power at ten million times the expenditure of electrical energy. Such costs meant nothing yet for they were still insignificant in relation to all the demands for electrical power, but then so were the rates of pollution low when industry began.

Yet the problem was probably not here, for the computers would become more efficient. As they grew smaller so, relatively, would their speed increase and their ability to store information. The problem was that the new brainpower was plastic, it would push to fill vacuums, press on to simulate what had hitherto been out of the range of simulation, occupy problems whose outer margins would be lost as the center was sucked into the binary system. All this excess brainpower! It could end by having as little to do with the real needs or natural balance of the earth as the delivery of ten automobiles to every citizen of the earth. Once delivered, people might of course use them, if only to crash them, but it would make for a most peculiar world – everybody living with ten cars – the structures of homes and streets would turn inside out, and the enterprising would look into the possibilities of twenty cars. If there had been a period in the history of culture when all human effort gave promise of being translated ultimately into money, now the time was conceivably approaching when the money of moneys would be here, the quintessence of number, the world reclassified by combinations of 1 and 0, nothing more. So had Leibnitz once attempted – in his lust to create a symbolic logic – to show that God was 1 and had created the world out of 0. Now another accounting was near. For if one could eventually define all existence by variations of 1 and 0, how easy would become the next step – to dissolve the world. Down all the steps of entropy we would flow.

If one were, however, to condemn the computer for this nightmare of a future, condemn it today for its automation, its new methods in banking, condemn its use by the Bar Association and the Stock Exchange and Boeing and the Bureau of Mines, the Cancer Society for research, condemn Control Data Corporation and General Dynamics, General Motors and General Electric, condemn computer analysis of the Dead Sea Scrolls, condemn its employment in the Department of Commerce and the Department of Defense, condemn Douglas Aircraft and Dow Chemical, du Pont and the election forecasts, condemn the FAA and freight trains shunted by program in and out of freight yards, condemn General Precision and Goodrich, Goodyear and Charles Goren, Hoffman Electronics and Hughes Aircraft, IBM 704 and IBM 1401, condemn the Industrial Advertising Research Institute and the Internal Revenue, International Air Transport and Johnson's Wax, Kresge Eye Institute and Lincoln Laboratory, the Public Library and Lockheed Aircraft, McGraw Hill and Merrill Lynch Pierce Fenner and Smith, Minneapolis Honeywell, Monsanto, NATO and the National Cash Register – the names go by like sounds in a coffeehouse poem – the U.S. Navy and the National Bureau of Standards, Ohio State, Patent Office, Philco, Phillips Petroleum, Radcliffe, the Rand Corporation, Rockefeller Institute and the Sara Lee Bakeries, the Signal Corps and Social Security, Southern Methodist and Sun Oil, TWA and UNESCO, Union Carbide, USC, the Upjohn Company, Wall Street, Westinghouse – does one condemn them for using ADAM and BINAC and BRAINIAC, CALCULO and CLASS, ENIAC, ERMA, ILLIAC, JOHNNIAC, LARC and MANIAC? Does one forbid MIPS, MOBIDIC and MUSE, RAMAC, RAYDAC, RECOMP and SAGE, or not permit STRETCH, UNIVAC, or VIDIAC, wait there is still DIDAC and EURATOM, FIELDATA, FINER and HIPO, HAYSTAQ and IRE, MEDLARS, BIAX and MIND, NADGE, NANWEP and NORAD, PLATO, TASCON and ARTOC. There is AUTOPROMPT, AUTOTAG, APT, AID and AIEE but maybe the point has been made. It will come as no surprise that without that detumescence of development in computers which saturated the electronics of the Fifties and the Sixties, without the five IBM 360/75 computers and the IBM 1460's on the floor of the Real-Time Computer Complex at Mission Control Center there would have been no trip to the moon, and indeed no Instrument Unit on Apollo-Saturn. Rockets would still be punching holes in clouds, for without a digital computer no trajectories could be calculated. The simple work of addition, subtraction, multiplication and division to calculate a trip to Venus would take one man eight hundred years. Yet now there were supercomputers to calculate the journey in thirty seconds. That plastic brainpower could turn over every straw in the haystack to find the needle. The rush to extermination or apocalypse was being accelerated by every computer on earth.

v

Of course, the computer on the Lem was nowhere so vast. The brain capacity of computers designed in the same period was roughly proportional to their volume, and this computer, identical to the one on the Command Module, was two feet long, one foot high, and six inches wide. It weighed seventy pounds and consumed seventy watts and had a vocabulary of 38,916 sixteen-bit words, a way of stating that the Guidance Computer could take information in the form of ones and zeros in any variety up to sixteen units in a row. That made one word among 38,916 words. Hordes of navigation data and engine data had thus been laid in for reference; also, a series of programs which had been designed to keep the Lem on its trajectory by issuing precise engine and thruster commands. Another of its functions was to keep the Inertial Measurement Unit or gyroscope in constant alignment with the stars. It was capable to a small degree of correcting malfunctions in its own system, and it could also compute navigation information and display it to the astronauts, or answer their questions, through a combination of Display and Keyboard called a DSKY which had lighted numbers in a small frosted window to show answers, offer information, or blink out alarms. In turn, the astronauts communicated with the computer through keys in the DSKY which they punched like adding machine tabs to begin computer actions. The keys went from the number 0 to 9, there was a key for plus and minus, and other keys marked VERB, NOUN, CLEAR, PRO, KEY REL, ENTR and RSET.

The Guidance Computer was the major part of the Primary Guidance and Navigation Section (whose acronym was PGNS) and so was sometimes called Pings to distinguish it from AGS, the Abort Guidance Section, a smaller computer and inertial system ready to take over if Pings failed. Ags could hardly fly the mission, it was there for backup in case the Guidance Computer failed and abort was necessary to return the Lem to the Command Module. So the success of the landing on the moon would depend on Pings' speed in processing the variety of information it would be absorbing, and its ability to issue commands as the Lem proceeded down. Coming into the maw of its calculations would be questions the astronauts put in on the DSKY, and programs they ordered it to initiate; coming in from other sources and sensors would be all the evidence of altitude change, velocity change, and fuel consumption. Since the rendezvous radar would be sending and receiving signals from the separated Command Module, so too would rendezvous radar be offering data to Pings on the change of distance and the relative position of the two ships. In addition, the landing radar would also be entering its evidence in the Guidance Computer, and to increase confusion the landing radar was bound to disagree with Pings, for it would be bouncing signals off the literal moon ground, rather than estimating like Pings where the ground ought to be on the basis of past observations and calculations. Therefore the computer would be having to make last-minute adjustments at a great rate. But here is Aldrin on the matter:

Suppose the computer tells us we're at thirty-two thousand. We give that to the radar and the radar comes back and says the computer is a liar; we're at twenty-eight thousand. The computer goes into a sulk and says it will split the difference; call it thirty thousand. The radar takes that and says no you don't, and besides you're down to twenty-seven now. We can keep narrowing the difference during the automatic part of the descent, until we get down to about five hundred feet, and at some point around about there we have to take over manually. That computer isn't going to dodge a boulder for you.

The Pings obviously would have enough to do. If it had almost thirty-nine thousand words, most of them were already fixed. Only two thousand could be applied to a transient problem. In the welter of functions which might descend upon it, beleaguered as a short-order cook when he has lost a series of orders at the rush hour, the Pings had the power to give priority to particular functions, set other functions aside, or even cease calculations for a moment if an error were made, or a discrepancy could not be accounted for, or information was simply coming in faster than it could be processed. The Pings was presumably an effective computer for its size, but weight was crucial, and so it was squeezed in its tasks. It bore the same relation to the computers at Mission Control that a dollar bill bears to a thousand dollar bill. As limited, relatively, were the services it could purchase. Of course the descent from beginning to end could have been computed on the ground from the tracking of the huge antennae in the Manned Space Flight Network, and in fact it was, but to be dependent on Mission Control for the landing was too dangerous: the telemetered data could always fail. Static, or any other blocking of the signal, could cut out crucial instants of data. Often did. Besides, the ground was a quarter of a million miles away, and information therefore took more than a second to arrive; if it took even an instant to process and send back, three seconds could elapse in the interval, a catastrophe in a situation where the controls on the Lem were rapidly going wrong. So the limited seventy-pound Guidance Computer with its 38,916 words was going to be the chief pilot for the first part of the trip down, its indispensable function to calculate the ratio between the fuel which still remained and the miles yet to descend. Without a functioning computer those rates were next to impossible for a pilot to estimate, since the Lem would be traveling at several thousand miles an hour when it began its descent and settling at only several feet a second by the end. And it must be remembered that one could not fly the Lem down, which is to say one did not have the fuel to take it through circles, allow it to descend, then rise again, then take a long turn. There would be some maneuvering available at the end, but for the majority of the descent, few were the adjustments to be made with the trajectory. The Lem was still riding on the effects of the momentum which had first brought it to the moon, riding like a ball thrown into a canyon. One might nudge it a degree or two from side to side, one might brake its fall a little more or less, but essentially its momentum was its capital and it was traveling on a braking ballistic curve toward the moon ground – it was not going to wheel and circle and soar and turn and negotiate a descent – not in a lunar vacuum. So little changes of attitude or braking thrust were large in importance, and a computer could estimate their effect on fuel far more rapidly than a man. Yet, given its limited capacity and variety of tasks, it was obvious that one burden of preparation at NASA was to make certain Pings never got into the condition of a quarterback to whom everybody is talking in the huddle when the time-outs are used up and the clock is running into the end of the game.

Who could be more aware of such difficulties than the flight controllers in the Mission Control Room? The functions of a computer were always in as much danger of going awry as society is in danger of some final collapse into crime; experts and chains of experts were forever at work on electronic species of crime detection – they were vigilant to search for sneak circuits in the computer, circuits which if exercised at the same time would interreact in ways no one had foreseen. There were millions of switches and billions of permutations: complex electrical circuits were like patients with modern viruses – after prescribing several medicines, the doctor could never know which, if any, had triggered the cure, and just what in fact the patient had suffered from. As medicine had grown more complex, so the defined edges of all diseases had frayed and shaded into one another. Something like that was forever in danger of happening with computers as their tasks grew more complex and the nature of their response came closer to Webster's definition of thought: 'to form in the mind, to exercise judgment.' As computers developed mysterious malfunctions, computer detectives came into existence with more and more developed instincts for where sneak circuits might occur, where functions could intertwine and paralysis crop up in whole areas of an operation. There was a checklist the astronauts had to go through with every instrument on the Lem, and for the Pings the checklist was detailed. It engaged dozens of switches which produced hundreds of combinations of circuits which in turn offered thousands of possibilities for sneak circuits. For months, Staff Support rooms and every corporation associated with Guidance Control had worked on the anticipation of where and how sneak circuits might occur. No one, however, could be certain – to anticipate every whim of a computer was equal to foreseeing the steps of a virgin whose heart was nymphomaniac – electricity could not run through a multiplicity of circuits without creating a wake of electromagnetisms and interferences whose results were occasionally so bizarre a flight controller could be known to give assent, with unhappy expressions on his face, to the possibility that a computer operator charged with psychic tension on an extraordinary day could also have his effect on the malfunction of the computer. At any rate, the checklist, product of months of work and months of preventive detection, was gone through by the astronauts while circling the moon, and after every item was fulfilled correctly, every switch properly thrown, a sneak circuit was still not avoided. In the no-man's-land of electrical hegemony, sneak circuits resided at the very edge of thought. 'If we had been supersmart,' said the Flight Director Gene Kranz on another day, 'we could have picked up the possibility.'

Collins had said, 'The most dangerous items are the ones we've overlooked.'

Yes, it was not possible to anticipate everything. Not by time, not by cost, not even by space. More than sneak circuits were sitting upon the descent. It was for example impossible to use a real rendezvous radar in the simulator, for how could one maintain a real situation for hours on earth in which two moving vehicles might be anywhere from two feet apart to two thousand miles apart and moving with velocities which varied from inches a second to thousands of miles an hour, no, a project close to the size of the Space Program itself might be necessary to create a real equivalent of rendezvous radar, and yet it was only necessary to make certain that the amount of data coming in by rendezvous radar and landing radar would not overload the Guidance Computer. Therefore, a conceivable set of received signals was written up in electrical code and piped as rendezvous and landing radar data into the computer during these simulations. No overload occurred. Not in the simulations.

Still, there was hardly a sense of security at Mission Control. Decisions would have to be taken on the floor in the course of the descent. Order after order would have to be given to the astronauts on whether to continue with the mission, or abort. Each engineer working a console on the floor of the Mission Operations Control Room was an expert on the limits which could be negotiated by elaborate equipment if things began to go wrong, but the decision to say GO or NO GO might have to be taken in some arena of crisis where the answer would not be clear. So each man on that floor knew he could enter a stricken instant, a cauldron of adrenalin, a failure of nerve which could lay a shadow upon all the hours of his life: an order to abort the mission which later proved to be unnecessary, or an injunction to go ahead which resulted in death would have to leave an isolation of the soul. Suicide could be the neighbor at one's elbow for many a year. No wonder technicians at NASA so often had hands clammy to the touch.

Therefore, the man who would be Flight Director during Eagle's descent to the moon, Gene Kranz, had taken his flight controller's crew through a series of planning sessions the month before, a field seminar which could well have been termed the Engineering of Emergency Situations, for Flight Control proceeded to trace out the connotations of every alert and every alarm the computer could show. How quickly, they inquired, would a particular situation deteriorate into disaster after the first alarm was sounded? What might be the symptoms of each deterioration? what were the remedies? what the partial remedies? for which functions could Mission Control substitute for Pings, given those few seconds of lag? which were the functions Mission Control could not support? And then – no small matter – into which consoles, and before whose desks would these alarms terminate? Who would be required to make each final decision on the floor of the room? What would be the feasible criteria to encourage one to go ahead through a minefield of alarms, yes, under which conditions could the torpedoes be damned? It was more intricate to answer than to ask. Not all factors were known. Some variables would have to be estimated, some guesses taken. The Lem would be up at fifty thousand feet when the descent began. From that distance a pilot could hardly fly it to the ground, for he could not compute the fuel consumptions quickly enough, and the Lem was going to burn some large part of eighteen thousand pounds of fuel on the last leg down. So a pilot could probably not take it in from thirty thousand feet, nor from twenty nor ten, probably not even from ten. At NASA the official guideline was two thousand feet. If the computer ceased to function above two thousand feet, the directive was to abort. Kranz had a private estimate of seven thousand feet. Tall, rough, hard-driving, looking like one of those lean hard-running quarterbacks who play in Southwest conferences, he was built for touchdown, he was looking to go all the way: these sessions were geared to find modes to bring Eagle in if necessary on half a computer, and yet bring her down safely so that she did not find herself one hundred feet in the air with no fuel left, no time to abort, and a crash to the moon ground, first lunar explorers dead or, worse, left alive on a craft too damaged ever to ascend again. At Mission Control they must all have felt as if they were breathing through oxygen masks when the hour of descent approached.

vi

'How does it look?' asked the Capcom.

'The Eagle has wings.'

Eagle came back over the hill an intimate distance apart from Columbia. Pads and updates vibrated on the radio waves while the moon passed below. Each ship made little adjustments for trim, two spacecraft passing sixty-odd miles above the craters of the moon.

COLUMBIA: *We're really stabilized, Neil. I haven't fired a thruster in five minutes . . .*
I think you've got a fine-looking flying machine there, Eagle, despite the fact you're
upside down.

ARMSTRONG: *Somebody's upside down.*

From the view of the Command Module, the Lem was floating with its legs in the air. Collins took a photograph here. The Lem swims toward us out of some darkness of space, a tropical fish with extraordinary red-gold antennae and a lead-gray skin to its body. The wrapped pads on its feet shine like gold in the sun. 'You guys take care,' said Collins. By radio across the hundred feet of gap, Armstrong said, 'See you later,' and Columbia gave a burp to its motors and pulled out a few feet a second until it was almost a quarter of a mile in front.

'Going right down U.S. 1, Mike,' said Armstrong. An earlier flight of astronauts had given this name to the route, and some shade of the loneliness of driving a highway at night must have been in the voice, for they were onstage, the curtain had at last gone up – there was a glare to the footlights. Into that glare they would walk. Only it must have been more like some dream of theater within theater, as if to step before an audience would transport one to still another theater, the stage was dividing, the walls turned out, they were on a new stage, an entrance further within – some sense of their passage into an isolation within the isolation must have come upon them in the trip up from earth, and now alone in the Lem going down Highway One; in another half-revolution they would begin to descend, another stage would appear, the footlights would shine upon the floor of the moon. Was there time in all the rapt transmissions of future data – lunar surface data pads already on the way up – was there time to taste the dimensions of their new anxiety, its pleasures of risk, its throttled fear like the sensuous tremorings of a fall in a dream? Through aisles of quiet fear the psyche would descend.

But nervous they were. Nervous they all were, Eagle, Columbia, Capcom, and the controllers of the flight. The landing was now two hours off, and Collins called down that he had unexplained roll thruster activity. 'I may have bumped the hand control.' A little later Capcom was stumbling on instructions. 'Give us a mark when you're at seven miles – I mean seven-tenths of a mile.' Collins started to call 'Houston,

Apollo,' and quickly corrected to 'Houston, Columbia.' They went over the hill, the Loss of Signal occurred. The flight controllers in Mission Operations stood around in little groups. Now Eagle was firing her thrusters on the far side. Her bottom flipped forward like a runner sliding spikes first, she ignited her descent motor and let it burn for a half-minute – 29.8 seconds. That braking of her momentum would slow her into a smaller swoop about the moon. Now her orbit would be no longer a rough circle but an ellipse fifty-seven miles by eight and a half miles above the surface, and when she swung down to the lowest reach of her orbit, some fifty thousand feet from the Sea of Tranquility beneath, a decision would be made; GO or NO GO. Either she would fire her motors again to slow her speed still further and thereby begin that descent where her motors continued to burn until her legs touched the ground, or if something seemed amiss she would pass, and motors not fired, would fling out into another orbit fifty-seven odd miles by eight and a half, perhaps to descend next time around.

The viewing rooms at Mission Control had begun to fill. Every notable and high official in NASA was now there, Dr Thomas Paine and Rocco Petrone, Von Braun and Debus, Chris Kraft, Sam Phillips, Seaman and Low, John Houbolt. There was George Mueller and Dr Gilruth and John Glenn, Deke Slayton, and Cernan, Conrad, McDivitt, and Lovell, and others, and then others. Name an astronaut. He was there. The high society of NASA was a group as closed to superficial penetration as a guild of Dutch burghers in the Seventeenth Century – no one but the men in that room would ever begin to know the novels and dramas of conflict, the games of loyalty, and what captures and frustrations of power had played back and forth among these men in the last ten years – it was another of the great novels of the world which would never be written. And was the world a little more polluted for that?

Since they were also a most sophisticated audience, they were now to go through a half hour of excruciating theater, that high theater of symbolic languages and masks and incantations where no word is familiar to the uninitiate and the motions of the actors are communicated from behind a screen. An observer unfamiliar with the technical terms of the subject or stranger to the small reserved gestures of the flight controllers on the floor would have gathered nothing of the dramas developing – merely the tension which constricted the air of the room.

There would have been enormous tension in any case – ten years of work would now be concentrated in an hour – but as Eagle came over the hill for the last time, gliding down in its reduced orbit to the point of decision at fifty thousand feet, so the radio began to give trouble:

CAPCOM: *Columbia, Houston. We've lost all data with Eagle. Please ask him to re-acquire to high gain. Over.*

COLUMBIA: *Eagle, this is Columbia. Houston would like you to reacquire on the high-gain antenna.*

COLUMBIA: *Eagle, did you copy Columbia?*

CAPCOM: *Eagle, Houston. Did you call?*

The Capcom was Charley Duke, the man who had been Capcom during the lunar rendezvous of Apollo 10, a thirty-three-year-old astronaut who had been accepted for the program as late as the spring of '66. Armstrong had asked him to be Capcom for Apollo 11. 'I would have liked to say,' Duke later remarked, 'that I was on a crew and wouldn't have time to do it, but I wasn't on a crew.' So Duke was the Capcom even if he had never been up in space.

Later he would talk about the moment their radio went bad.

It always happens that when we have the critical revolution or the critical pass, we have lousy communications. It just seems like that's our luck. The data kept dropping out. I said to myself, 'Oh, no, here we go again,' because we had a mission rule that said we needed adequate communications and data from the spacecraft before we would commit to powered descent.

Yet Eagle continued. The decision to land would not be taken until they were at fifty thousand feet. If the radio was still bad they could take another orbit. So they went on. Sometimes data came, sometimes it was out again. One can hardly divine the tension from their dialogue:

COLUMBIA: *Eagle this is Columbia. Houston lost you again. They're requesting another try at the high-gain.*

CAPCOM: *Eagle, Houston. We have you now. How do you read? Over.*

ALDRIN: *Loud and clear . . . I know what the problem was there. It just started oscillating around in yaw.*

CAPCOM: *Roger, we'll work on it.*

PUBLIC AFFAIRS OFFICER: *Aldrin is referring to the Lem steerable antenna. That comment about the oscillations.*

The steerable antenna, sensitive as a finger wet for the wind, the steerable antenna charged, for all one knew, with some unknown charisma, had been oscillating, and no one was about to be certain just why.

ALDRIN: *Did you copy the star – I mean the sun check, Charlie?*

CAPCOM: *That's affirmative. We did, Buzz. Out . . .*

CAPCOM: *Eagle, Houston. We recommend if you yaw 10 right, it will help us on the high-gain signal strength. Over.*

PUBLIC AFFAIRS OFFICER: *Coming up on five minutes to ignition. Gene Kranz getting a GO-NO GO for descent.*

CAPCOM: *Eagle, Houston. If you read, you're a GO for powered descent. Over.*

COLUMBIA: *Eagle, this is Columbia. They just gave you a Go for powered descent.*

CAPCOM: *Columbia, Houston. We've lost them on the high gain again. Would you please – we recommend they yaw right ten degrees and reacquire.*

COLUMBIA: *Eagle, this is Columbia. You're a GO for a PDI and they recommend you yaw right ten degrees, and try the high gain again.*

COLUMBIA: *Eagle, you read Columbia?*

ALDRIN: *Roger, read you.*

COLUMBIA: *Okay.*

CAPCOM: *Eagle, Houston. We read you now. You're GO for PDI. Over.*

ALDRIN: *Roger. Understand. AELD control circuit breaker. Second Gimbal AC, closed.*

vii

Good communication was reacquired just three and a half minutes before ignition. Mission Control could take air into their lungs again. It is the first cliché of tense activity. We do not breathe, as if we are afraid to alter the benevolent dispositions of the universe by the evil emanations of our heart. Or is it rather the thought of freezing one's existence at the instant so that everything good in the heart can be deposited to the credit of the protagonist we watch? The questions are without answer, but then as equally are many serious questions about the curious functioning of a radio. They could talk – Eagle to Columbia – from as little as ten feet away; they could call across a quarter of a million miles of space. Either way, their communications could fail, or stagger, or crackle into a caterwauling of static – some irritated zone of universe always between. When one abstracted one's knowledge of radio the knowledge ended in one final abstraction: All solids, liquids and gases could be reduced to structured and structureless arrangements of molecules, and radio seemed to be the communication from one kind of activated structure – sometimes no more than a crystal which was scratched – to another structure, and the message, or the resonance, or the wave went out in all directions through any medium of fluids or gases or near-voids of structureless molecules. It suggested that it was in the nature of structures to address each other. It was not as if a transmitter shot a direct communication on a line to a unique spot, or on direct lines to millions of receivers, it was rather that a particular message went out to be added to all the other messages resonating in the atmosphere and – however it happened – in the vacuums of space as well, and so every structure on earth, and in space, was forever being passed through myriads of invisible messages which pervaded atmosphere and space, any one of them to be picked up if the equipment of the structure were appropriate to the task.

If one took into account that besides radio, and all other electromagnetic transmissions, there were also transmissions of nature to be recognized in the antenna and signal-makers of insects, and half-recognized in the whiskers of cats, the homing instincts of birds, the wheeling of schools of fish, the God-knows-what of bacteria and viruses adapting at high speed to antibiotics which would kill them, if one stretches credulity so far as to permit some astral plane or marketplace in sleep where the souls of the living and/or the dead could speak well, what a devil of a soup of numberless messages careening in all directions of the universe, what a bath of meanings never glimpsed in every molecule of breath to pass each tree in the forest of the nostrils of the nose, what a small glimpse in the static of all the lost messages which snarl at the passage of favored communications – the wonder is that radio works at all, and not that there are other messages transmitted by other means which are sufficiently close in kind to distort reception.

Any man who lived with a radio was an unspoken student of the psychology of machines and so was likely to believe – even if he would confess it only on a torture rack – that coming so near to the moon, that incomparable dish antenna of the stars, five-billion-year-old instrument of universes back of universes, who was to think the radio would not act up, who was ready to swear his life was so clean that if he were a NASA notable, seated in that room at Mission Control, he did not have to hold his breath, for space held no curses, and he had violated no taboo nor intervening message? They held their breath. The moon was coming ever nearer, the power descent was begun, from eight and a half miles up it was begun, and who had not thoughts at that instant of the inexplicable? Did Collins in his Command Module now advancing at greater and greater distance ahead of Eagle, now off alone on thirty hours of continuing orbit before he would ever see them again, if ever he did – did Collins think of the night before the launch when a violent thunderstorm had struck the plains south of Houston and an old oak on his front lawn had been uprooted? Was it lightning or deluges of rain? Good cheer for Collins that he had no savage mind. And Cernan at Mission Control or Armstrong in the Eagle – did they think of that mysterious and still unexplained moment when the Lunar Module of Apollo 10, just thirteen miles above the moon, went – against all the controls of its systems – automatic, when it took off on its own flight plan, and Cernan, surprised to the seat of his meat, gave vent to a curse all audible over the mass media of America. 'Son of a bitch,' Cernan had cried out. And indeed the Lem had had its moment of acting like a child of passion. Or did Armstrong think of the night before, when the Lem had tried to point to the moon? Boxed in their bulky pressure suits, tied in and swaddled like Eskimo children in baskets, all moves bulky, always in fear of rapping a bank of switches with the insensitive surface of their suits – 'You're so clumsy and there's so much force required to move inside the suit,' Aldrin had said, 'that everything is WHAM! I could bump right into you and maybe I wouldn't know it' – constrained in vision, they began their powered descent. The motors of the descent stage were fired. Once again they braked. Once again the reduction of their speed began to bring them down from high velocity to low. Now they came below the orbital parameter of fifty thousand feet. They were committed. They would land, or they would crash, or they would abort and return to Columbia, but they could not try it again.

It is worth a moment to consider the complexities of the descent. The Lem, we recall, had started at a speed of thirty-six hundred miles an hour when it left the Command Module and would come hovering to the lunar soil at three feet a second or less. Its first burn had braked its speed enough to reduce its orbit, but this was relatively a small reduction. Now at fifty thousand feet it would go from approximately three thousand and more miles an hour down to six hundred miles an hour at 13,500 feet, a steady forceful braking of the motors for two hundred and fifty miles. Conceive of them on this long brake, two hundred and fifty miles with their rocket motors burning in front of them in order to take their speed down to no more than the speed of a jet plane, and that is but the beginning. They have lost but five miles of altitude in this period of braking – their route has been near to parallel to the curve of the moon – they will descend yet another half-mile and cover another ten miles before their horizontal speed is relevant to the speeds of an automobile, sixty miles an hour, fifty miles an hour, there, slowed, ready for the last descent, they are at high gate, seven thousand feet, not a mile and a half above the moon. Yet all this while, for all of this trip, from a point forty-five thousand feet up, down to a high gate at seven thousand feet, there in their bulky-wham suits, they have been riding on their backs, feet forward, eyes looking up through their windows, so that they see nothing of the moon ground, but instead are staring up at the earth. What a curious position for descent. But the landing radar is on the other side of the Lem (it would require a huge outrigger to install it in a position more comfortable for them) and the landing radar is now their eyes as it bounces signals off the moon and returns them estimates of altitude. So they will ride from forty-five thousand feet to seven thousand feet flat on their backs, unable even to see the moon. Yet there has been additional cause for worry already. At the beginning of powered descent, at fifty thousand feet, before they turned over, they had been lying face down, looking out backward on the Sea of Tranquility as it receded through their windows, and their first mark of orientation, the crater Maskelyne W, did not appear on time and in the place where it should be. A fatefully long second passed. A second at three thousand miles an hour is nearly a mile, another second passed. Were they lost? Maskelyne W came under the window two full seconds and then a little more late. They were already two miles off the mark. At forty-five thousand feet, motor burning all the way in front of them, the Lem rolled them over on their backs. The moon was now invisible, the earth above their window – their hands reached up to the instrument panel overhead, they floated, held in position by cables and restraints for there were, we may remember, no couches in the Lem. How must the earth have looked in the quiet anxiety of the moment. Their toes were leading them to the moon, their eyes were looking at the earth – would they ever see it again, or was this the hour of signing off? So descended the Lem, weird unwieldy flying machine, vehicle on stilts and never before landed, craft with a range of shifting velocities more than comparable to the difference from a racing car down to an amphibious duck, a vehicle with huge variations in speed and handling as it slowed, a vehicle to be flown for the first time in the rapidly changing field of gravity of the moon, going from weightlessness to one-sixth gravity, and one-sixth gravity had never been experienced before in anything but the crudest simulations, and mascons beneath, their location unknown, their effect on moon gravity considerable, angles of vision altering all the time and never near to perfect, the weight of the vehicle reducing drastically as the fuel was consumed, and with it all, the computer guiding them, allowing them to feel all the confidence a one-eyed man can put in a blind man going down a dark alley, and when, at the moment they would take over themselves to fly it manually, a range of choices already tried in simulation but never in reality would be open between full manual and full computer. Armstrong could, if he wished, control the attitude of the Lem and its forward progress, but allow the computer to manage the rate of descent. The only difficulty was that the hand controls of the Lem were curious indeed. To his right was a short thick red pistol grip like a small club of a joy stick, and it could be cocked left and right, forward and back, or be turned like a screwdriver to translate separate commands for roll, pitch, and yaw – it had even a push-to-talk switch, and this control (Attitude Controller Assembly) by pitching the Lem forward or back, could serve to increase or retard the forward speed once the ship was slowed for the final descent; it could also twist her, sidle her, waddle her, or cant her into new positions for skimming over boulders. What a swivel, what phallus!

At the left hand was a control which altered the speed of descent, a Thrust-Translater Controller, a glorified toggle switch which changed the rate of descent one foot per second every time one clicked it. Armed with this set of instruments, they would come in over ground never seen closely before, in an unfamiliar field of gravity, flying a ship which had never been landed, there to come down on four legs rather than wheels, all the while directing their vehicle with two hand controls as difficult of manipulation as some sophisticated version of patting the head with one hand while rubbing a circle on the belly with the other. All of this to be done in the bulky-wham suit.

It was not easy. No question of that. Many of the simulations on the ground had ended in mock crashes or mock aborts. Not all problems were solvable on the spot. Only Armstrong in fact was trained to take the Lem all the way in, and he had the memory of smashing the LLRV. But let us listen to an account of the difficulties by Aldrin.

Neil will take it down. He has the controls on his side. I don't have them on my side. There's a throttle on my side, but I don't have this rate of descent control, and I don't have this redesignation capability . . . But during the actual landing there is a fairly neat division of labor. Neil will be looking more and more out the window with his hand on that stick. He's not able to look much at the displays on board. This is where we have to have a finely tuned team-work, so that Neil gets the information he needs to transfer whatever he sees into something meaningful. I'll have to relay this information. If Neil has to take his attention away from looking out the window, look down to the keyboard and then back again, this is wasteful.

There were items sufficient for tension, yes. It was the measure of the trip to the moon that the more one knew of the difficulties, the more intense was the anxiety. Now we can add to the failures of communication and the difficulties of flying the Lem that extraordinary moment in the descent when the alarm light went on and the number 1202 blinked on the Display and Keyboard.

viii

We can go back to the point not four minutes after ignition when the Lem turned over on its back and the astronauts approaching the moon looked across a quarter of a million miles to earth. The comments are laconic, the difficulties are doubtless still ahead, but the radio is giving trouble again. They have passed through a minute where their remarks are garbled.

CAPCOM: *Eagle, Houston. You are GO. Take it all at four minutes. Roger, you are GO – you are GO to continue power descent.*

ALDRIN: *Roger.*

PAO: *Altitude 40,000.*

CAPCOM: *And Eagle, Houston. We've got data dropout. You're still looking good.*

ALDRIN: *PGNCS. We got good lock on. Altitude light is out. Delta H is minus 2900.*

CAPCOM: *Roger, we copy.*

ALDRIN: *Got the earth right out our front window.*

How powerful must have been such a sight to draw comment when the radio is working poorly, and a garbled remark can throw confusion back and forth. Perhaps it suggests some hint of happiness that they are finally at the beginning of the entrance to the last tunnel.

In the next moment Aldrin's voice speaks. 'Houston, you're looking at our Delta H program alarm?'

'That's affirmative,' replies the Capcom. 'It's looking good to us. Over.'

Aldrin's voice calls out, '1202, 1202.' It was an alarm from the computer – 'Executive Overflow' was its title. What a name! One thinks of seepage on the corporation president's bathroom floor. In fact it meant the computer was overloaded, and so unable to perform all its functions. In such a case the computer stops, then starts over again. It has recalculated its resources. Now it will take on only the most important functions, drop off the others. But what a moment at Mission Control! They have worked on this alarm in the day Kranz devoted to emergency situations. They know that if 1202 keeps blinking, the activities of the computer will soon deteriorate. The automatic pilot will first be lost, then control over the thrust of the engine, then Navigation and Guidance – the pilots will have to abort. In fifteen seconds it can all happen.

Picture Aldrin on his back looking up at the DSKY. 'Give us the reading on the 1202 program alarm.' It is his way of saying, 'Is it serious?'

Thirty seconds go by. Duke speaks up the quarter of a million miles. 'Roger, we got . . .' Pause. 'We're GO on that alarm.'

Kranz has been quizzing his Guidance officers and his Flight Dynamics officers. It is a ten-second roll call, and each one he queries says GO. The words come in, 'GO. GO. GO. GO.' The key word is from Guidance Officer Stephen G. Bales. It is on his console that the 1202 is also blinking. But they have been over the permissible rate of alarm on which they can continue to fly a mission, and the 1202 is coming in not that fast – the Executive Overflow is not constant. So Bales' voice rings out GO. Listening to it on a tape recorder later, there is something like fear in the voice, it is high-pitched, but it rings out. In the thirty seconds between Aldrin's request for a reading and the reply that they were GO, the decision had been taken.

Capcom: 'Eagle, Houston. We'll monitor your Delta H.' Mission Control would take over part of Pings. Data from the landing radar would no longer be fed to the computer on the Eagle, it would be sent uniquely to the ground. Eagle would be able to continue with the computer eased of some of its burden. But if telemetry failed again, and Pings could not reassume the burden, they would still have to turn back.

So they were obliged to proceed through the long braking burn with their minds on thoughts of Abort all the way, their eyes on the instrument panel, Armstrong's eyes as much as Aldrin's searching for a clue to what had caused the overload, six hundred dials and switches to consider and eliminate in blocks and banks – a hopeless activity: the cause of the malfunction might reveal itself on no dial, and in fact could derive from the load of the rendezvous radar in addition to the landing radar, and that introduced still another fear, for the rendezvous radar would have even more work on the trip up from the moon. As Duke said: 'Here we are with a computer that seems to be saturated during descent and my gosh, we might be asking it to perform a more complicated task during ascent.' And periodically Kranz queried his FIDO, his GUIDO, his TelCom and his EECOM, and the voices came back, 'GO, GO, GO, GO,' and Eagle descended, now at twenty-one thousand feet, now at sixteen thousand, thirteen thousand five hundred. 'Stand by,' said the Capcom, 'you're looking great at eight minutes.' Altitude ninety-two hundred feet.

In another twelve seconds they were at high gate. Now their horizontal speed was very low, their thrust was reduced, and they began to sit up from their near horizontal position. Slowly the astronauts' legs inclined with the legs of the Lem toward the ground, slowly their heads came up. And as they did, so returned a view of the moon through the bottom of the window now that the Lem inclined back toward the vertical. There were markings on their windows, horizontal lines to line up against the horizon of the moon and so serve to pinpoint the area where they would land, but as the ground of the Sea of Tranquility came into sight from four thousand feet, from three thousand feet, as their angle shifted and more and more of the moon was visible through that slanted window, so to their distress was the terrain unrecognizable. Quick glimpses showed no landmarks, no particular craters on the flat desert of Tranquility which might bear familiar relation to the photographs and charts Apollo 10 had prepared for them, and Armstrong memorized, no, now they were in another place, whether four miles or fourteen away from the selected landing site was impossible to tell. And at that instant a new alarm blinks on the computer. 'Twelve alarm,' says Aldrin, '1201.' The Capcom answers, 'Roger. 1201 alarm.'

When nothing further is said, Aldrin's voice comes in again. 'We're GO. Hang tight. We're GO. Two thousand feet . . .'

Capcom: 'Roger.'

Eagle: 'Forty-seven degrees.'

Capcom: 'Eagle looking great. You're GO.'

And the flight controllers at Mission Control are screwed to the parameters of the consoles, the roll calls come in for GO. There is always attention to Bales' answer to Kranz's query. 'GO,' Bales' voice will pipe out against the alarms.

Kranz is a leader. He is a man who gives others the feeling that they are about to go through the door together into the stadium where they are each going to play the best game of their life. Kranz, like Slayton, has the look of a man who had lived for years in space, and Bales is a young engineer with a large round face and large horn-rimmed glasses, destiny sits on him with a moist touch, but the limits of decision had been clarified that June morning a month ago and Bales had done the work to separate a total crisis from a partial crisis on 1202 and 1201, and now, distinguishing the differences – no time to ask for confirmation – calls GO. The mission continues.

But the astronauts continue without the division of labor Aldrin has specified as tidy. They come down toward the gray wife of the earth's ages with their eyes riveted to the instruments. There are no more than peeps and glimpses of the oncoming ground. At two thousand feet, Armstrong finally leaves the dials, studies the view from the window. He can land by hand if necessary now.

But he cannot locate himself. No landmarks are familiar. From one thousand feet up it is apparent that Eagle is headed for a wretched crater with a boulder field of rocks and 'the rocks seemed to be coming up at us awfully fast.' Fast and mean are those rocks accelerating to the eye like the zoom of a camera down to the ground. 'The clock runs about triple speed in a situation like that.' And now Aldrin is calling out relevant computer and instrument readings. Here they come at Armstrong! '750, coming down at 23.' That was altitude seven hundred and fifty feet, rate of descent twenty-three feet per second. '700 feet, 21 down. 600 feet, down at 19. 540 feet, down at 30 – down at 15.' In the tension Aldrin has miscalled a number for an instant. '400 feet, down at 9.' Something is now garbled. Then the voice again. It is quiet, it is almost sad. '350, down at 4. 330 feet, 3½ down. We're pegged on horizontal velocity.'

They are drifting horizontally over the boulder fields, skittering like a water bug debating which pad it will light on. Armstrong has taken over all of attitude control and part of throttle control – his commands are now inserted in with computer commands. The descent rate has reduced from ten feet a second to three feet, then only a foot a second. They hover, Armstrong searching for a spot in the boulder field, 'because I'm sure some of the ejecta coming out of such a large crater would have been lunar bedrock, and as such, fascinating to the scientists. I was tempted, but my better judgment took over . . .' No, he is not rushing in. He has gone through the computer alerts, the loss of all landmarks, has descended into that narrow field of vision where the horizon of the moon is always near, now drops down toward those boulders, hovers and skims, he has not had his recurring dreams as a boy night after night without tutoring the synapses of his growth into a thousand simulations of deliberate entrance into a dark space.

At last there is a place, 'the size of a big house lot' between craters and a boulder field, and drifting almost done, they see the shadow of their Lem slanting across the moon ground like a giant prehistoric bird of destiny and '200 feet,' says Aldrin, '4½ down, 5½ down. 160, 6½ down, 5½ down, 9 forward. 5 percent. Quantity light. 73 feet. Things looking good. Down a half. 6 forward.'

'Sixty seconds,' says the Capcom. That is the limit of their fuel.

'Lights on,' returns Eagle. Now their landing lights burn down on the sunlit moon ground to beam through the dust, and now comes the dust. At thirty feet above the ground, a great amount blows out in all directions like an underwater flower of the sea and the ground is partially visible beneath as if 'landing in a very fast-moving ground fog,' and the fuel gauges almost empty, and still he drifts forward. 'Thirty seconds,' calls out the Capcom for warning. And in a murk of dust and sunlight and landing lights, the Eagle settles in.* Contact lights light up on the board to register the touch of the probes below her legs. Aldrin's voice speaks softly, 'Okay, engine stop. ACA out of detente. Modes control both auto. Descent engine command override off. Engine arm off. 413 is in.'

Capcom: 'We copy you down, Eagle.'

Armstrong: 'Houston, Tranquility Base here. The Eagle has landed.'

Then was it that the tension broke for fifty million people or was it five hundred million, or some sum of billions of eyes and ears around a world which had just come into contact with another world for what future glory, disaster, blessing or curse nobody living could know. And Armstrong and Aldrin, never demonstrative, shook hands or clapped each other on the back – they did not later remember – and back at Mission Control, Charley Duke said, 'Roger, Tranquility, we copy you on the ground. You've got a bunch of guys about to turn blue. We're breathing again. Thanks a lot.'

And Kranz, who had issued every order to Duke, and queried his

---

* Mr Earle E. Spamer corrects this point by informing me there are no landing lights on the LEM. Lights on 'is to indicate that the 60-seconds-of-fuel-remaining light' has just gone on.

controllers in a voice of absolute calm for the entire trip down, now tried to speak and could not. And tried to speak, and again could not, and finally could unlock his lungs only by smashing his hand on a console so hard his bones were bruised for days. But then if his throat had constricted and his lungs locked, his heart stopped, he would have been a man who died at the maximum of his moments on earth and what a spring might then have delivered him to the first explorers of the moon. Perhaps it is the function of the dream to teach us those moments when we are GO or NO GO for the maximum thrust into death. They were down, they were on the moon ground, and who could speak?

'All right,' said Kranz. 'Everybody settle down, and let's get ready for a T-1 Stay–No Stay.'

7

A Sleep on the Moon

They were down, they had landed – the decade in which the project had been conceived was the decade in which it was achieved. What an achievement! Hannibal had taken his legions across the Alps, Cortez had conquered the mighty armies of the Aztecs with less than a thousand conquistadors, Castro landed in Cuba with eighty-two men, lost seventy in ambush on the beach, and five days later, hiding in the jungle, said to the survivors: 'the days of dictatorship are numbered.' Now Apollo 12 had landed on the moon, for if Apollo 11 was down, Apollo 12 could not be far behind. A new age of man had probably begun: just conceive of those purifications of discipline and cooperation that ordinary technicians working for most ordinary if immense corporations had shown. What an abstention from intrigue, treachery and betrayal had been forged in the million links of the chain.

Yet the question remained. Apollo was the god of the sun, so NASA did not fail to use his name for the expedition to the moon. Was the voyage of Apollo 11 the noblest expression of a technological age, or the best evidence of its utter insanity? It was the question which would dog Aquarius into the tenderest roots of his brain. So any spirit of impatience which would now have the astronauts open the hatch may as well burn its fumes – six hours are to go by before a man's boot comes off the Lem and puts the mark of sneaker cleats on the moon. In the meantime, moon conceivably all quivering in its finest registers from the four landing feet of the Lem, the question persists – are we witness to grandeur or madness? So the mind casts back to the source of Project Apollo and the birth of NASA, back into the warlock's pot of high politics in that hour after American prestige drooped in the eyes of the world because the Russians had put Sputnik into orbit and Soviet technology was in some ways at least superior to American. Eisenhower complained. If he had signed the bill which created NASA in 1958, he still did not know, he would announce, why everyone was thus inflamed about space. And in 1961, on the way out, in his last message to Congress a few days in January before John F. Kennedy would be inaugurated, Ike gave the word that he would not advise any extension of space flight after Mercury, no, not unless – and here came Elixir of Eisenhower – not unless 'further experiment and testing' gave go to good reason. Can we be certain it was altogether out of Eisenhower's reach to realize that a vain young President from an opposing party, consumed with the large desire to do things his way, was not going to take the ex-President's advice? But then we never give credit to Eisenhower for being so sly a man he was not even aware of it himself.

If it is natural to assume Kennedy was sympathetic to a moon shot when Eisenhower was publicly opposed, Kennedy was also a man with a regard for consensus; a national poll showed fifty-eight percent of the public sample were opposed to spending forty billion dollars in a race against the Russians to the lunar sphere. Kennedy was therefore not about to make a push. Not then. After the Bay of Pigs, however, the national desire may have moved up to the stars; certainly, after Alan Shepard's flight in Mercury-Redstone 3, public opinion took a full shift. It shifted with a rush. All the while, Johnson had not been chairman of the Senate Space Committee and head of the National Aeronautics and Space Council for nothing, no, and had not personally picked James E. Webb to be head of NASA for no reason, nor looked upon fast-expanding Houston without plans for even more rapid expansion. It is not hard to conceive of how fast the tip of his tongue could flicker in and out of Kennedy's ear when public opinion began to turn. Indeed the President had to agree with the Vice President about something. And Space was congenial to Kennedy, congenial to the metaphor of a new frontier, congenial to his love of doughty ships and mysterious seas; besides, the brightest of his economists were fond of the absurdity of space. The most advanced of them had come to the conclusion that economics was a phenomenon which resisted planning – the Soviet Union had been giving its unwilling demonstration of this point for thirty years: yes, economies were most interesting when they developed off the target. It was as if one did not achieve a career in the theater by studying how to act but by mountain-climbing or road racing. The effort, not the technique, seemed to prepare the result, or if not the intended result, another equally desirable. The voyage to the moon came to be seen by many of Kennedy's liberal advisers as the most imaginative way to prime an economic pump in a time of relative prosperity and no war. Who cared if it succeeded? When the decade was out, there might be no moon rocks in the hand, but a new kind of prefabricated housing could well have been discovered instead, or some new fuel for automobiles. Plastics were bound to accelerate themselves into products and industries not yet conceived. Advances in metallurgy and electronics would inspire huge new plants of higher education, boom them up from swamps newly filled by the exudations of the cities. The technological age, bloated with waiting, was ready to burst on the landscape of America. That technological age would solve all the old problems – so declared the confidence of Kennedy's elite. What better for the symbol of a new age than a landing on the moon before the decade was out? It was a blind push, equal to the hot sobs of the Oklahoma land grabber who plunked down his marker and said, 'Mah land runs from this stone to that tree,' before he even knew if he had bottom lands or water. Before the decade was out. Why? Because the trip to the moon had to serve as the embodiment of a new vision and visions are obliged to be neat. Kennedy, like many an enterprising young man before him, knew the best approach to large and complex mysteries was to plunge your hands into the short hair. 'This is a new ocean,' he said, 'and I believe the United States must sail upon it.'

ii

So Project Apollo was born. On a landscape of machination, economic cynicism, warlockery, the maneuvers of Lyndon Johnson's gravyboat navy, and out of John F. Kennedy's profound respect for the dynamics and mechanics of new romantic ages, was Project Apollo born. She would prime a pump for the Government until the war in Vietnam would rip the handle out of her grasp, she would gild the balance sheet of many a corporation until the war in Vietnam would cause a crisis of priorities for the work of corporations, she would give work to hundreds of thousands, and did, until she had to let many of them go; she would create towns and counties; she would be handmaiden, then wife, to the computer; and her spin-offs had hardly begun, she would still change the age; but she had failed until now to become a vision of technology which would put light in the eyes of every poor man. She had failed to become a church for the new age. She was only a chalice for the wounded bewildered heart of the Wasp, a code of honor for corporation executives hitherto bereft of pride. She set electronic engineers and computer programmers to dreaming of ways to attack the problems of society as well as they had attacked the problems of putting men on the moon. But she had not unified nor purged that accompanying world of feverish development and pollution which had kept pace with her pure endeavor.

In this hour they landed on the moon, America was applauding Armstrong and Aldrin, and the world would cheer America for a day, but something was lacking, some joy, some outrageous sense of adventure. Strong men did not weep in the street nor ladies copulate with strangers. Any armistice to any petty war had occasioned wilder celebrations. It was almost as if a sense of woe sat in the center of the heart. For the shot to the moon was a mirror to our condition – most terrifying mirror; one looked into it and saw imitations of a final disease. But probably it will take the rest of the trip to absorb the remark.

Armstrong and Aldrin were to do an EVA that night. EVA stood for Extravehicular Activity, and that was presumably a way to describe the most curious steps ever taken. It is one thing to murder the language of Shakespeare – another to be unaware how rich was the victim. Future murders stood in the shadow of the acronyms. It was as if on the largest stage ever created, before an audience of half the earth, a man of modest appearance would walk to the center, smile tentatively at the footlights, and read a page from a data card. The audience would groan and Beckett and Warhol give their sweet smiles.

So, now, on that moon (whose name was *Mond* in German, even if *monde* was the word for 'world' in French, now on that moon called *maan* and *maane* in Dutch and Danish) a set of television activities would soon begin. The astronauts would go out from the Lem and set up a TV camera and the American flag, they would lay out three experiments, obtain two tubes of moon soil, fill two boxes of rocks, speak to the President, take photographs, and go back to their Lem. Never would they be more than a hundred feet from the ship. Out of the variety of activities available, they would go through a schedule about as attractive to watch as an afternoon of qualitative analysis in a college lab. Fact, they would not even perform the experiments – they would merely lay them out.

Somewhere in the center of NASA was the American disease: Focus on one problem to the exclusion of every other. When Communism had been the problem, nothing had existed for national policy but anti-Communism. Now, ever since the fire of Apollo 204, there had been only one idea at NASA. Get men to the moon by the end of the decade and get them back. If drama had to be sacrificed, rid the situation of drama. If scientific investigations would hamper a smooth flight, restrict scientific investigations. A narrowness of vision, constricted by the panic which followed the Apollo fire, lost all register of the true complexity of the event. Propriety had always been the natural soup of any engineering society; now NASA propriety spoke of super-congelations of moral lard. No curse, omen, oath, scar or smell could intrude on the landscape, no revel, no voice, and no unnecessary chancing of human life. It was not that anybody wanted the blood of astronauts any more than they desired the death of bullfighters, auto racers or boxers, it was that NASA had come to believe that if Apollo 11 resulted in death, all space investigation was gone, whereas in fact the irony was that the world, first sacrifices in outer space paid, would have begun to watch future flights with pain and concern. It was not that anybody was about to argue for the taking of unnecessary chances, it was more as if some of the precautions became absurd and so expensive that the mind had to engage in brute thoughts of what the money would mean in all the other places one could name.

Say, it was even that the most necessary experiments were not made. The most cumbersome element in all the flight was the Pressure Garment Assembly, the bulky-wham suit. It was a danger in itself, for it was all too easy, as Aldrin had noted, to press against a whole bank of buttons. It had been designed to protect the astronauts in the vacuums of space, and it consisted of fifteen layers of plastic material, a suit as hard as the material on a fire hose – it took prodigies of strength to move it and intricate adjustments of grace and balance – it is yet worth describing in detail – and this suit, which left the astronauts about as much coordination as a two-year-old in three sets of diapers, was designed to protect them against puncture. Yet no one knew what puncture might mean in space. It was uncertain whether exposure of the flesh to a vacuum would result in quick death or supportable injury. Still, no canary was brought along to be partially exposed, its death or its ability to survive for a time carefully noted. It is a horror, of course, to let a bird die on the moon – ecology hard upon us, who can jeer at the claim of the antivivisectionists that something divine might not also die, but animals are used everywhere in every closet of behavioral psychology labs, rabbits go out of their minds in electrode-laden harnesses before they are finally fed to the town dump, so what would it have cost NASA except an incision in their hypocrisy if the first critical experiments had been taken to see if partial or temporary exposure to a vacuum was feasible. If so, the suits could be made more agreeable – the explorations more adept.

Well, such experiments were for other occasions. The astronauts, in preparation for the EVA, had been garbed in EMU, the Extravehicular Mobility Unit, or moon-walk space suit complete. EMU consisted first of three envelopes, a liquid cooling garment, a pressure garment, a thermal-and-micrometeoroid garment, plus a Portable Life Support System for oxygen, plus a backup system, a radio, a waste-management system, a maintenance kit (to work on their own life-supporting machinery), a set of special visors for their helmets, and the ubiquitous biomedical belt. The total weight was one hundred eighty-three pounds – on earth the astronauts staggered around in the EMU when they were able to move at all. On the moon weight of the suit and pack would be more like thirty pounds, but it still had to feel as much like a spaceship as a space suit, and indeed with a quadrant of rocket thrusters, a computer, and a bag of food-injector, there would have been little to keep the astronauts from going into orbit for a revolution or two. Six hundred pounds of fuel weighing only one hundred pounds on the moon might have done the job!

That was a joke which in another year or two could prove less of a joke. In any case, the EMU was designed to provide an environment in which a man might live while walking through the vacuum of space and the radiations of the solar wind plus the cosmic rays – a suit designed to be bulletproof as well against the occasional micrometeoroid. Most meteoroids were smaller than buckshot, but they traveled at a speed of forty-five thousand miles an hour. (Still, the chance of getting hit was only one in ten thousand – no army was going to be stopped by that.) More to the point: the EMU would also protect against the heat of the sun in the full lunar day when temperatures rose clear above boiling to 243 degrees Fahrenheit on the moon soil. So the suit had to be insulated against the heat of a fall on the seared ground.

We can imagine what a construction was involved. Early that morning, over nine hours before their landing, back in the Command Module, they had taken off their constant-wear garments, been naked for once after all these days, and as quickly had put on the liquid cooling garment, which was a long-sleeved pair of long johns with two layers of Beta cloth, one on either side of a network of tubing through which cooled water would pass – the liquid cooling garment was not dissimilar in appearance to a scuba diver's rubber formfitting suit. It was estimated that their labor on the moon pushing the total EMU around would be equal to the exertions of a man shoveling sand, or snow, or sawing wood. So cooling was critical. In that sealed envelope, that man-shaped spaceship surrounding their skin, even a little too much overheating could fog their visor, overload all purge systems.

Next came the Pressure Garment Assembly, a submarine chamber complete in itself. Putting that on over the liquid cooling garment, they would be safe in flight if there were any leaks or punctures in the Lem. The Pressure Garment Assembly was in fact a true space suit itself, the one indeed they wore for the launch; this PGA consisted of a helmet, gloves, and the torso-and-limb suit. PGA had a 'comfort' layer of nomex, then a neoprene-coated pressure bladder, finally a nylon restraint layer. The bladder pressed against the skin of the astronaut, pressed all over their bodies with a pressure about a quarter of the earth's normal atmosphere. Without this bladder, it was assumed that blood might boil in the veins or skin pop open from the internal pressure of the body, which from birth had been stressed to maintain itself against normal atmospheres on earth – in fact no one knew exactly what would happen if a few square inches of skin were suddenly exposed by a tear in the suit to the moon vacuum, no one could even be certain whether the skin would burn, for the heat of the sun might be less than 243 degrees Fahrenheit even a foot above the moon soil since there would be no air to trap the heat. In any case, once in the pressure garment suit, the effect was equivalent to moving about in a man-shaped balloon. There were restraints on the bladder of course, subtle hoops and rings at all the joints, not altogether unlike the joints in armor; the gloves, in fact, to enable bending of the knuckles, were reminiscent of jousters' gauntlets, but a balloon it was still, and movements were necessarily deliberate in order not to work at further compressing air already overcompressed in a joint. Needless to say the Pressure Garment Assembly had a plenitude of valves, faucets, plugs, taps, and redundancies. There were four gas connectors, two for oxygen in, two for oxygen out, a water interface between the Portable Life Support System (PLSS) which was worn on the back and the liquid cooling garment underneath, an electrical connector to provide communications, instrumentation and power interface from PLSS to PGA. There was also a urine transfer collector (with valve) to avoid depressurizing the pressure garment. Also pressure gauges, pressure relief valves and seven pockets for data cards, penlight, sunglasses, lanyard, scissors, checklist and utility.

This was the space suit for wear inside the cabin of the Command Module or Lem during launch, docking, undocking or through any period when the hatch door had to be open. For the walk itself, they would add still another covering, the Integrated Thermal Micrometeoroid Garment, a formfitting cover which was laced over the pressure garment and consisted of a plastic armor, proof against heat and meteoroids, composed of two layers of nylon, seven layers of Beta Kapton space laminate, and an outer cloth of Teflon-coated Beta, Teflon being the synthetic with which nonchar frying pans are coated.

The helmet was a transparent polycarbonate shell of Lexan with a sealing ring for attachment to the collar of the suit. For the moon walk two extra visors were attached, a lightly tinted inner visor to reduce glare and fog, and an outer visor in appearance like a fencer's mask with a gold film on its surface to reflect solar radiation. When lowered, one could not see the face, but the gold film obstructed no vision from within and altered no colors.

Finally, there was the PLSS and the Oxygen Purge System, which had a half hour of oxygen supply as backup for the PLSS. Both were packs, the Oxygen Purge sitting on the PLSS, and both were hefted in a special fiber-glass shell molded to fit the back. The PLSS contained four hours of oxygen supply and a VHF transmitter and receiver to enable the astronauts to address each other, or by relay to connect themselves to the S-band on the Lem and thereby be able to speak to the ground. There were batteries in the PLSS, and oxygen supply, and ventilation, and a cooling system for the water in the liquid cooling garment, also a remote control system worn on the chest with a camera attached. It had a fan switch, a pump switch, a space suit communicator switch, a volume control for voice, an oxygen quantity indicator, five status indicators, and an interface to switch over to the oxygen purge system. Slung around the neck just beneath the helmet, it was as if a slab of the instrument panel on the Lem had been brought along.

So equipped, in an outfit bulkier than a diving suit, an enormous pack on their back, and heavy lunar overshoes, they were ready to go out on the moon. But we have leaped over their activities in the hours between; it will come without surprise that the astronauts have hardly been idle.

iii

They had landed, there was jubilation in Mission Control, and a moment of fraternization between Armstrong and Aldrin, but in fact they were actually at work in the next instant. No one knew what would await them – there were even theories that most of the surface of the moon was as fragile as icing on a cake. If they landed, and the moon ground began to collapse, they were ready to blast off with the ascent stage even as the descent stage was sinking beneath. But no sound of crumbling came up through the pipes of the legs, no shudder of collapse. A minute passed. They received the order to Stay. The second Stay–No Stay would be on them nine minutes later, and they rushed through a checklist, testing specific instruments to make certain they were intact from the landing. The thirty-odd seconds of fuel they still had left when they touched down was vented from the descent stage, a hissing and steaming beneath the legs like a steed loosing water on icy ground. Verbs and Nouns were punched into the DSKY. Now came the second Stay. There would not be another Stay–No Stay until the

Command Module had made a complete revolution of the moon and would be coming back toward them in good position for rendezvous. So, unless some mishap were suddenly to appear, they had at least another two hours on the satellite. It was time to unscrew their gloves at the wrist and take them off, time to unscrew their helmets at the neck, lift them off.

They gave their first description of the landing, and made a few general remarks about the view through the window, the variety of rocks. But there was too much work to look for long. After a few comments on the agreeableness of lunar gravity, after a conversation with Columbia and mutual congratulations, they were back at the computer. Now, in the time before the next Stay–No Stay, they had to simulate a countdown for a planned ascent and realign the Inertial Measurement Unit, that is, determine the vertical line of moon gravity, and install its index into the Inertial Measurement Unit, then level the table and gyroscope from which all navigation was computed. Star checks were taken. Meanwhile, Armstrong was readying the cameras and snapping photographs through the window. Now Aldrin aligned the Abort Guidance Section. Armstrong laid in the data for Program 12, the Powered Ascent Guidance. The Command Module came around again. The simulated countdown was over. They had another Stay. They powered down their systems.

In the transcript the work continues minute after minute, familiar talk of stars and Nouns, acronyms, E-memory dumps, and returns to POO where Pings may idle. They are at rest on the moon, but the dialogue is not unencumbered of pads, updata link switches and noise suppression devices on the Manned Space Flight Network relay.

Then in what is virtually their first pause in better than an hour on the moon, they request permission to do their EVA early, begin in fact in the next few hours rather than take a halt to sleep. For days there had been discussion in every newspaper of the world whether the astronauts could land on the moon and a few hours later go to sleep before they even stepped out of the Lem; now the question has been answered – they are impatient to go.

CAPCOM: *We will support it.*
ALDRIN: *Roger.*

CAPCOM: *You guys are getting prime time TV there.*
ARMSTRONG: *Hope that little TV set works, but we'll see.*

Now the astronauts stopped to eat and to relax. Over the radio came the dialogue of Mission Control talking to Collins in orbit overhead. Around them, through each pinched small window, were tantalizing views of the moon. They could feel themselves in one-sixth gravity. How light were their bodies. Yet they were not weightless. There was gravity beneath them, a faint sensuous tug at their limbs. If they dropped a pencil, it did not float before drifting slowly away. Rather, it dropped. Slowly it dropped, dropped indeed at the same leisurely speed with which Apollo-Saturn had risen off its launching pad four and a half days ago. What a balm for the muscles of the eye! One-sixth of earth gravity was agreeable, it was attractive, it was, said Aldrin, 'less *lonesome*' than weightlessness. He had, at last, 'a distinct feeling of being somewhere.' Yes, the moon was beneath them, hardly more than the height of a ten-foot diving board beneath them – they were in the domain of a presence again. How much like magnetism must lunar gravity have felt.

ALDRIN: *This is the Lem pilot. I'd like to take this opportunity to ask every person listening in, whoever and wherever they may be, to pause for a moment and contemplate the events of the past few hours, and to give thanks in his or her way.*

In the silence, Aldrin took out the bread, the wine, and the chalice he had brought in his Personal Preference Kit, and he put them on the little table in front of the Abort Guidance Section computer. Then he read some passages from the Bible and celebrated Communion.

A strange picture of religious intensity: there is of course no clue in Aldrin's immediate words – they are by now tuned to precisely what one would expect.

'I would like to have observed just how the wine poured in that environment, but it wasn't pertinent at that particular time. It wasn't important how it got in the cup. It was important only to get it there' – and not spill, we may assume, this most special blood of the Lord. 'I offered some private prayers, but I find now that thoughts, feelings, come into my memory instead of words. I was not so selfish as to include my family in those prayers at the moment, nor so spacious as to include the fate of the world. I was thinking more about our particular task, and the challenge and the opportunity that had been given us. I asked people to offer thanks in their own way, and it is my hope that people will keep this whole event in their minds and see beyond minor details and technical achievements to a deeper meaning behind it all, challenge, a quest, the human need to do these things and the need to recognize that we are all one mankind under God.'

Yes, his recollections are near to comic in their banality, but one gets a picture of this strong-nosed strong-armed gymnast in his space suit, deep in prayer in the crowded closet space of the Lem, while Armstrong the mystic (with the statue of Buddha on his living room table) is next to him in who knows what partial or unwilling communion, Armstrong so private in his mind that when a stranger tried to talk to him one day on a bus, he picked up a book to read. There, before his partner, Aldrin prayed, light lunar gravity new in his limbs, eyes closed. Can we assume the brain of his inner vision expanded to the dimensions of a church, the loft of a cathedral, Aldrin, man of passions and disciplines, fatalist, all but open believer in predestination, agent of God's will, Aldrin, prodigy of effort on Gemini 12, whose pulse after hours of work in space had shot up only when he read a Veteran's Day message to the ground. Patriotism had the power of a stroke for Aldrin and invocation was his harmony. Tribal chief, first noble savage on the moon, he prayed to the powers who had brought him there, whose will he would fulfill – God, the earth, the moon and himself all for this instant part of the lofty engine of the universe, and in that eccentric giant of character, that conservative of all the roots in all the family trees, who now was ripping up the roots of the ages, that man whose mother's name was Moon, was there a single question whose lament might suggest that if the mission were ill-conceived or even a work of art designed by the Devil, then all the prayers of all good men were nothing but a burden upon the Lord, who in order to reply would be forced to work in the mills of Satan, or leave the prayers of his flock in space. Not likely. Aldrin did not seem a man for thoughts like that, but then his mind was a mystery wrapped in the winding-sheet of a computer with billions of bits.

iv

Later, Armstrong would say, 'That first hour on the moon was hardly the time for long thoughts; we had specific jobs to do. Of course the sights were simply magnificent, beyond any visual experience that I had ever been exposed to,' and Aldrin would describe it as 'a unique, almost mystical environment.' In fact, there is an edge of the unexplained to their reactions. Their characteristic matter-of-fact response is overcome occasionally by swoops of hyperbole. And to everyone's slight surprise, they were almost two hours late for their EVA. Their estimate of time was off by close to fifty percent. For astronauts that was an error comparable to a carpenter mistaking an eight-foot stud for a twelve-foot piece. If a carpenter can look at a piece of wood and guess its length to the nearest quarter-inch, it is because he has been working with lengths all his life. Equally, people in some occupations have a close ability to estimate time.

With astronauts, whose every day in a simulator was a day laid out on the measure of a time-line, the estimate of time elapsed had to become acute. Armstrong and Aldrin had consistently fulfilled their tasks in less time than was allotted. Now, curiously, they fell behind, then further behind. There were unexpected problems of course – it took longer to bleed the pressure out of the Lunar Module than had been anticipated, and the cooling units in the backpacks were sluggish at first in operation, but whether from natural excitement and natural anxiety, or an unconscious preoccupation with lunar phenomena so subtle that it is just at the edge of their senses, any extract from the transcript at this point where they are helping to adjust the Portable Life Support System on each others' backs shows real lack of enunciation. Nowhere else do the NASA stenographers have as much difficulty with where one voice ends and another begins.

> TRANQUILITY: *Got it (garbled) prime rows in.*
> TRANQUILITY: *Okay.*
> TRANQUILITY: *(garbled)*
> TRANQUILITY: *Let me do that for you.*
> TRANQUILITY: *(Inaudible)*

TRANQUILITY: *Mark I*

TRANQUILITY: *(garbled) valves*

TRANQUILITY: *(garbled)*

TRANQUILITY: *Okay*

TRANQUILITY: *All of the (garbled)*

TRANQUILITY: *(garbled) locked and lock locked.*

TRANQUILITY: *Did you put it –*

TRANQUILITY: *Oh, wait a minute*

TRANQUILITY: *Should be (garbled)*

TRANQUILITY: *(garbled)*

TRANQUILITY: *Roger. (garbled)*

TRANQUILITY: *I'll try it on the middle*

TRANQUILITY: *All right, check my (garbled) valves vertical*

TRANQUILITY: *Both vertical*

TRANQUILITY: *That's two vertical*

TRANQUILITY: *Okay*

TRANQUILITY: *(garbled)*

TRANQUILITY: *Locked and double-locked*

TRANQUILITY: *Okay*

TRANQUILITY: *Miss marked*

TRANQUILITY: *Sure wish I would have shaved last night.*

PAO: *That was a Buzz Aldrin comment.*

The hint is faint enough, but the hint exists – something was conceivably interfering with their sense of order. Could it have been the lunar gravity? Clock-time was a measure which derived from pendulums and spiral springs, clock-time was anchored right into the tooth of earth gravity – so a time might yet be coming when psychologists, not geologists, would be conducting experiments on the moon. Did lunar gravity have power like a drug to shift the sense of time?

Armstrong was connected at last to his PLSS. He was drawing oxygen from the pack he carried on his back. But the hatch door would not open. The pressure would not go low enough in the Lem. Down near a level of one pound per square inch, the last bit of man-created atmosphere in Eagle seemed to cling to its constituency, reluctant to enter the vacuums of the moon. But they did not know if they could get the hatch door open with a vacuum on one side and even a small pressure on the other. It was taking longer than they thought. While it was not a large concern since there would be other means to open it – redundancies pervaded throughout – nonetheless, a concern must have intruded: how intolerably comic they would appear if they came all the way and then were blocked before a door they could not crack. That thought had to put one drop of perspiration on the back of the neck. Besides, it must have been embarrassing to begin so late. The world of television was watching, and the astronauts had exhibited as much sensitivity to an audience as any bride on her way down the aisle.

It was not until nine-forty at night, Houston time, that they got the hatch open at last. In the heat of running almost two hours late, ensconced in the armor of a man-sized spaceship, could they still have felt an instant of awe as they looked out that open hatch at a panorama of theater: the sky is black, but the ground is brightly lit, bright as footlights on the floor of a dark theater. A black and midnight sky, yet on the moon ground, 'you could almost go out in your shirt-sleeves and get a suntan,' Aldrin would say. 'I remember thinking, "Gee, if I didn't know where I was, I could believe that somebody had created this environment somewhere out in the West and given us another simulation to work in."' Everywhere on that pitted flat were shadows dark as the sky above, shadows dark as mine shafts.

What a struggle to push out from that congested cabin, now twice congested in their bulky-wham suits, no feeling of obstacle against their flesh, their sense of touch dead and numb, spaceman body manipulated out into the moon like an upright piano turned by movers on the corner of the stairs.

'You're lined up on the platform. Put your left foot to the right a little bit. Okay, that's good. Roll left.'

Armstrong was finally on the porch. Could it be with any sense of an alien atmosphere receiving the fifteen-layer encapsulations of the pack and suit on his back? Slowly, he climbed down the ladder. Archetypal, he must have felt, a boy descending the rungs in the wall of an abandoned well, or was it Jack down the stalk? And there he was on the bottom, on the footpad of the leg of the Lem, a metal plate perhaps three feet across. Inches away was the soil of the moon. But first he jumped up again to the lowest rung of the ladder. A couple of hours later, at the end of the EVA, conceivably exhausted, the jump from the ground to the rung, three feet up, might be difficult in that stiff and heavy space suit, so he tested it now. 'It takes,' said Armstrong, 'a pretty good little jump.'

Now, with television working, and some fraction of the world peering at the murky image of this instant, poised between the end of one history and the beginning of another, he said quietly, 'I'm at the foot of the ladder. The Lem footpads are only depressed in the surface about one or two inches, although the surface appears to be very very fine-grained as you get close to it. It's almost like a powder.' One of Armstrong's rare confessions of uneasiness is focused later on this moment. 'I don't recall any particular emotion or feeling other than a little caution, a desire to be sure it was safe to put my weight on that surface outside Eagle's footpad.'

Did his foot tingle in the heavy lunar overshoe? 'I'm going to step off the Lem now.'

Did something in him shudder at the touch of the new ground? Or did he draw a sweet strength from the balls of his feet? Nobody was necessarily going ever to know.

'That's one small step for a man,' said Armstrong, 'one giant leap for mankind.' He had joined the ranks of the forever quoted. Patrick Henry, Henry Stanley and Admiral Dewey moved over for him.

v

Now he was out there, one foot on the moon, then the other foot on the moon, the powder like velvet underfoot. With one hand still on the ladder, he comments, 'The surface is fine and powdery. I can . . . I can pick it up loosely with my toe.' And as he releases his catch, the grains fall back slowly to the soil, a fan of feathers gliding to the floor. 'It does adhere in fine layers like powdered charcoal to the sole and sides of my boots. I only go in a small fraction of an inch. Maybe an eighth of an inch. But I can see the footprints of my boots and the treads in the fine sand particles.'

Capcom: 'Neil, this is Houston. We're copying.'

Yes, they would copy. He was like a man who goes into a wrecked building to defuse a new kind of bomb. He talks into a microphone as he works, for if a mistake is made, and the bomb goes off, it will be easier for the next man if every detail of his activities has been mentioned as he performed them. Now, he released his grip on the ladder and pushed off for a few steps on the moon, odd loping steps, almost thrust into motion like a horse trotting up a steep slope. It could have been a moment equivalent to the first steps he took as an infant for there was nothing to hold onto and he did not dare to fall – the ground was too hot, the rocks might tear his suit. Yet if he stumbled, he could easily go over for he could not raise his arms above his head nor reach to his knees, his arms in the pressure bladder stood out before him like sausages; so, if he tottered, the weight of the pack could twist him around, or drop him. They had tried to shape up simulations of lunar gravity while weighted in scuba suits at the bottom of a pool, but water was not a vacuum through which to move; so they had also flown in planes carrying two hundred pounds of equipment on their backs. The pilot would take the plane through a parabolic trajectory. There would be a period of twenty-two seconds at the top of the curve when a simulation of one-sixth gravity would be present, and the two hundred pounds of equipment would weigh no more than on the moon, no more than thirty-plus pounds, and one could take loping steps down the aisle of the plane, staggering through unforeseen wobbles or turbulence. Then the parabolic trajectory was done, the plane was diving, and it would have to pull out of the dive. That created the reverse of one-sixth gravity – it multiplied gravity by two and a half times. The two hundred pounds of equipment now weighed five hundred pounds and the astronauts had to be supported by other men straining to help them bear the weight. So simulations gave them time for hardly more than a clue before heavy punishment was upon them. But now he was out in the open endless lunar gravity, his body and the reflexes of his life obliged to adopt a new rhythm and schedule of effort, a new disclosure of grace.

Still, he seemed pleased after the first few steps. 'There seems to be no difficulty in moving around as we suspected. It's even perhaps easier than the simulations . . .' He would run a few steps and stop, run a few steps and stop. Perhaps it was not unlike directing the Lem when it hovered over the ground. One moved faster than on earth and with less effort, but it was harder to stop – one had to pick the place to halt from several yards ahead. Yes, it was easier once moving, but awkward at the beginning and the end because of the obdurate plastic bendings of the suit. And once standing at rest, the sense of the vertical was sly. One could be leaning further forward than one knew. Or leaning backward. Like a needle on a dial one would have to oscillate from side to side of the vertical to find position. Conceivably the sensation was not unlike skiing with a child on one's back.

It was time for Aldrin to descend the ladder from the Lem to the ground, and Armstrong's turn to give directions: 'The shoes are about to come over the sill. Okay, now drop your PLSS down. There you go. You're clear . . . About an inch clearance on top of your PLSS.'

Aldrin spoke for future astronauts: 'Okay, you need a little bit of arching of the back to come down . . .'

When he reached the ground, Aldrin took a big and exuberant leap up the ladder again, as if to taste the pleasures of one-sixth gravity all at once. 'Beautiful, beautiful,' he exclaimed.

Armstrong: 'Isn't that something. Magnificent sight out here.'

Aldrin: 'Magnificent desolation.'

They were looking at a terrain which lived in a clarity of focus unlike anything they had ever seen on earth. There was no air, of course, and so no wind, nor clouds, nor dust, nor even the finest scattering of light from the smallest dispersal of microscopic particles on a clear day on earth, no, nothing visible or invisible moved in the vacuum before them. All light was pure. No haze was present, not even the invisible haze of the finest day – therefore objects did not go out of focus as they receded into the distance. If one's eyes were good enough, an object at a hundred yards was as distinct as a rock at a few feet. And their eyes were good enough. Just as one could not determine one's altitude above the moon, not from fifty miles up nor five, so now along the ground before them no distance was real, for all distances had the faculty to appear equally near if one peered at them through blinders and could not see the intervening details. Again the sense of being on a stage or on the lighted floor of a room so large one could not see where the dark ceiling began must have come upon them, for there were no hints of gathering evanescence in ridge beyond ridge; rather each outline was as severe as the one in front of it, and since the ground was filled with small craters of every size, from antholes to potholes to empty pools, and the horizon was near, four times nearer than on earth and sharp as the line drawn by a pencil, the moon ground seemed to slope and drop in all directions 'like swimming in an ocean with six-foot or eight-foot swells and waves,' Armstrong said later. 'In that condition, you never can see very far away from where you are.' But what they could see, they could see entirely – to the depth of their field of view at any instant their focus was complete. And as they swayed from side to side, so a sense of the vertical kept eluding them, the slopes of the craters about them seeming to tilt a few degrees to one side of the horizontal, then the other. On earth, one had only to incline one's body an inch or two and a sense of the vertical was gone, but on the moon they could lean over, then further over, lean considerably further over without beginning to fall. So verticals slid and oscillated. Rolling from side to side, they could as well have been on water, indeed their sense of the vertical was probably equal to the subtle uncertainty of the body when a ship is rolling on a quiet sea. 'I say,' said Aldrin, 'the rocks are rather slippery.'

They were discovering the powder of the moon soil was curious indeed, comparable in firmness and traction to some matter between sand and snow. While the Lem looked light as a kite, for its pads hardly rested on the ground and it appeared ready to lift off and blow away, yet their own feet sometimes sank for two or three inches into the soft powder on the slope of very small craters, and their soles would slip as the powder gave way under their boots. In other places the ground was firm and harder than sand, yet all of these variations were to be found in an area not a hundred feet out from the legs of the Lem. As he explored his footing, Aldrin sent back comments to Mission Control, reporting in the rapt professional tones of a coach instructing his team on the conditions of the turf in a new plastic football field.

Meanwhile Armstrong was transporting the television camera away from the Lem to a position where it could cover most of their activities. Once properly installed, he revolved it through a full panorama of their view in order that audiences on earth might have a clue to what he saw. But in fact the transmission was too rudimentary to give any sense of what was about them, that desert sea of rocks, rubble, small boulders, and crater lips.

Aldrin was now working to set up the solar wind experiment, a sheet of aluminum foil hung on a stand. For the next hour and a half, the foil would be exposed to the solar wind, an invisible, unfelt, but high-velocity flow of noble gases from the sun like argon, krypton, neon and helium. For the astronauts, it was the simplest of procedures, no more difficult than setting up a piece of sheet music on a music stand. At the end of the EVA, however, the aluminum foil would be rolled up, inserted in the rock box, and delivered eventually to a laboratory in Switzerland uniquely equipped for the purpose. There any noble gases which had been trapped in the atomic lattice of the aluminum would be baked out in virtuoso procedures of quantitative analysis, and a closer knowledge of the components of the solar wind would be gained. Since the solar wind, it may be recalled, was diverted by the magnetosphere away from the earth it had not hitherto been available for casual study.

That was the simplest experiment to set up; the other two would be deployed about an hour later. One was a passive seismometer to measure erratic disturbances and any periodic vibrations, as well as moonquakes, and the impact of meteors in the weeks and months to follow; it was equipped to radio this information to earth, the energy for transmission derived from solar panels which extended out to either side, and thereby gave it the look of one of those spaceships of the future with thin extended paperlike wings which one sees in science fiction drawings. In any case it was so sensitive that the steps of the astronauts were recorded as they walked by. Finally there was a Laser Ranging Retro-Reflector, an LRRR (or LRQ, for LR-cubed), and that was a mirror whose face was a hundred quartz crystals, black as coal, cut to a precision never obtained before in glass – one-third of an arc/sec. Since each quartz crystal was a corner of a rectangle, any ray of light striking one of the three faces in each crystal would bounce off the other two in such a way that the light would return in exactly the same direction it had been received. A laser beam sent up from earth would therefore reflect back to the place from which it was sent. The time it required to travel this half-million miles from earth to moon round trip, a journey of less than three seconds, could be measured so accurately that physicists might then discern whether the moon was drifting away from the earth a few centimeters a year, or (by using two lasers) whether

Europe and America might be drifting apart some comparable distance, or even if the Pacific Ocean were contracting. These measurements could then be entered into the caverns of Einstein's General Theory of Relativity, and new proof or disproof of the great thesis could be obtained.

We may be certain the equipment was remarkable. Still, its packaging and its ease of deployment had probably done as much to advance its presence on the ship as any clear priority over other scientific equipment; the beauty of these items from the point of view of NASA was that the astronauts could set them up in a few minutes while working in their space suits, even set them up with inflated gloves so insensitive that special silicone pads had to be inserted at the fingertips in order to leave the astronauts not altogether numb-fingered in their manipulations. Yet these marvels of measurement would soon be installed on the moon with less effort than it takes to remove a vacuum cleaner from its carton and get it operating.

It was at this point that patriotism, the corporation, and the national taste all came to occupy the same head of a pin, for the astronauts next proceeded to set up the flag. But that operation, as always, presented its exquisite problems. There was, we remind ourselves, no atmosphere for the flag to wave in. Any flag made of cloth would droop, indeed it would dangle. Therefore, a species of starched plastic flag had to be employed, a flag which would stand out, there, out to the nonexistent breeze, flat as a slab of plywood. No, that would not do either. The flag was better crinkled and curled. Waves and billows were bent into it, and a full corkscrew of a curl at the end. There it stands for posterity, photographed in the twists of a high gale on the windless moon, curled up tin flag, numb as a pickled pepper.

Aldrin would hardly agree. 'Being able to salute that flag was one of the more humble yet proud experiences I've ever had. To be able to look at the American flag and know how much so many people had put of themselves and their work into getting it where it was. We sensed – we really did – this almost mystical identification of all the people in the world at that instant.'

Two minutes after the flag was up, the President of the United States put in his phone call. Let us listen one more time:

'Because of what you have done,' said Nixon, 'the heavens have become a part of man's world. And as you talk to us from the Sea of Tranquility, it inspires us to redouble our efforts to bring peace and tranquility to earth . . .'

'Thank you, Mr President. It's a great honor and privilege for us to be here representing not only the United States, but men of peace of all nations . . .'

In such piety is the schizophrenia of the ages.

Immediately afterward, Aldrin practiced kicking moon dust, but he was somewhat broken up. Either reception was garbled, or Aldrin was temporarily incoherent. 'They seem to leave,' he said to the Capcom, referring to the particles, 'and most of them have about the same angle of departure and velocity. From where I stand, a large portion of them will impact at a certain distance out. Several – the percentage is, of course, that will impact . . .'

Capcom: 'Buzz, this is Houston. You're cutting out on the end of your transmissions. Can you speak a little more forward into your microphone. Over.'

Aldrin: 'Roger. I'll try that.'

Capcom: 'Beautiful.'

Aldrin: 'Now I had that one inside my mouth that time.'

Capcom: 'It sounded a little wet.'

And on earth, a handful of young scientists were screaming, 'Stop wasting time with flags and presidents – collect some rocks!'

vi

There were as many as one hundred and fifty laboratories waiting for rocks. Five hundred of the world's best selenologists and geologists were preparing experiments to measure the age of the moon, its trace elements, stable isotopes, rare gases and particle tracks. The mineralogy of the satellite would be studied and its magnetic and electrical properties; the thermoluminescence, compressibility and elastic wave of lunar material would be examined as well; the soil mechanics, the solar radiation, the analysis of organic compounds, even the simple search for paleontological traces of once-living material were all going to be explored in the products of the two boxes of rocks the astronauts would bring back. There was a time-line on the EVA – they had hardly more than two hours on the moon, and a string of chores to complete, duties sufficiently numerous for them to keep plastic cards with a list of reminders taped to the wrist of their suits. Now, after almost an hour of Armstrong's oxygen supply had already been used up, now sensing that they have once again fallen behind in their sense of time, Armstrong began to work at filling the first rock box. Using a scoop, he picked up those rocks and fragments he thought most unusual or most significant, a gross collection selected as quickly as possible and inserted and packed in the rock box, which was immediately sealed with a vacuum-type seal. Later, they hoped to have time to fill the other box. There were plans to document the second collection of rocks, photograph them with a stereoscopic camera as they were picked up, for the position of rocks could offer significant clues to a trained geological eye. But they were already half an hour behind, and the Lem had still to be given a thorough photographic examination, the seismometer and the LRQ had yet to be deployed, and the core samples of moon dirt were still waiting to be obtained. The careful documentation of the rock gathering finally suffered. We might assume there was finally too much stimulation, too much near-familiar and subtly unfamiliar phenomena to be absorbed. The rocks themselves were full of unexpected variation. Some were as ordinary as cinders in an ash dump, others seen in the spectroscopic photographs are without dimension. One does not know if it is a photograph of a three-inch fold of rock or the full study of a ledge and a cave.

So they hopped around, prodigies of discipline, soldiers of caution. Standing on the edge of craters even six feet deep, the shadow was so dark they could have been looking down an air shaft to the cellar of the moon. Such phenomena must have teased the powerful fortifications of their common sense. Armstrong was to repeat over and over that the moon was friendly, the moon was hospitable. They were to say it again and again. It was presumably to the advantage of NASA that the moon be friendly to justify the outlay of those billions of bucks; never spend money on an ingrate! although in fact, the assumption was about as deep in real knowledge of public opinion as any thought the colored races of the world would be soothed by Muzak. How much more the world might have honored the exploit if the moon aroused anguish, awe and terror. No matter! It was friendly, it was beguiling. Afterward, Armstrong found himself describing again and again the mysterious properties of color on the moon. The terrain, by his description, was tan if one looked along it in the direction of the sun; it was the same tan if one turned completely around and stared at the land beyond one's shadow. But to right and left, at either side, the colors were darker, not tan but brown. Directly beneath one's feet, or looking at soil in one's hand, the color was dark gray, sometimes even black.

Then there were other phenomena. One is mentioned by Armstrong only once. He looks at his shadow on the ground and . . . 'Down-sun through a very, very light gray, light gray color, a halo around my own shadow, around the shadow of my helmet.' Yes, immediately after they had landed, they had spoken of how interesting were the colors. Twenty minutes later, immersed in routine, the colors were matter-of-factly described as tan and gray. Yet later there are halos, and color has become a function of the vector along which one looks. Aldrin, in turn, was having his troubles with soil mechanics. Aldrin, familiar with shovel and pick, was driving core tubes into the ground, but the resistance was a phenomenon. To drive the tube down even six inches, he had to use his hammer with such force that the top of the core tube was dented. The soil may have been loose to the touch of his foot, but it was almost as firm as rock just a few inches down. Still, it supported nothing. Difficult as it was to drive a tube into the ground, yet the tube would not stand. 'It was a unique, almost mystical environment.' Yes, soil mechanics like light, altered its properties with a change in direction. Perhaps the friendliness of the moon was also a matter of the direction in which one faced. A turn of the body in a dream can drop us into the long slide of a nightmare.

But now the time had come for them to pick up the rock boxes, pick up the foil of the solar wind, pick themselves up and the tendrils of their attention and reenter the Lem. Their EVA – it is now just after midnight – is done. They climb up the ladder, they close the hatch, bring up the pressure in the cabin. Now they take off their PLSS and their overshoes, and remove their helmets. They sniff. There is a pronounced odor in the cabin. The moon dust they have brought back on their suits now smells like gunpowder to Aldrin, and like wet ashes to Armstrong. The moon has a smell.

vii

Still, the EVA has been a gross disappointment. With every effort by Aquarius to find an edge of the sinister in this first expedition to the peculiar soil of the moon, the astronauts are obviously as equally determined – they must certainly be employed by NASA! – to make the moon a playground of the future. Tranquility Base! 'The moon was a very natural and very pleasant environment in which to work,' Aldrin reports after what excessive expenditure of BTU's it has taken a strong man like himself to drive a narrow pipe all of six or eight inches into the ground when the flight plan had called for twelve. Then the dented hard-smacked stick wouldn't even stand up. 'I was sure,' said Armstrong, 'it would be a hospitable host. It had been awaiting its first visitors for a long time.' The logic was impeccable. 'Come into my house, Joey Namath,' said the eighty-year-old spinster, 'We got a welcome for you, my sister and I.' No, sentiments conceived in buildings with windowless walls were pushing upon the very perceptions of the explorers. They could not help but like the moon – cold curse of their employers if they did not. Still, they were now back in the Lem; their EVA was done. They had only to open the hatch one more time, throw out some pieces of equipment, lock up again, go to sleep, and in the morning they would be ready to depart, perhaps never to return again. Can they conceivably have felt cheated? They would be a mirror to the sentiments of the world. If the moon was not sinister, then NASA was heir to a chilling disease, for they had succeeded in making the moon dull, the moon, that planet of lunacy and harvest lovers, satellite unlike any other moon in the solar system, the plane of its orbit even canted at an angle to the plane of the earth and the sun – no other moon could make such a claim – and besides the moon had properties of light so mysterious as to suggest that a shift of direction might be its equivalent for a passage of time, since a turn of the head could alter the mood of its colors from the afternoon. A step into shadow was a visit to night.

They were, however, not near to contemplation for this hour, nor for the next. Duties pressed to check over the systems on Eagle and prepare for the jettisoning of extra equipment. There was film still left in the magazines and additional pictures to take through the windows. It was the middle of the night in Houston but still young in the two-week-long day of the moon. The sun glared on the ground like the sun of the Southwest on desert wastes. Programs 8 to 13, relating to the ascent from the moon, were sent up from earth, and the Environmental Control Canister was changed. Their heart rates during the EVA were reported. Aldrin had reached a peak of 125, Armstrong had gone as high as 160 at the end, a very high rate, even higher than the rate at which his heart was beating during the descent.

But languor, or disorientation, or some intoxication of the moon kept delaying progress. The time-line in the flight plan called for them to jettison the Portable Life Support System packs an hour and fifteen minutes after they first closed the hatch, but now two and a half hours had gone by and still they were not ready to depress cabin pressure and open the hatch, get rid of the packs. Fatigue or some indifference to time had slowed them up again. Capcom asked, 'Do you have a time estimate for us until you're ready to start cabin depress? Over.'

'Fifteen minutes maybe?'

That would put them seriously back of the latest revision on schedule. They had been almost two hours late starting the EVA, they had lost an additional half hour during the EVA. Altogether, it was astonishing. Men who were never late had somehow consumed four extra hours in the last nine.

Deke Slayton was on the mike at Mission Control, 'I guess you guys know that since you're an hour and a half over the time-line, and we're all taking a day off tomorrow, we're going to leave you. See you later.'

Armstrong: 'I don't blame you a bit.'

Slayton: 'That's a real great . . . I really enjoyed it.'

Armstrong: 'Thank you. You couldn't have enjoyed it as much as we did.'

Slayton: 'Roger. It sure was great. Sure wish you'd hurry up and get that trash out of there, though.'

Now they sped up. It took them seven minutes instead of the estimated fifteen to begin depressurization of the cabin, and almost immediately the PLSS, the lithium hydroxide canisters, and the armrests from the Lem were thrown out. Fatigue, bemusement, and the rest, they must still have watched in fascination as the objects sailed away on a long lazy throw to the ground. Even the PLSS, which on earth would have dropped like a loaded barrel, floated off as far as a diver arching out from the high board.

Capcom: 'We observe your equipment jettison on TV and the passive seismic experiment reported shocks when each PLSS hit the surface. Over.'

Armstrong: 'You can't get away with anything any more, can you?'

Yes, this was the year in which for the first time a naked couple had fornicated on the New York stage in a play called *Che!* and man had landed on the moon. If hippies had fantasies of being measured and taped during extraordinary experiences, it was in fact the astronauts whom it was done to.

Capcom: '. . . a magnificent job up there today.'

Aldrin: 'Thank you very much. It has been a long day.'

Capcom: 'Yes, indeed. Get some rest there, and have at it tomorrow.'

The faintest trace of the Elizabethan had at last entered the language of the event. 'Have at it tomorrow.' It was almost three in the morning, but Aldrin responded, 'Did you-all come up with any other solution that we might try to the mission-timer problem?'

That was worked out in the next ten minutes, then there was a consumables update, then a list of ten questions relating to observations they had made on the moon that day. 'We can either discuss it,' said the Capcom, 'a little later on this evening, or sometime later in the mission. It's your option. How do you feel?'

'I guess we can pick them up now.'

So for the next fifteen minutes, at the end of this twenty-one-hour day in which they had entered the Lem, prepared it, separated from the Command Module, gone into a separate orbit, landed, and done a first walk on the moon, they were still in harness, still looking for more work. The ten detailed questions they chose to answer at three in the morning were on the position of their Lem, the depth of their digging, the rays beneath the descent-engine bell, the driving of the core tubes, the stroke of the landing struts and the possible existence of hills on the horizon which might block the solar rays during sunset, those questions and others. It was almost three-thirty in the morning when the astronauts finally prepared for sleep. They pulled down the shades and Aldrin stretched out on the floor, his nose near the moon dust. Armstrong sat on the cover of the ascent engine, his back leaning against one of the walls, his legs supported in a strap he had tied around a vertical bar. In front of his face was the eyepiece of the telescope. The earth was in its field of view, and the earth 'like a big blue eyeball' stared back at him. They could not sleep. Like the eye of a victim just murdered, the earth stared back at him.

viii

It used to be said that men in the hour of their triumph knew the sleep of the just, but a modern view might argue that men sleep in order to dream, sleep in order to invoke that mysterious theater where regions of the unconscious reach into communication with one another, and charts and portraits of the soul and the world outside are subtly retouched from the experience of the day. If this is so, what a gargantuan job of ingestion had fallen upon the unconscious mind of the two astronauts, for the experience of their world would not include the moon. Deep in a state of exaltation and exhaustion, tonight – it is now four in the morning – was hardly the time to embark upon the huge work of a dream which could begin to feed into the wisdom of the unconscious those huge productions of the day behind. If their senses had been witness to sights and sensations never experienced before, their egos were also in total perturbation, for on the previous day their names had been transported to the eternal moonlight of the ego – they were now immortal. It is not so easy for men to sleep on such a night, for they know their lives have been altered forever – what a dislocation of the character's firm sense of itself; in fact it is precisely in the character of strong egos that they are firmly rooted. Now, they are uprooted and in a state of glory. What confusion! A disorientation of the senses and coronation of the ego are the problems to be approached this night by the dream, and that while lying in the most uncomfortable positions possible in the foreign skin of a pressure garment while the temperature grew chilly inside that stiff sack, for even with water circulation down to a minimum, the suits insisted on cooling their tired heat-depleted bodies.

How indeed to go past the threshold and enter the great chamber where the kingdoms of sleep will greet them with a revel equal to the hour, no, no man in a state of exhaustion would dare to chance the rigors of a powerful night of dreams, for important decisions which may shape the future can be decided on bad nights by the poor artwork of dreams not sufficiently energetic or well enough conceived to offer the unconscious a real depth of answer – which is why perhaps insomnia is self-cycling, for, too tired or fearful to engage in serious dreams one night, we are even more exhausted for the next, and do not dare to sleep. Who will be the first to swear that deep contracts of the soul are never sworn to in the darkest exchanges of slumber, or failure too quickly accepted in the once ambitious hearts of exhausted men? No, the astronauts were in no shape to sleep. Just so quickly ask a computer to work when its power supply is erratic, its mechanical parts need oiling, and it has just been instructed to compute all further trajectories as if earth gravity did not exist and the moon were motionless before the sun!

Yet, all of this – this kaleidoscope of impressions, this happiness of the heart, sore and tender and merry in the very pumping of its walls, these thrifty ducts of aviator's love for family, children and mate open now if ever open, the patriotic incandescence of a dialogue with the President, the delivery of a job promised to the team – these tributaries of happiness carrying the uprooted tree of the ego and the mysterious house of the moon downstream in full flood are still only half of their inability to sleep. Few men could sleep with such happiness flowing unaccustomed in them, but now add a fear which has been kept in the vaults, a firm well-regulated natural concern of the executive mind, yet a fear not even primitive, but primeval in its uncharted depths: are they going to be able to ascend from the moon? It is one thing to shatter a taboo, it is another to escape the retribution which follows the sacrilege – where is the savage society whose folklore is not crowded with tales of the subtlety of every outraged curse? It is a fear which still lives in many an athlete and celebrity, in many an excessive modesty in any poor man who has found at last some luck, it is in the groveling of a dog at what comes next after he has won a fierce fight. The danger is always greatest just beyond victory – in some men that is a deeper belief than any other, for not yet at climax they can see themselves as deserving; once triumphant their balance has shifted, they know guilt, they are now not deserving. Well, whether the astronauts were deeply superstitious in this fashion or barely superstitious at all, we can be certain that any residual of this prime and hallowed fear would be awake in them tonight, for they were not by any measure yet free of the moon. In the morning, after all preparations were taken, a moment would still come when they would have to fire up their engines and lift off in the upper half of the Lem from the descent stage left behind. The ascent stage would rise if the ascent motor functioned – they were doomed if it did not, for all the redundancies of the equipment passed here through the bottleneck of one and only one piece of mechanism. There was no substitute for the ascent engine. Double tanks of fuel, and double tanks of oxidizer; double containers of helium to put pressure on fuel and oxygen; valves and cutoff valves in a plenitude of substitutions and alternate paths; but, finally, there was only one motor with only one throat and one bell, and that motor would have to flame up to 90 percent of its full thrust in the first three-tenths of a second after it ignited so that the ascent stage would lift and not settle back – no refinements to blast-off here!

There had been tests beyond measure on that motor, tests in vacuums and tests in fast-descending elevators to simulate lunar gravity; there had been refinements inserted within refinements to make certain that when a fuel spray of hydrazine and unsymmetrical dimethylhydrazine came through the injectors to meet a spray of nitrogen tetroxide the combustion would take place.

Since it had been designed for ignition in a lunar vacuum, so there was no air to feed the fires of the fuel. The equivalent of air was supplied by the nitrogen tetroxide, whose oxygen would sustain the burning hydrogen, whose oxygen would indeed ignite the hydrogen by merely meeting it. The very elements of water were here the elements of a fire so contagious it needed no match, merely the mating of the gases, a fire so explosive that a motor with a nozzle only two and a half feet across could lift the ascent stage and the men inside, lift them and fling them into orbit, all ten thousand pounds and more.

Still, this motor had never been fired before on the moon. It had been fired in vacuums, yes, but they were artificial vacuums on earth and not the pure vacuum of the moon where who knew what subtleties atomic particles, subparticles, and cosmic rays could present upon ignition? Nothing in any theory or working hypothesis of physics even began to suggest there was any reason why a moon vacuum should not prove the practical equal of an earth vacuum, but nobody could be certain, nobody could swear there were not unforeseen conditions which could inhibit the flame or cause it to flame out. Who knew the dispositions of fire on the moon when the air we breathed was also the stuff of fire, and hydrogen and oxygen could make water or electricity or fire? Yes, the real explanation of the flames remain as much a mystery as man's first hour in a forest after lightning struck a tree.

Primeval fears inspire primeval thoughts. There in the Lem, one body on a floor, the other with his heroic posterior carefully spotted on the very cover of that ascent engine which would lift them off in the morning, how they must have drifted on runs of happiness and rills of deep-veined fear. How easily they must have passed into large sleep-deprived inchoate thoughts of a world of men and women back on earth secretly wishing them well or ill, intervening in the long connected night of the world sleep with whatever gods or powers were sitting upon the ignition of the engine in the morning. So scourged and exalted, hovering on that ultimate edge of moral balance where one wonders if the sum of one's life has been for good or ill and if the morning will return a fair and just verdict, fingers crossed, ready fair enough to laugh or cry, dopey, exhausted, chilled, feverish no doubt with desire for morning to come, alert, high on the empty holes of the numbers they would punch into the computer in the morning, ears alert to the quiet pumps of the nocturnal Lem whose skin was baking in the moon morning heat, there in a cave of chilly isolation, happy, numb, and full of a fear of dreams, not knowing if their glory was to be doubled by the next night or if they would be at one among the martyred dead, yes, how were they to sleep and dare to dream when the future must look either to a transformation of the psyche, or a trip down the underground river of them all? Yet, if they were to die on the moon – was there an underground river there? or would they be forced, full strangers, to wander together, a queer last place for the mortal bones. Who, indeed, could sleep?

ix

Collins was listening to the pumps. He had been alone in space for fifteen hours, more than enough to go around the moon seven times, and
for seven times he had been alone on the far side, which was seven more
times than any other man had ever been alone there. Even the Public
Affairs Officer was ready to use superlatives for the occasion. 'Not since
Adam has any human known such solitude as Mike Collins is experiencing during this forty-one minutes of each lunar revolution when he's
behind the moon with no one to talk to except his tape recorder aboard
Columbia.' Shades of Krapp's Last Tape, one had another facet of Beckett's vision of the apocalyptic loneliness of the end; why, even the near
side of the moon was civilization in comparison to the far side. There,
out of contact with any voice on radio, alone in his Command Module,
rattling around in his new commodious dimensions with the center
couch removed and no other astronauts to break into his peace, Collins
and his machine were like a coin on edge in the universe, the acme of
technology, the acme of uprooted existence. Yet later Collins spoke
warmly of what he had come to call his 'mini-cathedral.' The tunnel
which connected up to the Lem was now like a bell tower; with the
couch removed, he had a center aisle; the instrument panel could serve
as nave and transept. Later he would protest that he was not lonely in
Columbia. 'I've been flying airplanes by myself for about seventeen
years, and the idea of being in a flying vehicle alone was in no way
alarming. In fact sometimes I prefer to be by myself.' Still, it is a protest.
He was obviously filled with uneasiness much of the time; his later
comments give more than one evidence of it. 'I figured that any chain
as long and tenuous as this had to have a weak link. Believe me, I spent
a lot of time worrying about that link. Could I be it? Could my training
have neglected some vital bit of information? Or had I been properly
exposed but simply forgetful?' A phrase later, he is stating that by launch
day he had convinced himself he had taken all the steps to prepare himself, at least within reason, but it is a remark without logical substance
against the larger fear of something loose and unknown in the material,
or something treacherous in himself. He had a curious job in the morning. If the Lem lifted without difficulty and came into its proper orbit,

the rendezvous would be simple. The Lem in fact would fire all the moves and do all the work to join him. He need merely stay in his orbit. If something went wrong with the Lem's ascent however, and it went into odd orbit or low orbit or if it had to fire off prematurely and therefore might be half the circumference of the moon away from him, why then he would have to make the maneuvers. Some might be difficult. Rendezvous could always prove deceptive, for one was able to approach another spaceship by accelerating one's speed but also by braking it. A reduction of speed reduced one's orbit and so reduced the time it took to circumnavigate a planet – one could thereby catch another spaceship. It was also possible, however, to catch up with the other spaceship by accelerating, provided one was certain of the rendezvous. If you missed the other ship, however, the price was to go rocketing off into a long and wasteful orbit; the amount of fuel they could spend in such maneuvers was limited. Moreover, rendezvous, if the circumstances were unusual enough, might have to be independent of Mission Control. Suppose immediate maneuvers were necessary on the far side of the moon. Then there would not be time to come around from Loss of Signal and get detailed programs from the ground – Collins might have to work out the computations on his own computer, or in a deteriorating situation make rough estimates without a computer and hope he could sight the Lem. It was an uneasy situation, reminiscent of the favorable odds on the ascent engine. By any logical or practical measure those odds had to be 100 to 1 or 1000 to 1 in favor of a good ascent; yet in those unspoken fears where wonder resides about the real nature of the universe, the unspoken odds were nearer to even.

So, too, with rendezvous was nothing likely to go wrong, yet Collins was obliged to live in the readiness for everything to go wrong. But if any disaster occurred to the others, how could he be certain afterward that some envy in himself had not triggered the result? It was not a happy position to be in, and Collins was the chief worrier of the three astronauts. Over and over the transcript is filled with his finicky insertions on the difficulties and small discrepancies of the equipment. Less of an engineer than Aldrin, less of a pilot than Armstrong, and in a flicker of doubt about his own body after an operation had been necessary to remove a growth from his spine (which if uncorrected

could have left him paralyzed) Collins was one of those men who nib-
bled on the details of routine programs and data loads the way people
who are bored at a party nibble peanuts. Rare was the astronaut who
could be happy as the pilot of the Command Module; Collins with his
quiet but wholly intent competitiveness was not going to be the first.
Besides Collins had suffered an added tension in the day which had just
passed. The Lem, during the excitement of the 1202 alarms, had failed
to land in any place Mission Control could locate precisely on the map.
Collins had spent the day searching the moon ground from sixty miles
up for some glint of the Lunar Module, but in five passes had had no
luck, not even in a painful eye-screwing search over one grid square
after another. While that was not great cause for anxiety since the Lem
could orient itself quickly enough once up in orbit, it was like a bad
augury. He was out of touch with the men he must meet tomorrow in
rendezvous.

Besides, there had been hordes of proper deportment required of
him throughout the day, bountiful copious congratulations to offer the
other astronauts for their glory. Oversweet may have been the recollection of his voice.

CAPCOM: *Columbia, this is Houston. Say something. They should be able to hear
something. Over.*

COLUMBIA: *Roger. Tranquility Base, it sure sounded great from up here. You guys did
a fantastic job.*

ARMSTRONG: *Thank you. Just keep that orbiting base ready for us up there now.*

What formal relations! It is as if the winning captain is patronizing
the losing captain. Collins' loyalties were certainly not tenuous – no
man could remain an astronaut without a strong sense of responsibility –
but Collins was being tried in a court of highest pressure. If there was
any joy to be alone in the Command Module, alone with himself in the
incommunicable regions of the far side of the moon, nearer then to the
messages of the outer galaxies than ever before, if Collins was free to
bathe in the pleasure of lonely thoughts like no man ever had before, if
Collins could even indulge the legitimate narcissism of the pilot who
lives in a machine which is an exquisite extension of his will, well,

Collins also had all the anxiety of listening to every tick in the murmur of the pipes, every slip and minuscule clutch of sound in that machine which transported him, and if ever a man felt the anxiety to think cheerful rather than evil thoughts for fear he would spring a glitch in the labyrinthine conjunctions of his machine, Collins had to be that man. There were unhappy precedents for being too long alone in space. Collins had to be aware, as all the astronauts were, of what had happened to Bonny, the monkey who had gone up in orbit around the earth on Biosatellite 3 and had come down nine days later, three weeks ahead of schedule, in an emergency splashdown. Bonny had been ill and died twelve hours later from causes which could not be determined. An unpleasant business. Bonny's death had been but a week before their own launch. Now, as much alone as Bonny, did Collins think for an instant of that trained fourteen-pound space animal who had become sluggish in weightlessness and ceased to perform his tasks? Did Collins wonder if the animal had sickened out of boredom, or out of the misery of its pure animal heart being unable to pick up a clue to where it might now be located in its weightless cage, or did Bonny begin to sicken and die because of some drear but most recognizable message its animal senses had received from space, some message too fine for the insulated nerves of man to receive? Yes, did Collins brood on Bonny? It was a long seven revolutions and then some more.

Yet Collins, at least, was able to sleep. Did he dream of Ags and Pings, of REFSMMAT, IMU and EMU? Did he descend to the first disorder of the dream with DSKY in hand and DAP and POO? To bore with one's brain into the hard-stuffed methods and modes of technology did one not also go back to those chaos-holds, those ledges of meaning that meaningless words provided in infancy as a set of arbitrary stations of sound which were somehow better and less chaotic than no sound? Acronyms! Collins slept.

X

And in Mission Control down on earth in the black reaches of the night, the Black team would normally have been clowning around. They were the Blacks, they were the lonely, they were the youngest engineers with the lowest seniority, who invariably worked the consoles while the astronauts were asleep. Theirs was the job with the least to do. Whether called Black because they worked through the night, or Black out of some brimming class humor at NASA, their job had the pleasure of frequent breaks for coffee and wild technological discussions about how much it would cost to build an actual and real superhighway U.S. Moon 1 – how many thousands or millions of Apollo-Saturns to send up to do it brute and direct, as opposed to how many if factories were built over lunar ground. On quiet nights they could put the question into an unoccupied computer and get back answers which opened other games. The computer on such nights was their farm animal washed and ready for picnic.

But the coffee was chilly coffee tonight, as cold and dank with anxiety as the plastic of their consoles. There were dungeons in the liberty of this moon-conquered night. The vault of silence in the Mission Operations Control Room (display for Eagle at rest; display for Columbia slowly crossing the screen) was there to offer a rebuke to any levity. In fact, the Black team was not even on. They would do work for the rendezvous later, on assignment out of turn – it was the plum which had been given them, even if the astronauts were to do most of the work.

xi

Two hours of sleep. One hour to sleep. And no real sleep for Armstrong and Aldrin. They are heroes, they are first among their peers, the knights of the silent majority, but they are suffering from insomnia. They have finally emerged onto the landscape of the modern novel. They emerge incidentally as promotion men as well.

PAO: *This is Apollo Control. Let's join the call to Tranquility Base.*

CAPCOM: *How is the resting, standing up there? Or did you get a chance to curl up on the engine can?*

ALDRIN: *Roger. Neil has rigged himself a really good hammock with a waist tether, and he's been lying on the hatch and engine cover, and I curled up on the floor.*

xii

At 12:53 on Monday morning, July 21, not twenty-two hours after they landed, with rendezvous radar put most carefully in Off position to avoid new program alarms, with hearts beating, fair to assume, and with minds wondering for the last occasion whether fire might not lose a vital property or two in the immediate domain of the moon – time, after all, was known to alter at the speed of light; with Pings loaded with every bit of data for Program 12, the Powered Ascent Guidance; with the simple ascent motor incapable of being run at anything other than full throttle they now put the master arm on, gave a last count to Mission Control, 'Forward 8, 7, 6, 5, Abort Stage! Engine Arm Ascent! Proceed!' and fired off from the moon.

The detonator cartridges exploded on time to separate the thorax of the Lem from the sac; the ascent stage rose, the descent stage remained. Just before that separation, all signal and electrical power between the two was sundered. Then the nuts and bolts joining the stages were also exploded. In the same fraction of a second, an explosive guillotine severed the connecting spine of wires, cables, and water lines between the two stages. The ascent motor flamed up to 90 percent of full thrust in three-tenths of a second, and with hardly more than a big jerk and a blast! and a Proceed! they rose off the moon in a wobbling climb, oscillating from side to side as their fuel sloshed in the tanks.

Aldrin: 'That was beautiful. 26 feet, 36 feet per second up. Be advised of the pitchover. Very smooth . . . very quiet ride. There's that one crater down there.'

PAO: '1000 feet high, 80 feet per second vertical rise.'

Later Armstrong would say, 'a beautiful fleeting final view of Tranquility Base as we lifted up and away from it.' Did they have the recognition at this instant that on another day there might be lunar cities under domes, and moondromes with their names? On climbed the ascent, up half a mile in the first minute, its direction no longer vertical, but tipping out, then pitching over toward the eventual curve of its orbit. Behind them was the memory of the blast-off, the Kapton and all the other loose-wrapped plastic insulations of the descent stage being blown in all directions, far out in the bulletlike trajectories of the moon, all that plastic, silver and gold debris, and behind them – first refuse of the first moon city – was already the handle of the rock contingency sample, the TV camera and its tripod, the staff for the solar wind experiment, the passive seismometer, the closeup camera, the Laser Ranging Retro-Reflector and its packing materials and brackets – there had been over a hundred brackets and they were now strewn on the moon ground – and there were two backpacks of the PLSS also left behind and overshoes, and tramped ground for a hundred feet around the descent stage, ten thousand prints of the marks of their boots on top of other marks of their boots, messy as a bivouac where troops have been milling in the rain. If men never came back, those marks might remain for millions of years. And the motionless waving of the flag.

But that was behind them, and their little wobbling ascent stage climbed up through its oscillations and out into the sea of space.

ALDRIN: *We're at 3000, 170 up, beautiful . . . 1500. 185.*

CAPCOM: *You are GO at 3 minutes.*

ARMSTRONG: *We're going right down U.S. 1.*

Rising right out of their dread; they were leaving the loneliest death in the world. If that ascent engine had not worked – there were no suicide capsules on the Eagle. They would not have needed them. When the frustration of being trapped on the moon proved too great, they would only have had to open the hatch, and remove their helmets. That could not have felt much worse than being a drowning man. But now they slung themselves down that track, pouches of fuel tanks carried like chaws of tobacco, one for each cheek of the ascent stage. On they came up into orbit.

CAPCOM: *Eagle, Houston, 4 minutes . . . everything's great.*

PAO: *Horizontal velocity approaching 2500 feet per second.*

ALDRIN: *Now we got – got Sabine off to the right now . . . There's Ritter out there. There it is right there. Man, that's impressive-looking, isn't it?*

CAPCOM: *Eagle, Houston, you're looking good.*

PAO: *One minute to go in the burn. 4,482 feet per second horizontal velocity.*

ALDRIN: *About 800 to go. 700 to go. Okay I'm opening up on the main shutoffs. Ascent feed closed. Pressure's holding good. Crossfeed on. 350 to go. Stand by on this engine arm. 90. Okay, off. 50, Shutdown . . .*

PAO: *Showing a perilune of 9.4 nautical miles, apolune of 46.7 nautical miles on the PNGCS. Shutoff velocity showing about 5,537 feet per second.*

In seven minutes and eighteen seconds after fires had been lit, they had consumed their fuel, turned off their ascent engine, and were coasting over a mile a second in an eccentric orbit around the moon, something like forty-seven miles from the surface at farthest point and nine and a half miles up at the nearest or perilune, at which point they had just arrived.

Capcom: 'Eagle, Houston . . . The whole world is proud of you.'

Armstrong: 'We had a lot of help down there.'

PAO: 'Flight Operations Director Chris Kraft commented that he felt like some five hundred million people around the world are helping push Eagle off the moon and back into orbit.'

During the ascent, they had monitored the heart rates. Armstrong had been hardly above normal, pegged at 90, a low figure for him. Aldrin had been up at 120. It was the only time in the flight that his heartbeat had been higher than Armstrong's. Is it possible that Aldrin was feeling a new sense of dread at the oncoming rendezvous? It would be a curious state to find oneself in after the worst technological moment has just been passed in a flying test.

xiii

To hold a gyroscope in the hand is to obtain an inkling of orbit. There is a sense of energy revolving in a powerful pattern, of rapid movement in some alliance with rest, for a gyroscope offers to the palm a sensation of high speed and high stability, as if all its activity is devoted to being precisely where it is.

So, now at high speed and yet in no more than a fast-moving corollary of the state of rest, the Lem is in orbit and the Command Module is in orbit, each traveling in elliptical rings around the moon, Columbia in the outer ring, which is close to a circle of sixty nautical miles in diameter, while Eagle is in the ellipse we have described of 46.7 by 9.4 nautical miles. It has taken a burn of 438 seconds in its ascent motor to reach this stage and it has consumed two thousand pounds of fuel and three thousand pounds of oxidizer in four tanks whose volume was each thirty-six cubic feet. Now it would go the rest of the way to joining the Command Module on small increments of velocity or small braking burns offered it by its four quadrants of thrusters and they are fed by two tanks of fuel and two tanks of oxidizer whose volume is no more than two cubic feet each, or taken all together, the four thruster tanks could fit into a cube with two-foot sides. The amount of energy capable of being released under fire by that much hydrogen and oxygen will be sufficient to close the gap which remains between Eagle and Columbia.

In a sense, the critical part of rendezvous has been finished already. The Eagle was only obliged to get into some kind of orbit. If its ascent motor had failed any time in the last sixty seconds of ascent, its thrusters would have been able to drive it the rest of the way into that first planned ellipse, and then the Command Module could have descended for rescue. Indeed, once Eagle succeeded in getting into any kind of orbit at all, the Command Module would be able to come down for it, but all orbits below six miles of altitude were dangerous indeed, for then Eagle would have no ability to clear mountains. Rendezvous maneuvers might have had to be speeded up. One could even conceive of a cinematic rescue with Columbia accelerating down toward the moon, slowing just long enough only for Armstrong and Aldrin to open the hatch, crawl outside, get a handhold on a quadrant of Columbia's thrusters and Sput! Sput! Columbia would be on her way up again as the Lem coasted into a lunar peak.

That was hardly the operation today. It proceeded smoothly. A little while after Eagle coasted up to her apolune of 46.7 miles above the moon, she fired her thrusters for a little more speed and was inserted into Concentric Sequence Initiation, a rough circle forty-five miles in diameter which was situated within the rough sixty-mile circle of the Command Module. Almost an hour later, halfway around from that position, came another burn, called the Constant Differential Height, whose purpose was to make certain that if both ships were traveling in concentric ellipses, the distances between them did not vary. (All this, well planned in advance, was readjusted in flight by measurements taken from the Inertial Measurement Unit, then calculated by the guidance computer, as well as by checks obtained from rendezvous radar.)

About thirty-eight minutes later, Eagle was ready to begin Terminal Phase Initiation. That was at a point where Columbia was thirty miles in front of her and seventeen miles overhead. Driving forward at a small increase of velocity, which closed distance at about a nautical mile a minute, on an angle 27 degrees above her local horizontal, Eagle swallowed the last gap in something less than forty-five minutes. The two ships came within view of each other in less than four hours and less than two revolutions from the time lift-off had occurred. Collins reported a great feeling of relief at seeing them come up toward him. 'I really got excited then because for the first time, it was clear they had done it. They had landed on the moon and got off again.' They had in fact come up all the way on their own power, with Columbia – power available not only in her thrusters but in her main propulsion motor – maintaining all the tools and options in reserve. It was more elegant to solve the problem with the lesser means. Besides, it left Columbia in possession of more fuel for the trip back.

Now came the last maneuvers. Little braking burns to put their velocity equal to one another were done in an operation called stationkeeping: now they could wheel through lunar space close enough to take photographs of one another. Let us listen to the transcript as they approach. It is very calm. What has happened to Aldrin's anxiety of a few hours ago?

ARMSTRONG: *Okay, Mike. I'll get – I'll try to get in position here, then you got it . . .*
I'm not going to do a thing, Mike. I'm just letting her hold in attitude HOLD.
COLUMBIA: *Okay.*
ARMSTRONG: *Okay, we're all yours.*
COLUMBIA: *Okay. Okay, I have thrusters D3 and D4 safetied.*
ARMSTRONG: *Okay.*
COLUMBIA: *I'm pumping up cabin pressures.*

The docking took place with a light touch. Collins never even felt the two ships meet. The probe of the Command Module slid into the Eagle's drogue. 'They're held together then,' Collins said, 'by three tiny capture latches, and it's almost like tiny little paper clips holding together two vehicles, one of which weighs thirty thousand pounds, the other five thousand. It's a tenuous grasp. To make the combination rigid you fire a little gas bottle that activates a plunger which literally sucks the two vehicles together. At this point the twelve capture latches fire mechanically and you are held together very strongly. That's the hard dock.'

Just before that moment 'all hell broke loose.' It was Collins' remark, there on the transcript, but he has no recollection of saying it. As he fired the charges, there was an abrupt, shocking and 'abnormal' oscillation. The ships began to yaw from side to side at a rapid rate. What an instant for Armstrong – did the memory of the sun flashing through the window of Gemini 8 come back to him? What a thunder for Aldrin after the mishaps with the computer on the day before, what a stroke of doubt for Collins at where the mistake could be. 'All hell broke loose.' Hell was when the unforeseen insisted on emerging. Shivering and quivering, the ships slapped from side to side.

Well, it lasted for 'eight or ten rather dubious seconds,' while Collins and Armstrong worked to get back in line with one another, and all the while the automatic retract was working and they finally came together with a big bang and were docked 'and it was all over.'

It was essentially all over. The chores were done, the suits were vacuumed of moon dust, the tunnel was opened, and they met and shook hands, no comment recorded for posterity, and then passed the rock boxes through to Collins who handled them 'as if they were absolutely jam-packed with rare jewels which' he adds, 'in a sense they are.' Now the hours were spent on the details of final housekeeping aboard the Lem, in the final transfer to the Command Module, the repositioning of the probe and the drogue. Soon the Lem would be jettisoned. The two ships would separate and Eagle would ebb away at a few feet a second. Once it was out of sight, it would never be seen again. Still, an essential part of it would certainly expire in full view of their recording instruments for the primary cooling system, which kept the computer from overheating, was disconnected before they left. As the hours passed, the computer on the Eagle kept sending data, but the signal became weaker and weaker. Finally it died. Pings with her sneak circuits, Executive Overflows, and DSKY was dead. Long before that, they had fired up their Service Propulsion Motor on the back of the moon, and the

Command and Service Module had come over the hill with velocity sufficient to throw her out of lunar gravity and out on her long way home. Pings on the Lem was not in fact to die until after midnight when thousands of miles away the crew was settling in for a real sleep.

The trip back had begun. They had sixty hours to spend in modest work, in a repetition of chores, and in a great deal of thought. It is doubtful if they brooded too long on the wild gyrations before the hard dock. Possibly there had been more concern at Houston. For the Lem was a machine of machines, a beauty of a beast which had never seen work on earth. She had habits like a horse who was crazy once a year. She had taken off on Cernan for a rampant little trip, she had dipped this crew toward the moon, then, as if magnetized by magnetisms never quite measured before, charged with some sense of person from the remarkable components of her crew, or by her contact with the moon, charged perhaps by some sensitivity to the difference between the men in herself, and the man in the Command Module she soon would meet, she had quivered, or the Command Module had quivered, or both had quivered, space machine to moon machine, quivered like magnets which approach and gyrate when suspended on a string. Something happened up there which no one could explain, something once again had stirred the hairs in the secret cave. The psychology of machines was a whisper in the dark and sultry Houston night.

xiv

It is almost sixty hours before reentry for the astronauts, and their return will be without events of the largest scope. Collins will grow a mustache, and Mission Control will report that the crime rate in Italy was at a low for the year on the night they walked the moon. A girl will be born in Memphis who is christened Module McGhee, and a boy named Greg Force will fix a bearing on the huge antenna in Guam because he is the only one whose arm is small enough to reach into the hole. The Capcom on duty the following night will mistake the moon for the earth on a murky television screen and the astronauts will have the experience of seeing the earth and moon looking equal in size out opposite windows. Slowly the earth will grow in the window. Blue she will gleam and brown and gray and silver and rose and red. Her clouds will cover her like curls of white hair, her clouds will turn dark as smoky pearls and the lavender of orchid, her clouds will be brown and green like marsh grass wet by the sea, and the sea will appear beneath like pools of water in the marsh grass. The earth will look like a precious stone, blue as sapphire, blue as a diamond, the earth will be an eye to look at them in curious welcome as they return. They have been as far as Achilles and Odysseus, as far as Jason who sailed to meet the argonauts, far as Magellan and Columbus, they have been far. And their fingertips are smooth from plastic, their lungs are leather from days of bottled gas. What does an astronaut give up of the ultimate tastes to travel so far? We are back to Aquarius moldering on flatlands not far from the sea.

PART III

The Age of Aquarius

The Hanging of the Highwayman

There was a melancholy to the end of a century. The French, who were the first to specify a state for every emotion, would speak of the *fin de siècle*. It was the only name to give his own mood, for Aquarius was in a depression which would not lift for the rest of the summer, a curious depression full of fevers, forebodings, and a general sense that the century was done – it had ended in the summer of 1969.

If he had had his extraordinary night of insomnia in Houston, and had thought his way out, well, that was just for one night. The woes of these hot weeks sat upon him as he slept and as he worked. He was used to writing in moods so bad he could assume he was passing through a swamp at midnight – some of his best work had come out of periods worse than this, and some of his worst efforts had emerged from hours which had been too pleasant. It was almost as if he had to suffer while working in order to come closer to exercising some more ultimate faculty of judgment. It was a terror to write if one wished to speak of important matters and did not know if one was qualified – sometimes the depressions helped to give sanction to the verdicts taken. It was not so unreasonable. The question is whether it is better to trust a judge who travels through his own desolations before passing sentence, or a jurist who has a good meal, a romp with his mistress, a fine night of sleep, and a penalty of death in the morning for the highwayman.

To write was to judge, and Aquarius may never have tried a subject which tormented him so.

ii

He had come home the day before the astronauts came back to earth. There were splashdown parties promised all over Houston, and he was tempted to remain, for there would be portraits in plenty to paint of Texas drinking and poolside brawls, but he was also in a panic to get back and start work – his first deadline was not three weeks off. Each extra day at this end could be a reprieve at the other. Still much in the middle of the event, mental digestions churning, he returned to the bosom of his family.

The house was sour; the milk gave every intimation that it had curdled. His wife and he were getting along abominably. They had had hideous phone calls these last few weeks while he was away. Several times, one or the other had hung up in the middle of a quarrel. It was impossible to believe, but they each knew – they were coming to an end. They could not believe it for they loved their two sons as once they had loved each other, but now everything was wrong. It was sad. They had met on a night of full moon, and would end in the summer of the moon. Sometimes his wife seemed as if deranged by Apollo's usurpation of the moon. She was extraordinarily sensitive to its effects; she was at best uneasy and at worst unreachable when the moon was full. Through the years of their marriage Aquarius had felt the fullness of the moon in his own dread, his intimations of what full criminality he might possess, had felt the moon in the cowardice not to go out on certain nights, felt the moon when it was high and full and he was occasionally on the side of the brave. And she was worse. Call her Pisces for the neatness of the scheme. Beverly, born sign of Pisces. She was an actress who now did not work. An actress who does not work is a maddened beast. His lovely Pisces, subtle at her loveliest as silver, would scream on nights of the full moon with a voice so loud she sounded like an animal in torment. They were far and away the noisiest house on the street.

It hardly mattered in Provincetown. That was the land of the free. At the very tip of Cape Cod, a fishing town curled around a spiral of land whose sand dunes separated the bay and the sea, it was a town of Portuguese and Yankees in winter, of artists, faggots, hippies, bikers, debs, dikes, off-course jets, groupies, and beefed-up beer-drinking tourists from Jersey in summer, not to mention hordes of middle-class professionals with progressive views and artistic liens. An isthmus of quiet in the calm months, it was no island of the mind in July, no, it was the Wild West of the East, and it took forty-five minutes in the middle of August to drive a car half a mile down the one-lane main street. Marijuana was as available in Provincetown in the hideous hocks of summer as popcorn is plentiful in a drive-in movie in Iowa. Aquarius, of course, had none of it, not these years, not when working. He could not afford it. His brain was always lost for the following day. But now he had to work in its presence. There was hardly a dentist, a psychoanalyst, a townie, or a narco agent who was not turned on half the time, and the drinking parties among the most sedate began at five and ended at five with the dawn coming up his window on the bay, the gulls croaking their readjustments to all the twisted vertebrae of sand and sea. Stoned out of the very head of sensation, the summer populace was still groping and brooding and pondering its way down the gray and lavender beach in the red-ball dawn, sun coming over the water in one long shot of fire – Provincetown was the only place in the East he knew where the land spiraled so far around that you could see the sun rise out of the dunes in the east and set in water to the west. What a town! There was not one of his wrong and ill-conceived books he had not written in part here, and all of his good books as well, all of his books. He had learned how to work in summer if he had to, but one needed the skill of a contemplative who pitched his tent by a hot dog stand. So he hated his beloved Provincetown this summer above all.

It had been bad from the moment he was back. One of the early nights after his return, perhaps two or three days after splashdown, he had taken his wife and one of his best friends to a restaurant for dinner. The friend was Eddie Bonetti, a battered knob-nosed working writer out of South Boston, handsome as an old truck to those who knew him well, a small rugged prodigy of talent who had boxed a few fights professionally and been given working lessons in the gym by Willie Pep. Bonetti wrote poetry, perhaps he was the best working poet in Provincetown, certainly the best Aquarius had heard, and he had written a very good short novel about an old Italian making wine, a manuscript which was always on the edge of getting published by editors who were almost ready to put up with its brevity and its chastity – like many an Italian before him, Eddie Bonetti did not swear in print.

'Norman, I'm so fucking glad you're back,' he declared for the fourth time in his loudest voice five minutes after they sat down in the restaurant.

Bonetti stored his talent in many places. He had acted in two of Aquarius' movies, memorable in a small part in one, unforgettable in another. He had played an axe murderer who killed his wife after fifteen years of marriage; Aquarius was fond of saying that Bonetti was as good an actor as Emil Jannings for one night in his life. But that had been in passing. Bonetti also grew the best tomatoes in town, and had been known to play his flute to them in the middle of the night. Eddie was also capable while riding a bicycle down the main street (if he saw a friend driving behind him) of jumping his bike off the street across the sidewalk and into the bushes, where he would take a wild dive over the handlebars into the grass, just to give his friends the craziest laugh of the week. Bonetti could say, 'I'm worried about my heart,' and fall immediately on his back, there to wink at you. Bonetti was a prodigy of talent.

But he was drunk this night. He was drunk before the evening began. Because he had a big punchy sepulchral voice even in the quietest of times, it was booming everywhere tonight on his drinks. 'Fuck, Norman, I didn't know whenna fuck you were gonna get back,' he bellowed again in his best Savin Hill South Boston tones and the carnal communicant quavered like an organ pipe with a crazy nonstop overtone in the clean white tablecloth Wasp spa to which they had gone, an error of incomparable dimension, for Eddie in his dungarees and blue sweat shirt was as funky as the upholstery in the last used car on the lot. His clothes were in line for nine out of ten restaurants in town, but not where they were now – indeed Aquarius had picked it to obtain some afterthoughts on the moon shot. But Bonetti had a good century-old stiffening of his drunken proletarian senses when they walked in. No restaurant was going to put *him* down. So Aquarius, proud Aquarius, iconoclast of the last two decades, was obliged to act as a middle-class silencer, 'Will you keep your voice down,' he blasted in a hopeless murmur.

'Norman, this place is filled with drunken assholes. Fucking drunken assholes.'

'Eddie, I'll give you two to one you can't go through the meal without saying fuck.'

'Norman, I don't want to take your money.'

The bet was made. Eight dollars to four dollars. Before three minutes had passed, Bonetti had lost. Aquarius bet him again. Another two minutes and Eddie said, 'These shrimp are fucking good shrimp.'

Down eight dollars, his good mood cracked. Bonetti's wife was meeting them later. She worked as a waitress while Bonetti wrote – the lost eight dollars was now salt in his sores. Bonetti lived with his wounds. So he grew morose, and the meal took solid conservative steps. The Wasps at the neighboring tables recovered a few of the harmonies which had been blasted out of their bite. The sense of being stitched across the back by rays of displeasure abated. Aquarius did not know how many pinholes had been left in him, but the air in the restaurant was like the awful air of America on its perpetual edge, nihilisms gathering at the poles, dreams of extermination in all the camps. He looked at the Wasp at the adjoining table, a sturdy worthy with silver-rimmed glasses, red righteous ire in the flat red washes of his cheeks, the mottling of his neck. Two mature ladies with silver-rimmed glasses and silver curls and cones of marcel in their beauty parlor lacquer sat in court upon his specimen of the great unwashed, Bonetti, eating lobster right next to them. He felt suddenly as if he had betrayed Eddie – to calm him down was to leave him a target for every wild nihilism of the Wasp, that same laser of concentration and lack of focus on consequence which had taken us up to the moon.

Later, Aquarius was livid. At another place, listening to music, Bonetti's wife joined them, and he told her with keen cruelty, 'I hope Eddie bleeds over those eight bucks. He ruined the meal.' What he could not give voice to was a voice large and endless in its condemnations of himself and all the friends of his generation and the generations which had followed, an indictment of the ways they had used their years, drinking, deep into grass and all the mind illuminants beyond the grass, princelings on the trail of the hip, so avid to deliver the sexual revolution that they had virtually strained on the lips of the great gate. They had roared at the blind imbecility of the Square, and his insulation from life, his furious petulant ignorance of the true tremor of kicks, but now it was as if the moon had flattened all of his people at once, for what was the product of their history but bombed-out brains, bellow-ings of obscenity like the turmoil of cattle, a vicious ingrowth of informers, police agents, militants, angel hippies, New Left totalists, entropies of vocabulary where they would all do their thing – but 'thing' was the first English word for anomaly – an unholy stew of fanatics, far-outs, and fucked-outs where even the few one loved were intolerable at their worst, an army of outrageously spoiled children who cooked with piss and vomit while the Wasps were quietly moving from com-mand of the world to command of the moon, Wasps presenting the world with the fact after prodigies of discipline, while the army he was in, treacherous, silly, overconfident and vain, haters and despisers of everything tyrannical, phony, plastic and overbearing in American life had dropped out, goofed and left the goose to their enemies. Who among all the people he knew well had the remotest say on the quality of these lunar expeditions whose results might yet enter the seed of them all with concentrates worse than their collective semen already filled with DDT. An abominable army. A debauch. And he hated his good friend Eddie Bonetti for this, hated him for drinking at the post. 'You've been drunk all summer,' he felt like saying to him, 'and *they* have taken the moon.' Yes, there was a wild nihilism in his own army: the people were regurgitating the horrors of the centuries, looking to slip the curse out of their seed and into the air, while the curse reentered their seed through every additive in every corporate food. And on the other side, heroes or monsters, the Wasps had put their nihilism into the laser and the computer, they were out to savage or save the rest of the world, and were they God's intended? Looking at his drunken own, Aquarius did not know. He was one judge who would write willy-nilly out of his desolations this year.

2

'The World Is Bigger Infinitely'

To make everything worse, he was forced to see the end of the mission on television. He had applied months ago to cover the splashdown from the *Hornet*, but NASA and the Pentagon limited that number of reporters to a pool of three. So the end of the greatest week was seen by him in his living room in Provincetown, glaring at the television set – there was nothing to see. The sky was fogged. He was left to watch a succession of commentators. Since this came after days of watching TV in Houston, days of hearing the score of *Yellow Submarine* inserted behind old color movies of Gemini 6 and 7 doing a space ballet, he felt drowned. The flavonoids and the plasticoids had taken over. It was his name for TV men. He sometimes thought they came from a species which did not seem to have blood precisely, but some high concentration of haemoglobin-flavonoid in cryogenic plasma. If someone came on their show and cut off their arm or their nose, they would grow another one – plasticoid-flavonoid was a mutation which came from years of talking into microphones and passing on the remark somebody else had just handed you.

So he felt somehow deprived of the last beauties of the Command Module and the flight plan. Reentry was now the most predictable part of the mission and if, in relative terms, it was safe, still with that part of his brain which would insist on remaining a technological child of his century, he had to admire the splendors of the design for reentry. There would after all have been never a trip to the moon if there had not been a means to get back to earth discovered years ago – the atmosphere surrounding the planet offered the friction of a Carborundum wheel toward any object which approached from space. The heat generated was sufficient to consume everything but an occasional meteor. Yet a means had been evolved for safe reentry. Apollo 11 would come back at the speed it left, come back at seven miles a second, 25,000 miles an hour. The Command Module would separate from the Service Module back of it, and the Service Module would burn in the atmosphere to leave only the little cone ten and a half feet high, twelve feet ten inches wide, the mini-cathedral holding the three astronauts strapped in their couches, just ten and a half feet to come back out of three hundred and sixty-three feet, that alone to come back out of all that mighty ship of Apollo-Saturn which had first gone up. The Command Module would come skipping into the atmosphere on a carefully measured route, guided by its thrusters, which were controlled in turn by the computer, or in event of malfunction, by the men. Approaching base-first, its rounded circular base slapping into the atmosphere like a flat stone popping along the surface of a pond, it would sear a path through the sky from eighty miles up and fifteen hundred miles away from the site of splashdown, singeing through the outer air in an incandescent deceleration down to a horizontal speed of a few hundred feet a second, slow enough for a pair of drogue parachutes to open and turn its horizontal path down over to a line of descent. That would occur four miles up. At ten thousand feet, the drogue chute would be released and three little pilot chutes would deploy three larger parachutes, each of a diameter of eighty-three feet. They would slow the vertical descent from one hundred seventy-five miles an hour to twenty-two miles an hour, and the Command Module, swinging on suspension lines one hundred and twenty feet below her three canopies (which had previously been packed in a ring around her docking tunnel), would be deposited in the water in an area within a few miles of the carrier *Hornet*. Immediately, a built-in cutter would sever the parachute lines. If the Command Module ended upside down in the water – a position called Stable II – then three inflatable bags in the forward compartment, blown up by compressors on board, would proceed to float the cone over so she was riding on her base. Swimmers, dropped from approaching helicopters, would attach a flotation collar, and bring up a raft. The astronauts would emerge from the hatch. After decontamination procedures they would be lifted in a sling to the helicopters and brought to the *Hornet*. It was neat. It had been as carefully worked out as the deployment of the Navy of Recovery over thousands of miles of the Pacific. Still, the foundation of all reentry remained the heat shield at the base of the Command Module, nothing but an epoxy resin, a species of phenolic plastic injected in a honeycomb screen. It was not even three inches thick at its widest but it would bear a reentry temperature of 5000 degrees Fahrenheit, hotter than the fiercest kiln, hotter than the melting point of all metals but tantalum and tungsten and they in alloy with anything but themselves would also have softened, so the heat shield of phenolic epoxy was a virtuoso piece of engineering. Indeed, it left no residual problem of cooling down from high temperatures once velocity had slowed. The material heated in chips which turned white-hot, charred, melted, and then flaked away leaving no ember behind, but only a chip of fresh material to be charred in its turn. The epoxy once gone, the spaceship was through reentry as well, its brazed-steel honeycomb heat shield back of that char layer no hotter than 600 degrees at touchdown, and the interior of the Command Module remained at 75 degrees. Yet on the way down, the spacecraft would gleam like a comet, a pale violet flame would flare behind it for hundreds of yards in a galaxy of molecules, a nebula of heat and light.

Well, none of that would he see. Not the flames of reentry on this cloudy day, nor the firing of the mortars on the Command Module to pop those first drogue parachutes far out beyond the turbulence of their immediate wake, no, nor would he have an eyewitness sense of the scene on board the *Hornet* – first instrument of rescue for the Wasp! – no, the news pool was providing all news of the *Hornet*. For the heat and energy of the reception one would have to depend on the atmosphere of the handouts from the press pool. They had accumulated every day in the mimeograph rack in the News Center at Houston, and he had taken a sheaf before he left. Now, he leafed through them while watching the plasticoids and the flavonoids on the TV screen, and President Nixon waiting upon the bridge of the *Hornet* with the Admiral. The Press Release, he decided, was an undiscovered literary form; indeed was it not the seed-bull of Camp?

ii

WUI-017 NASA-017 PRESS PD FROM WUI PRESS CENTER USS HORNET 14TH LOS ANGELES TIMES FROM CHARLES HILLINGER ABOARD THE USS HORNET

ABOARD THE HORNET – THE SHIP'S SKIPPER APPEARED ON TV STANDING BEFORE A CHART OF THE MID-PACIFIC WITH A POINTER IN HAND . . .

'EACH OF US MUST BE DOUBLY VIGILANT IN PERFORM-ING OUR DUTIES ON JULY 24,' DECLARED CAPT. CARL J. SIEBERLICH, 48, WHOSE HOME ASHORE WITH HIS WIFE, TRUDY, AND THREE CHILDREN IS IN THE MIRALESTE AREA OF PALOS VERDES PENINSULA, CALIF.

'THE THREE MOST IMPORTANT PEOPLE IN THE WORLD AT THAT TIME IN HISTORY WILL COME ABOARD OUR SHIP. WE ARE THE LAST LINK IN A VERY IMPORTANT CHAIN.'

GLUED TO 110 TELEVISION SETS SCATTERED THROUGH-OUT THE 20 DECKS OF THE 894-FT FLATTOP WERE NEARLY ALL 2,115 OFFICERS AND MEN OF THE HORNET (EXCEPT THOSE STANDING WATCH) AND MOST OF THE 107 NASA OFFICIALS AND CIVILIANS.

SEAMEN IN BERTHING SPACES LAY IN BUNKS WATCH-ING AND LISTENING TO 'THE OLD MAN' AS HE TALKED ABOUT RECOVERY PLANS AND PRESIDENT NIXON'S SCHEDULED VISIT.

OFFICERS IN THE WARDROOM GATHERED AROUND THE TV THERE. THE CAPTAIN WAS HEARD AND OBSERVED IN SHOPS, OFFICES, MESS ROOMS AND SLEEP-ING QUARTERS.

HE REPORTED TO HIS CREW THAT EVERY EFFORT WAS BEING MADE FOR ALL ABOARD TO WITNESS SPLASH-DOWN AND SUBSEQUENT RECOVERY, TO VIEW THE PRESIDENT AT CLOSE RANGE.

'THE PRESIDENT'S PRESENCE ABOARD THE HORNET,' SAID THE SKIPPER, 'ADDS GREATLY TO THE NATIONAL IMPORTANCE OF THE EVENT. HIS BEING HERE FOCUSES ATTENTION ON HOW IMPORTANT THIS ACCOMPLISH-MENT REALLY WILL BE.'

SOME ABOARD THE HORNET THAT HISTORIC DAY WILL HAVE THE DUBIOUS DISTINCTION OF BEING ON HAND FOR THE CULMINATION OF MAN'S GREATEST

ADVENTURE OF ALL TIMES, YET NOT SEE ANY OF THE EXCITEMENT.

ONE SAILOR IS SERVING 30 DAYS IN THE SHIP'S BRIG FOR THIEVERY. HE'LL MISS IT.

ANYBODY ILL IN SICK BAY AND UNABLE TO GET UP WON'T SEE IT.

LT. COMDR. ROBERT SCHMIDT, 41, CHIEF ENGINEERING OFFICER WHOSE HOME IS IN SPRING VALLEY, CALIF., AND 80 OF HIS 'SNIPES' – BOILER TENDERS, MACHINIST MATES AND ENGINE ROOM ELECTRICIANS – WILL BE AT WORK DEEP DOWN IN 'HELL'S HALF-ACRE' (THE MAIN CONTROL AREAS, FIRE AND BOILER ROOMS).

'I'M GOING TO HAVE MY MOST EXPERIENCED MEN DOWN THERE WITH ME,' SAID COMDR. SCHMIDT. 'SURE, IT'D BE GREAT TO BE TOPSIDE AND SEE THE BIG SHOW.

'BUT THE ENGINE ROOM GETS PAID FOR RUNNING THE SHIP. IT'S A THRILL FOR US TO HAVE THE OPPORTUNITY TO PLAY A VITAL ROLE IN THE RECOVERY OF THE FIRST MEN ON THE MOON.

'THOSE THREE GUYS WILL BE UP THERE BUSTING THEIR REARS TO DO A GOOD JOB. THE LEAST WE CAN DO IS RECIPROCATE IN KIND. THEY'RE DOING THEIR JOB. WE CAN DO OURS.'

IT'S A CASE OF MISSING BEING AN EYEWITNESS TO THE RETURN OF THE FIRST MEN FROM THE MOON BY BEING THE CREAM OF THE CROP IN THE ENGINE ROOM . . .

DURING THE RECOVERY EXERCISE THAT BEGAN AT 2 A.M. MONDAY, SQUALLS, WIND AND HIGH SEAS CAUSED A DELAY IN RECOVERING A REPLICA OF THE COMMAND MODULE.

SHARKS CAUSED A BRIEF SCARE AS SWIMMERS SCRAMBLED BACK ONTO THEIR RAFT UNTIL THE SHARKS DISAPPEARED.

BUT THERE WAS NO SWEAT ABOUT THE TIME FACTOR IN RECOVERING THE MODULE. 'WE ARE IN NO HORSE RACE OUT HERE,' SAID DR DONALD STULLKEN, HEAD

OF THE NASA SPACE RECOVERY TEAM, WHO DESCRIBED THE RUN-THROUGH AS A SUCCESS.

THE SHIP'S TV, 'HORNET TELEVISION CHANNEL 3,' FAR OUT OF RANGE FROM COMMERCIAL STATIONS, FEATURES NIGHTLY MOVIES, 'LOCAL' TALENT AND SPECIAL MESSAGES SUCH AS THE ONE FROM THE CAPTAIN.

THERE'S ALSO A RADIO STATION STAFFED WITH HORNET DISC JOCKEYS; A SHIP'S NEWSPAPER, 'THE STRAIGHT SKINNY' (NAVY JARGON FOR 'GENUINE TRUTH') EDITED BY JOURNALIST 2ND CLASS ROBERT REED, 24, OF YUBA CITY, CALIF.

MANY OF THE CREW HAVE THEIR OWN PERSONAL RECORD PLAYERS FOR ENTERTAINMENT, EVEN PERSONAL POPCORN POPPERS.

POPCORN IS ALSO A POPULAR ITEM AT 5 CENTS A BAG AT THE GEEDUNK (SODA POP) STAND.

A SMOKE SHOP FEATURES CIGARETTES AT 11 CENTS A PACK. THERE'S A CANDY STORE, CLOTHING STORE AND A SMALL-SIZED DEPARTMENT STORE. THERE'S EVEN AN ESCALATOR THAT RUNS FROM THE HANGAR DECK TO THE FLIGHT DECK.

THE SHIP HAS A DIAL TELEPHONE SYSTEM BOASTING MORE THAN 1,000 PHONES – A TELEPHONE FOR EVERY OTHER MAN ON SHIP'S COMPANY.

JOGGERS RUN AROUND THE FLIGHT DECK WHEN AIR OPERATIONS ARE SECURED. VOLLEYBALL AND BASKETBALL ARE PLAYED ON THE HANGAR DECK.

FOR THE BULL'S-EYE IN THE PACIFIC THIS TIME, THE ASTRONAUTS WILL HAVE A FLOATING CITY WAITING TO GREET THEM ON THEIR JOURNEY BACK TO EARTH.

END OF MESSAGE SENT 1734/JC

ABOARD THE HORNET – PLANNING A FLIGHT TO THE MOON SOON? IF SO, BETTER GET YOUR RESERVATIONS IN AT THE EARLIEST FOR HELICOPTER 66.

NOT THAT YOU'D BE FLYING TO THE MOON IN 'OLD 66.'

SIXTY-SIX IS THE 'CHOPPER' THAT PLUCKED

ASTRONAUTS FROM THE SEA FOLLOWING THE LAST TWO MOON SHOTS AND WILL DO THE JOB JULY 24 WHEN APOLLO 11 SPLASHES DOWN IN THE PACIFIC . . .

SIXTY-SIX FLIES SEA RESCUES BACK IN CALIFORNIA AND KEEPS IN SHAPE FOR HER PRIME MISSION TO SEEK OUT AND DESTROY, IF EVER NECESSARY, ENEMY SUBMARINES . . .

'OLD 66' IS THE PROPERTY OF HELICOPTER ANTISUBMARINE SQUADRON FOUR – A BUNCH OF FLYBOYS THAT CALL THEMSELVES THE BLACK KNIGHTS . . .

WHEN YOU WATCH THE BIGGEST TV SHOW IN HISTORY NEXT THURSDAY, YOU'LL SEE 'OLD 66' BIG AS LIFE HOVERING OVER THE SPACESHIP AFTER SPLASHDOWN.

THEN YOU'LL SEE THE ASTRONAUTS HOISTED UP INSIDE 'OLD 66' IN A BILLY PUGH RESCUE NET. FINALLY CAMERAS WILL PICK UP THE CHOPPER AS SHE FLIES THE ASTRONAUTS TO THE HORNET'S FLIGHT DECK AND DISAPPEARS ON AN ELEVATOR TO THE HANGAR DECK.

WHO KNOWS, 'OLD 66' MAY GET AN EMMY FOR HER PERFORMANCE.

ASSOCIATED PRESS
FROM RALPH DIGHTON ABOARD THE USS HORNET

THERE ARE AT LEAST SIX NOT-TOO-IMPLAUSIBLE WAYS APOLLO 11 ASTRONAUTS COULD CONTAMINATE EARTH WITH LUNAR MATERIAL – AND SOME ENVISION SITUATION AS DRAMATIC AS A MOVIE SCRIPT.

ANY ONE OF THEM COULD CONCEIVABLY TEAR HOLE IN NATIONAL AERONAUTICS AND SPACE ADMINISTRATION'S QUARANTINE PROCEDURES, MOST OF WHICH ASSUME ASTRONAUTS BE PICKED UP BY THIS CARRIER, THE PRIMARY RECOVERY SHIP.

SUPPOSE EMERGENCY IN FLIGHT FORCED COMMAND MODULE TO RETURN TO EARTH AT A POINT FAR FROM RECOVERY FORCES IN PACIFIC OR ATLANTIC. RESCUE

THEN BE UP TO NEAREST VESSELS OR 'SHIPS OF OPPOR-
TUNITY' AS NASA CALLS THEM.

CAPTAIN OF SUCH A SHIP BE FACED WITH SOUL-
SHAKING DILEMMA. EVEN IF HE COULD REACH
SPACECRAFT IN TIME, WOULD HE DARE BRING ASTRO-
NAUTS ABOARD?

WOULD ASTRONAUTS ALLOW HIM TO DO IT?

RECOVERY SHIPS CARRY LARGE QUANTITIES OF GERM-
KILLING SOLUTIONS WITH WHICH ASTRONAUTS AND
SPACESHIP ARE SCRUBBED. THEY HAVE DOCTORS AND
TECHNICIANS THOROUGHLY TRAINED IN DECONTAM-
INATION PROCEDURES.

A 'SHIP OF OPPORTUNITY' PROBABLY WOULD NOT.
TRAINED MEN AND QUANTITIES OF DISINFECTANT
COULD BE AIR-DROPPED, BUT WOULD TAKE HOURS.

WHAT IF FLIGHT-ABORTING EMERGENCY WERE DIS-
ABILITY OF AN ASTRONAUT – AN ILLNESS OR INJURY
REQUIRING IMMEDIATE ATTENTION?

– 'THAT WOULD BE ROUGH,' DR WILLIAM CARPENTIER,
NASA PHYSICIAN WHO BE ISOLATED WITH ASTRONAUTS
DURING 21-DAY QUARANTINE AFTER RETURN FROM
MOON, SAID FRIDAY IN INTERVIEW. 'A LOT OF THOUGHT
WOULD HAVE TO GO INTO SUCH A DECISION, BUT I'D SAY
THE CAPTAIN WOULD HAVE TO TREAT THE ASTRONAUTS
LIKE CARRIERS OF AN INFECTIOUS DISEASE AND KEEP
THEM IN QUARANTINE UNTIL HELP ARRIVED.'

WUI007 NASA 007 PRESS PD FROM WUI PRESS CENTER
ABOARD USS HORNET 13

COOK-UPI

ABOARD USS HORNET – A NASA OFFICIAL SAID SUNDAY
IT WAS HIGHLY UNLIKELY THERE COULD BE A BREACH
OF THE QUARANTINE PROCEDURES ON THE HORNET
FOR THE APOLLO 11 ASTRONAUTS – AND THAT IF IT
OCCURRED WHILE PRESIDENT NIXON WAS ABOARD HE
PROBABLY WOULD BE OUT OF JEOPARDY ANYWAY . . .

'HE (THE PRESIDENT) IS ALWAYS SEPARATED

ADEQUATELY SO WE COULD GET HIM COMPLETELY OUT
OF JEOPARDY,' STULLKEN SAID. 'I DON'T EVEN SEE WHY
WE HAVE TO DISCUSS THIS THING,' HE SAID, 'IT'S ONE OF
THOSE REMOTE POSSIBILITIES. WE MIGHT JUST AS WELL
WORRY ABOUT GETTING A TSUNAMI (TIDAL WAVE) DOWN
HERE.'

iii

There in the fever of summer, his children restive at his side, Aquarius
watched Nixon chat through a microphone while the astronauts smiled
at him from the rear window of their new trailer (called the Mobile
Quarantine Facility) on board the *Hornet*. Dressed in dark-green biological isolation garments, they looked rested and pleased. Nixon was
fair in mood and genial indeed.

> NIXON: *Neil, Buzz, and Mike. I want you to know that I think I'm the luckiest
> man in the world. And I say this not only because I have the honor to be President
> of the United States, but particularly because I have the privilege of speaking for
> so many in welcoming you back to earth. I could tell you about the messages we
> received in Washington. Over one hundred foreign governments, emperors and
> presidents and prime ministers and kings, have sent the most warm messages that
> we have ever received. They represent over two billion people on this earth, all of
> them who have had the opportunity through television to see what you have
> done . . . I called the three of, in my view, three of the greatest ladies and most
> courageous ladies in the whole world today, your wives. And from Jan and Joan
> and Pat, I bring their love and their congratulations. We think that it is just won-
> derful that they could have participated at least through television in this return;
> we're only sorry they couldn't be here. And also, I've got to let you in on a little
> secret – I made a date with them. (Laughter). I invited them to dinner on the
> thirteenth of August, right after you come out of quarantine. It will be a state din-
> ner held in Los Angeles. The governors of all the fifty states will be there, the
> ambassadors, others from around the world and in America. And they told me that
> you would come too. And all I want to know – will you come? We want to honor
> you then.*
>
> ARMSTRONG: *We'll do anything you say, Mr President. Just anything.*

NIXON: *One question, I think, all of us would like to ask. As we saw you bouncing around in that boat out there. I wonder if that wasn't the hardest part of the journey. Was that – did any of you get seasick?*

ARMSTRONG: *No, we didn't. And it was one of the harder parts, but it was one of the most pleasant, we can assure you.*

NIXON: *Yes, well, I just know that you can sense what we all sense. When you get back now – incidentally, have you been able to follow some of the things that happened when you've been gone? Did you know about the All-Star game?*

ARMSTRONG: *Yes, sir. The Capsule Communicators have been giving us daily news stories.*

COLLINS: *They daily post us.*

NIXON: *Were you American League or National League?*

ARMSTRONG: *I'm a National League man.*

ALDRIN: *I'm nonpartisan, sir.*

COLLINS: *That's right.*

NIXON: *There's the politician in the group, right?*

ARMSTRONG: *We're sorry you missed that game.*

NIXON: *Yes, well – you knew that, too.*

ARMSTRONG: *We hear that –*

NIXON: *The rain –*

ARMSTRONG: *The rain. Well, we haven't been able to control the weather yet, but that's something we can look forward to as tomorrow's challenge.*

NIXON: *Right, right. Well, I can only summarize it because I don't want to hold you now. You have so much more to do. And gee, you look great; do you feel as good as you look?*

ARMSTRONG: *Oh, we feel just perfectly, Mr President . . .*

NIXON: *Well, just let me close off with this one thing. I was thinking as you know, as you came down and we knew it was a success, and it had only been eight days, just a week, a long week. But this is the greatest week in the history of the world since the Creation. Because as a result of what happened in this week, the world is bigger infinitely, and also as I'm going to find on this trip around the world and as Secretary Rogers will find as he covers the other countries and Asia, as a result of what you've done the world's never been closer together before. And we just thank you for that. And I only hope that all of us in government, all of us in America, that as a result of what you've done, we can do our job a little better. We can reach for the stars just as you have reached so far for the stars. We don't want to hold you any longer. Anybody have a last request? How about promotions, do you think we could arrange something? (Laughter).*

ARMSTRONG: *We're just pleased to be back and very honored that you were so kind as to come out here and welcome us back, and we look forward to getting out of this quarantine –*

COLLINS: *Great.*

ARMSTRONG: *– and talking without having glass between us.*

NIXON: *And incidentally, the speeches that you have to make at this dinner can be very short. And if you want to say fantastic or beautiful, that's all right with us. Don't try to think of any new adjectives; they've all been said. And now, I think incidentally, that all of us who – the millions that are seeing us on television now, are seeing you, would feel as I do that in a sense our prayers have been answered, and I think it would be appropriate if Chaplain Plirto, the chaplain of this ship, were to offer a prayer of thanksgiving. If he would step up now. Chaplain, thank you.*

CHAPLAIN: *Let us pray. Lord God, our Heavenly Father, Our minds are staggered and our spirits exultant with the magnitude and precision of this entire Apollo 11 mission. We have spent the past week in communal anxiety and hope as our astronauts sped through the glories and dangers of the heavens. As we try to understand and analyze the scope of this achievement for human life, our reason is overwhelmed with abounding gratitude and joy, even as we realize the increasing challenges of the future. This magnificent event illustrates anew what man can accomplish when purpose is firm and intent corporate. A man on the moon was promised in this decade. And though some were unconvinced, the reality is with us this morning, in the persons of astronauts Armstrong, Aldrin, and Collins. We applaud their splendid exploits and we pour out our thanksgiving for their safe return to us, to their families, to all mankind. From our inmost beings, we sing humble, yet exuberant praise. May the great effort and commitment seen in this project, Apollo, inspire our lives to move similarly in other areas of need. May we the people by our enthusiasm and devotion and insight move to new landings in brotherhood, human concern and mutual respect. May our country, afire with inventive leadership and backed by a committed followership, blaze new trails into all areas of human cares. See our enthusiasm and bless our joy with dedicated purpose for the many needs at hand. Link us in friendship with people throughout the world as we strive together to better the human condition. Grant us peace beginning in our own hearts. And a mind attuned with goodwill toward our neighbor. All this we pray as our thanksgiving rings out to thee. In the name of the Lord, amen.*

COLLINS: *Amen.*

The anthem was played. The astronauts stood at attention. It had ended. It was done. Armstrong's face looked remarkable. Never as at this hour on television had it had so much of the shriven and scourged look of that breakfast food face which smiles in innocence at us from every billboard. A truly American saint. Of course, the Devil has power to assume a pleasing shape. Aquarius thought of that moment eight days before when Armstrong, carrying his Portable Life Support System, had passed through the fenced-off throng of passionately adoring photographers and journalists at Cape Kennedy and entered a van to drive off to the launch. Then, his head in his helmet, he had had the hard flat-eyed egocentric look of a kitten, eyes hardly cracked, who will be someday a cat. It had been a moment to suggest that in the mysteries of Armstrong's makeup, there might be a bona-fide devil in one soul if a saint in the other – assume he was twin-souled, yes – and if Aquarius had a glimpse of him as a mystic, he could see him now again as a cat-technician who would tamper with the rain. 'Haven't been able to control the weather yet, but . . . tomorrow's challenge.'

iv

He did not know if he learned any more when word came back to him of the splashdown parties. Like a true journalist he was on the phone for full sessions with his informants, and the accounts had that essential wonder which speaks of the exaggeration of the journalist overcome by the exaggeration of events. In deference to the mission it had been a quiet week in Nassau Bay until splashdown – night after night it was as if no one connected with NASA dared to get too drunk for fear of fudged responses in the morning. A few hours after splashdown, however, the parties began; they had begun in effect from the moment technicians from the Staff Support rooms began to fill the Mission Operations Control Room, and people wet cigars and waited for the astronauts to come in on the helicopter and land on the carrier, and when they did, little flags came out and were waved in everyone's hand. The aisles jammed between the consoles with scores of personnel who now were crowding in the door.

From there, parties spread in all directions. Out through the computer-designed suburbs around the Manned Spacecraft Center spread the celebrations, and up the highway to Houston. There was a large and formal ball in Houston that night at the Marriott Hotel from seven to nine, put together by the twenty-five main contractors in the Apollo Program, North American, Grumman and General Electric to lead the rest, a huge orchestration and libation with pâté de foie gras, pigs in blankets, shrimp and eggs and olives, and ice carvings on the tables of antelope, pumpkin and dolphin tails, plus two thousand guests, the cream (selected by the twenty-five corporations) of nabobs from NASA, king contractors, and bona-fide River Oaks Houston. It was a proper party, and the bar closed at nine-thirty. There were even ladies wearing red-white-and-blue Ed White scarves (autographed by every astronaut) which were sold by astronauts' wives to make money for the 'Ed White Memorial Fund.'

Word was out, however, of another party which had begun in the Nassau Bay Motor Inn, the motel off NASA Highway Number 1 with the round red velvet beds where Aquarius had stayed weeks before. There everybody was welcome – $1.50 bought barbecue beef and drinks were $1.15 if you did not bring your own. Three thousand people came not in beards and not with sideburns, rather in short-sleeve shirts with neckties, the ladies in cocktail dresses, scarce were the ladies in décolletage and miniskirts and pants suits – it was a regiment of office workers, engineers, technicians, secretaries, and people wandering in off the highway, the sun burning the pool until nightfall, then the night itself with all of young NASA-land driving into the great trough of all-out recreation, rebel yells finally tearing the Texas air. At seven-thirty in the evening two men threw a blond into the pool. A man followed immediately. The heats of the party were on. A go-go dancer got up on the diving board and worked to the gut rhythm of a band called the Astronauts – six Blacks. The Blacks were finally at NASA. Men climbed up the diving board, went flying past the go-go girls and into the pool, beer cans followed, and broken whisky bottles, chairs and shoes and pieces of clothing, bodies thumped in with the splat of mortars, and toilet paper was slung over the bushes and the lawn. A bouncer with a fire extinguisher went prowling the corridors in the main body of the inn looking for teeny boppers who had jammed the elevator. It went on until four in the morning.

Listening to his informants, Aquarius had a pang for not being there, as if some knowledge more revelatory than the rest might have come his way, some better sense of what resided in the computer men of the windowless walls, but he did not really know that it would have mattered any more than being on the *Hornet*. What did it matter finally if one were anywhere but on the moon for this story? God or Devil at the helm – that was the question behind the trip, and any vulgarities or fine shows of spirit on the good carrier *Hornet*, any verdict decided by the detritus in the pool on the morning after, would hardly reveal the core of the event. That core was buried in the nerve ends of everyone's life. One might as well judge the event from an armchair, for a species of apocalypse was upon us. This was, after all, repeat, the year in which a couple had fornicated on the stage and we had landed on the moon, this was the decade in which we had probed through space, and who knew which belts of protection had been voided and what precisely they had protected. A revolution was in the air which could overthrow every living establishment, an organization of society was also building which might march men daily through aisles monitored by computer probes, there to measure the individual deviations and developments of the night. That was equally on its way. We had contracted for a lunar program in 1961 and what a decade had followed! The times were loose, and no scientist alive could prove that the moon was wholly a dead body any more than they could show that death was a state of being totally dead. Teddy Kennedy's car went off the bridge at Chappaquiddick with Mary Joe Kopechne and the hopes of the Democratic Party went with them as a proper end to a period which had begun with the suicides of Ernest Hemingway and Marilyn Monroe; the younger brother of Martin Luther King was found dead in his swimming pool the day after men walked on the moon. And Provincetown was like a province of the moon in these days of a moon-crazy summer through which he was obliged to work, marriage with Pisces foundering around him, Provincetown, where Eugene O'Neill had lived in the dunes and Anna Christie's father cursed 'that ole debbil sea.' Did the seagulls call these bronze mornings of August dawn with a special fright across the long flats? He could not rid himself altogether of the thought that the moon might be a resting place for more than the hounds of the tide. Perhaps the mysterious magnetosphere had been designed to hold back all those streams of ignoble dead who did not deserve the trip, perhaps the belts of protection were now being voided in all of afterlife, and so anomalies were rising from hell – he was obliged to wonder if man had finally become a cancer in the forms of the Lord. Yet, equally, the full-ness of the moon in Provincetown these nights after the landing was more radiant with lunacy than ever. What if the moon had been draw-ing us to her for years, what if the plastic amphitheaters of NASA were nothing less than the intimations of her call? It was obvious that if he were without compass to the designs of the Father, then of course he had no clue to the nature of the moon: she could be a disguise of Heaven or as easily the Infernal Shades. For another man, such thoughts might have been dangerous, as dangerous as for Aquarius to drive a sports car with a loose wheel down a mountain road, but it was his pro-fession to live alone with thoughts at the very edge of his mental reach. If brooding over unanswered questions was the root of the mad, how-ever, and sanity was the settling of dilemmas, then with how many questions could one live? He would answer that it was better to live with too many than too few. Rave on, he would. He would rave on.

3

A Burial by the Sea

The strain of the summer did not abate. He went back to Houston in the middle of August to see the astronauts at a press conference when they came out of quarantine, and they looked astonishingly the same as they had in the last conference eleven days before they left for the moon. He had long held the theory that experts were men who had the least sensitivity to their subject and so experienced the smallest difficulty in memorizing a huge number of facts concerning their topic. He had only to think of some of the sports writers, literary critics and pornographers he had known, to be confirmed in his thesis. Now he wondered if that was why the astronauts were the first experts in walking on the moon.

Back in New York next evening, he was again a student of TV as he watched the dinner party President Nixon gave for the crew of Apollo 11 and several thousand NASA men, contractors and guests.

When Nixon got up to speak, Aquarius' host switched on a projector loaded with a color film of a boy and girl making love. But the television set was made to serve as the screen. Soon a vagina fluttered butterfly wings over the nose and mouth of the speaker. The laughter that came up from the toils and locks of the company's hard-hearted plumbing was close to apocalyptic. With astonishment Aquarius found himself laughing as hard as the rest. Jokes at the expense of Nixon usually bored him. If Aquarius thought Nixon's most striking effect upon America was as a bloodletter who would reduce all passions, Aquarius was on the other hand not so certain that America had not needed a leech for its fever. From across a political divide, he admired what he had come to decide was Nixon's grasp on political genius – to be so unpopular and yet successful – that was genius! So Aquarius was bored with liberals who thought politics was equal to loathing Nixon. But the sight of that young and wide-open pussy fluttering back and forth over the dish antenna of Richard Nixon's endlessly inquiring face touched off some explosion of frustrations in all of them, battered, bewildered, dislocated New Yorkers roaring now like college kids doubly in love with themselves for the success of the prank.

Back in Provincetown, however, marriages were breaking up as fast as tires blowing in a long race. The most astonishing couples – a man and woman, for example, who had been married unhappily and most tenaciously for twenty years – were breaking up. He counted at least five such surprising dissolutions where one, or at most two, might be par for a warm season. He didn't know if all those marriages had ended because the principals felt ridiculous before the serious actions of men in other places this summer, or whether the marriages had smashed on the outraged waves of some unmeasurable radiation from the roiled invisible waters of Tranquility Base.

As if answer to the moon landing, the Woodstock Music Festival came and went, and four hundred thousand children sat in the rain for two days and nights and listened to rock music, the electronic amplifications elevating the nerves beneath the fingernails of the musicians to the Holy of Holies; Sharon Tate, three friends, and an employee, plus her unborn baby seven months in the womb, were murdered in a guttering of blood all over the walls of her jewel-box of a dwelling in the whimpering Hollywood hills. He felt no shock in further weeks to come when Manson and his family were arrested, for like many novelists Aquarius had a few stray powers of divination, and had projected a novel two years before about a gang of illumined and drug-accelerated American guerrillas who lived in the wilds of a dune or a range and descended on Provincetown to kill. A year later, parts of four girls had been found in a common grave in Truro eight miles away. They would bring to trial a young man from the town who was steeped by report in no modest depths of witchcraft. Yes, drugs to expand consciousness were detonating the banks of fires burning beneath these hundreds of years, and Provincetown was country for witches: here the Pilgrims had landed, here first in the weeks before they moved to Plymouth, Provincetown was the beginning of America for Americans, an immense quadrangle of motel to prove it now on the ground where the Pilgrims first sailed around the point, anchored, and rowed an explorer's boat to shore.

ii

His friends, the Bankos, buried a car as Labor Day approached. They had purchased a heap for the summer, purchased it with a request of the salesman that he sell them a piece of well-used automobile which would manage to survive through August and into the first weeks of September, but it died before Labor Day was on them, bearings gone, valves gone, oil pan cracked, and broken crankshaft – it was gone. Something in the mood of the summer brought every neighbor in for the burial. The sculptor Jack Kearney became the master of the rites, and poets living near became sacramental officers of the day. Friends came with drinks, while Harold McGinn, local contractor with bull-dozer and earth digger, was there to scoop a hole six feet by eight feet by eight feet deep. A rope was put up to hold the neighbors and children from cavorting too near the abyss. And the car, a two-tone sedan of apricot and cream of a long-gone year with mourns of chromium now pitted by salt air and eight years of sun, such faded vehicle, was pushed back by the pallbearers, Aquarius among them, to land with its rear bumper, trunk and differential in the hole and its hood to the sky. The bulldozer leaned it up to a near vertical, and the pallbearers shoveled in earth and tamped sand at the base of the hole. Children ate cake and candy. A boy dressed in the black robes of a Byzantine priest read somber verses from Virgil, the Latin passing like a wash of coagulants over the car still settling in its half-buried grave, and Heaton Vorse in a cape and long-brimmed loose-hinged hat read from the Song of Solomon, sounds of mirth going up as the lines fell like hoops on the promontories of the apricot and cream Ford.

> I compare you, my love,
> > to a mare of Pharaoh's chariots.
> Your cheeks are comely with ornaments,
> Your neck with strings of jewels.

Vorse was the son of Mary Heaton Vorse, a lady radical who had par-ticipated in such events as the Paterson strike now fifty years gone, and Heaton Vorse had a long Yankee nose which virtually touched his plank of a chin. He read to the Ford:

Your navel is a rounded bowl,
That never lacks mixed wine.
Your belly is a heap of wheat,
    encircled with lilies.
Your two breasts are like two fawns,
    twins of a gazelle.
Your neck is like an ivory tower.
Your eyes are pools in Heshbon.

The crowd applauded, and Aquarius felt the proper warmth a funeral should evoke, a sorrow in the pit of merriment and the humor of the very sad – all these Provincetown neighbors out to applaud the burial of an old oil-soaked beast, and the Bankos circulated beer while children ran around the edge of the event, impatient for the ceremony to cease so that they might begin to paint the half of the auto protruding from earth. A child reached in through the open window and turned a switch. The windshield wipers went on in a flick. 'My God, it's not dead yet,' said a voice. But as if in a throe of its last effluents, the washers began to spurt a final lymph.

Eddie Bonetti read his poem, 'Duarte Motors giveth, and terminal craftsmanship taketh away.' Bonetti had worked all summer on a truck, finding the pieces he needed in the town dump, had worried over the Chevrolet manual for pickup trucks of the year of his buy the way a medical student in first-year anatomy goes through strings of flesh which might be nerves, Bonetti had lived with the mysteries of a working transmission through all of this summer, a dungeon of grease by evening to the groans of his pale bond Missouri wife, and Eddie read in his deep cockeyed booming voice, eloquent as the wind which announces a shift in the omens, played with his poetic humors, which moved ponderous phrases through turns of silver by the shift of weight, and his poem continued, honoring this buried friend, conceived in cynicism and sold in exhortation on the floor of Duarte Motors, agent of promises too huge for its fealty to the domains of work, too large for its embarkation back into the particulars of the soil. It was a heroic poem for the occasion (bound to have been printed if it had not been lost) and Aquarius, finding himself drunk unexpectedly on this afternoon of curious frolic, unable for once to resist the noise and calls of the last of summer and the ferments of the town, had come wandering out of his studio to attend half aghast, half sympathetic, to the idiocies of his friends – they would chop up a lawn mower to serve a salad.

The last of the poets, Walter Howard, was reading Numbers 16.

But if the Lord creates something new, and the ground opens its mouth and swallows them up, with all that belongs to them, and they go down alive into Sheol, then you shall know that these men have despised the Lord.

The children were out with brushes and paints, drawing figures, figure-drippings, and inchoate totems on the vertical roof of the car, and Kearney was limning the exposed bottom of the crankcase and chassis with lights of green luminescent pigment in slashes through the grease, hints of war paint – slowly the radiant ribbings of an insert's belly emerged from the dark and open works.

And as he finished speaking all these words, the ground under them split asunder and the earth opened its mouth and swallowed them up with their households. And all the men that belonged to Korah and all their goods.

Now the children were slinging paint through the open windows onto the vinyl of an old upholstery. Aquarius watched his wife at the other end of the lawn and knew again as he had known each day of this summer that their marriage was over. Something had touched the moon and she would never be the same. The sense of love as a balm for the vacuums of the day was departed from them – they were sealed from one another, a run of seven years was done, and his heart throbbed like a bruise in the thigh.

So he mourned the hour as well as any man would when his pains were not small, even mourned for the beast who cried out in Banko's half-burned Ford, mourned him like the skull of poor Yorick, and came back often in the next day and the next to watch Kearney the sculptor work with his torch and goggles to weld bumpers and angles of chromium into mandibles and legs while insect's antennae reached up in a mute's catalepsy to the sky. And they put floodlights at the base. The funeral had ended in an artifact for the summer of the moon in the East End of Provincetown not a hundred feet off the street which runs around the bay, not a half-mile in from the edge of town, Metamorphosis, titled by Kearney, a massive Yorick of half a Ford standing twelve feet high, first machine to die with burial in the land of the Pilgrims and the cod.

iii

And in those days, men will seek death and will not find it; they will long to die and death will fly from them.

That was from Revelations 8:7 and gives a clue to Aquarius' thoughts at the funeral. It was a day for more than a little to seek to die, for his work had him studying colonizations on the moon, conversion of oxygen from moon rocks, and cities of moon-based energy derived from radiations of the sun. Moon vegetables huge in size would grow in the reduced gravity of the field, the plants to thrive in hydroponic waters (also extracted from moon rocks) while algae proliferated in gardens of new-made atmosphere beneath a dome. The effort of these colonies would offer no less than the cheap manufacture on the moon vacuum of products of mass consciousness – electronics, communications, pharmaceuticals, yes, Sartre might be right and consciousness the conversion of Being to Nothingness, yes, the tools of the future mind seemed to be forged best in a vacuum – soon they would be orbiting rocket trains of cancer patients to take the cure in space, for the growth of malignancy was slowed apparently by radiation in weightless condition. Pain appeared at the thought of a new species of men born in lunar gravity, bodies grown in lunar gravity – what form would appear to their figures, pilot men of an electrical and interplanetary world which could speak across the ages of a failure of human potential, a smashing of mood, some loss of that other means of communication which once had lived in the carnal grasp of the roots of that earlier human so much closer to an animal in the ecological scheme, early human who had survived pregnancy, birth, first-year diseases, syphilis, loss of teeth – what a strength and substance to that earlier and lost human race, Aquarius brooded as the dirt flew in on the dead Ford, what a nice balance of food consumed and material used, equilibrium of lives, and deaths, and wastes in fair balance, as opposed to the oncoming world of parallel colostomies draining into the same main line, and the air of earth cities become carbon monoxide and lead, sulphur dioxide and ash, nitrogen oxide and other particulates of the noxious, earth staggering with sewages which did not rot, synthetics, aluminums, oils and pesticides, fertilizers, detergents and nuclear spews, acids and plastics and salt in the soil, cakes of suffocation in the rivers, hazes of nitrogen effluent to cut off the light from the sky, a burgeoning of artificials to addict the crops, another year of pollutions to choke the planet. And the population ready to double in four decades, no, less. One knew with the worst sense of bottoms disgorging into bottomless bottoms that if the military-industrial establishment was beginning to accept the idea that funds might be taken eventually from them and given over to the solution (or the barest hope of a solution!) to the critical symptoms of ecology, the nauseas of pollution, then the statistics presented to their private councils must have been incredible indeed. Was the end of the world at hand? Was that the message they now received? Io, Europa, Ganymede and Callisto! – we might be safer far on the moons of Jupiter. What did we know of what we did? Why the very organs of disease which once would kill a man were now delivered by surgeons to the womb of the open day, organs of disease reborn for an instant in a half-life, yes, cancer organs removed probably became the cancer communicants of ether yet unglimpsed.

Only a generation ago, they would have thought it was the essence of an insane heart to personify an organ, attribute a soul to the part, believe that a cancer of liver or cancer of lung was not extinguished so soon as its malignancy was removed. It would have been considered the core of psychosis to speak of the post-operative cancer communicants of the organ removed. Yet we were infants who tickled the navel of the moon while suffocating in the loop of our diaper. A line from a poem of Hemingway burned across the funeral festivities of the day.

> *In the next war*
> > *we shall bury the dead in cellophane*
> *The host shall come packaged*
> > *in cellophane.*

He broke up with his wife on Labor Day night and knew they would not be together for many a month, many a year, maybe forever. In the morning, after a night of no sleep, he was on a plane to Houston and the sifting of haystacks of technological fact for the gleam of a needle or a clue. And no computer named HAYSTAQ to serve as horse.

iv

It was a long September. He went back and forth between Houston and the mournful memories of the land of the Pilgrims and the cod. Pisces was away, and traveling. In the mend of Indian summer with the crowds gone and the rose-hip bushes in bloom on the dunes, their flowers artful as violet in a pearl, he bought a Land Rover for consolation and took long rides through lands of sand back of town, a corner of Sahara. In the bay, the flats at low tide heard the singing of the clams – dreams of glory at the majesty of oceans emerged in a sigh, a whistle, one could not quite hear the buried song of the clam. And the light dazzled across mirrors of inch-deep water and luminosities of glistening sand – he could almost have packed the literary equipment in for one good year of oil and gesso ground.

There were contracts however. Prose was never so much prose as when constructed with obligation. The more he visited Houston, the more he knew with what unhappiness is not automatic to tell that he might have blundered in accepting the hardest story of them all, for it was a sex-stripped mystery of machines which might have a mind, and mysterious men who managed to live like machines, and more than once in airplanes, high enough above the clouds to give a hint of other worlds in the gatherings and demarcations of airy attenuated farewell, he came to think again, as he had brooded again and again, on that simple conception of God as an embattled vision which had terrified him from the hour he first encountered the thought around one of the bends of marijuana fifteen years ago. Every other one of his notions had followed from that, for if God were a vision of existence at war with other visions in the universe, and we were the instruments of His endeavor just so much as the conflicting cells of our body were the imperfect instrument of our own will, then what now was the condition of God? Was He trapped in the wound of nature, severed from our existence as completely as the once exquisite balances of the shattered ecology? had that vision He wished to carry across the universe depended altogether upon human mind and flesh in sensuous communication with nature? had radio-by-machine been the cancer of communication? had the savage lived in a set of communions with the

invisible messages of nature which we had pulverized with our amplifiers? These days Aquarius carried Frazer's *Golden Bough* on long trips by
plane.

Bechuana warriors wear the hair of a hornless ox among their own hair because
the ox, having no horns, is hard to catch . . . a South African warrior who twists
tufts of rat hair among his own curly black locks will have just as many chances
of avoiding the enemy's spear as the nimble rat has of avoiding things thrown
at it . . . When you are playing the one-stringed lute, and your fingers are stiff,
the thing to do is catch some long-legged field spiders and roast them, and then
rub your fingers with the ashes; that will make your fingers as lithe and nimble
as the spiders' legs – at least so think the Galalereese. To bring back a runaway
slave an Arab will trace a magic circle on the ground, stick a nail in the middle
of it and attach a beetle by a thread to the nail, taking care that the sex of the
beetle is that of the fugitive. As the beetle crawls round and round, it will coil
the thread about the nail, thus shortening its tether and drawing nearer to the
center at every circuit. So by virtue of homeopathic magic the runaway slave
will be drawn back to his master.

It was the magic of savage metaphor, the science of symbol, it married spiders' legs to the music of the fingers and the useful frenzy of the
rat to the sensors in his hair. It made a wedding between the spiraled-in
will of insects forced to focus on a point of tether and the loss of any
will-to-escape in the slave. It was pretty, poetic and nonsensical, it was
nonsensical. Unless it were not. What if some real exchange between
insects, trees, crops, and grains, between animals and men, had lived
with real if most distorted power in the first hours of history? What if
that Vision of the Lord which had gone out to voyage among the stars
had obtained the power to be carried up by the artwork of a bounteous
earth exquisite in the resonance of all psyches in its field? – what if
radio, technology, and the machine had smashed the most noble means
of presenting the Vision to the universe?

What if God wrestled for the soul of man in some greased arena with
the Devil, who was now fortified by every emanation from baleful stars
beyond the sun – could that be so? What if God, losing cruelly here, and
yet gaining there, was in a combat just so crude as the counts of point in
a contest? What if, for the sake of a premise, one would assume that the

Devil was reconstructing nature with every electronic, plastic, surgery and computer and so had forced the Lord in desperation to descend into the earth and come back with His life in the grass of that most mysterious marijuana, a drug which made one aware of life in the veins at what severe price was not yet known? What if God, aghast at the oncoming death of man in man-deviled pollution, was finally ready to relinquish some part of the Vision, and substitute a vision half machine, and half of man, rather than lose all? What indeed if the Lord was allowing Himself to be consumed so that the angels and swine of His children who swallowed Him promiscuously each day and night on drugs were able to embark on journeys into the land of the dead, little journeys in through the first gates of the palace of death, and thus giving Himself to the children in the milk of their drugs was, yes, consumed by them each night and thereby relinquished the largest dreams of His future. Such thoughts were an agony of pain if one held them truly, for responsibility was then like a burning of blood, and the time of apocalypse was certainly near. A war of the millennia might yet rest on the shoulders of the young. What an abattoir of brain-splattered substance if they consumed their smack and left the world dirtier than when they began.

Or was it the Devil who had insinuated marijuana into every pot and every garage? Or was the Devil being consumed as well by computers and transistors, by agents of far-off stars? There was also an hour when questions trampled upon questions to leave the ground of thought as much a mire as the gray greasy boot-trampled soil of the moon. Sometimes he even thought that pot and hash and LSD had opened the way to the moon, for they might have voided the spiritual belts of real protection. Perhaps as the runaway slave came back to the master who kept the beetle on a string, so the drugged odysseys of inner space might have altered the zones of the outer. Again and again, staring out his airplane window he would say good-by to these thoughts and stare at clouds.

In several trips to Houston, he was like a man looking for the smallest sign. For the moon book which he had begun that summer idled now in the gap of Pisces' absence, and he did not know where to put his feet. One lifted a book like a boulder out of the mud of the mind, and his mind was a pit of wrenched habits and questions which slid like snakes. Where did you put your feet so that finally you might begin?

He found the answer at last in company with his favorite saying. 'Trust the authority of your senses,' Aquinas had said. He could repeat it again, for there was an object at last for his senses, there in the plastic vaults and warehouses of the Manned Spacecraft Center at Houston was a true object, a rock from the moon. Looking at it, answers came, answers strong enough to send him back to Provincetown for the fall and winter haul of his book, and a little of the spring. He finished in fact on a day when Apollo 13 was limping back to earth in wounded orbit with two fuel cells gone, its Lunar Module Aquarius never to reach the moon, yes, he finished in an hour when he did not know if the astronauts would return in safety or be lost, but he had written the ending in his mind long before; it came on the day he stood in quiet before that object from the moon, that rock which gave him certitude enough to know he would write his book and in some part applaud the feat and honor the astronauts because the expedition to the moon was finally a venture which might help to disclose the nature of the Lord and the Lucifer who warred for us; certainly, the hour of happiness would be here when men who spoke like Shakespeare rode the ships: how many eons was that away! Yes, he had come to believe by the end of this long summer that probably we had to explore into outer space, for technology had penetrated the modern mind to such a depth that voyages in space might have become the last way to discover the metaphysical pits of that world of technique which choked the pores of modern consciousness – yes, we might have to go out into space until the mystery of new discovery would force us to regard the world once again as poets, behold it as savages who knew that if the universe was a lock, its key was metaphor rather than measure.

Marvelous little moon rock. What the Devil did it say?

It was not so much. They led Aquarius through one back room after another, and up and down a stone stair or two. The week of exhibiting the rock at MSC was over – it was now on its way to the Smithsonian – and special favors were needed this particular afternoon to obtain a peek. But he reached a place at last he had been in months before, the room with the plate-glass window across its middle where magazine writers had hounded Armstrong until Armstrong confessed that man explored out as salmon swim upstream, and there on the other side of the glass was no astronaut today, but a small case vacuum-tight on the other side of the divide. He saw the lunar piece through not one glass but two, rock in a hermetically tight glass bell on the other side of another glass with still another hermetic seal. Yet she was not two feet away from him, this rock to which he instinctively gave gender as she – and *she* was gray, gray as everyone had said, gray as a dark cinder and not three inches across nor two inches high nor two inches for width, just a gray rock with craters the size of a pin and craters the size of a pencil point, and even craters large as a ladybug and rays ran out from the craters, fine white lines, fine as the wrinkles in an old lady's face, and maybe it was the pain of all these months of a marriage ending and a world in suffocation and a society in collapse, maybe it was just the constant sore in his heart as the blood pumped through to be cleared of love, but he liked the moon rock, and thought – his vanity finally unquenchable – that she liked him. Yes. Was she very old, three billion years or more? Yet she was young, she had just been transported here, and there was something young about her, tender as the smell of the cleanest hay, it was like the subtle lift of love which comes up from the cradle of the newborn, and he wondered if her craters were the scars of a war which had once allowed the earth to come together in the gathered shatterings of a mighty moon – there was something familiar as the ages of the bone in the sweet and modest presence of this moon rock, modest as a newborn calf, and so he had his sign, sentimental beyond measure, his poor dull senses had something they could trust, even if he and the moon were nothing but devils in new cahoots, and child of the century, Nijinsky of ambivalence, hanging man Aquarius, four times married and lost, moved out of MSC with the memory of the moon, new mistress, two feet below his nose, and knew he would live with the thought of a visit. All worship the new science of smell! It was bound to work its way through two panes of glass before three and a half billion more years were lost and gone.

PENGUIN MODERN CLASSICS

RABBIT, RUN
JOHN UPDIKE

'Brilliant and poignant ... By his compassion, clarity of insight, and crystal-bright prose, [Updike] makes Rabbit's sorrow his and our own' *Washington Post*

It's 1959 and Harry 'Rabbit' Angstrom, one time high school sports superstar, is going nowhere. At twenty-six he is trapped in a second-rate existence – stuck with a fragile, alcoholic wife, a house full of overflowing ashtrays and discarded glasses, a young son and a futile job. With no way to fix things, he resolves to flee from his family and his home in Pennsylvania, beginning a thousand-mile journey that he hopes will free him from his mediocre life. Because, as he knows only too well, 'after you've been first-rate at something, no matter what, it kind of takes the kick out of being second-rate.'

'Updike's punch is powerful' *Newsweek*

PENGUIN MODERN CLASSICS

LIBRA
DON DELILLO

'An unparalleled trip into the heart of America' *Observer*

Exploring the tale behind the six seconds that broke the back of the American century, *Libra* combines fact, hypothesis and fiction to portray the events leading up to the assassination of John F. Kennedy on 22 November 1963.

Powerful and eerily convincing, it chronicles Lee Harvey Oswald's odyssey from troubled teenager to a man of precarious stability searching for a role in history. When 'history' presents itself in the form of three disaffected CIA operatives, the scales are irrevocably tipped as Oswald finds himself caught up in events far beyond his control.

'This you feel is America, and the bad news starts here' *Sunday Times*

Penguin Modern Classics

A TRANQUIL STAR
PRIMO LEVI

'Works of the highest order of imagination … remind us why Levi is an indispensable writer' *Sunday Telegraph*

Primo Levi was one of the most astonishing voices to emerge from the twentieth century. This landmark selection of seventeen short stories, translated into English for the first time, opens up a world of wonder, love, cruelty and curious twists of fate, where nothing is as it seems. In 'The Fugitive' an office worker composes the most beautiful poem ever with unforeseen consequences, while 'Magic Paint' sees a group of researchers develop a paint that mysteriously protects them from misfortune. 'Gladiators' and 'The Knall' are chilling explorations of mass violence, and in 'The Tranquil Star' a simple story of stargazing becomes a meditation on language, imagination and infinity.

'We are blessed with this collection' *Herald*

PENGUIN MODERN CLASSICS

THE PRICE
ARTHUR MILLER

'A beautifully intelligent play about two brothers ... pinned in positions of flight from their own histories' *Observer*

Victor, a New York cop nearing retirement, moves among furniture in the disused attic of a house marked for demolition. Cabinets, desks, a damaged harp, an overstuffed armchair – the relics of a lost life of affluence he's finally come to sell. But when his brother Walter, who hasn't spoken to in years, arrives, the talk stops being just about whether Victor's been offered a fair price for the furniture, and turns to the price that one of them and not the other paid when their father lost both his fortune and the will to go on ...

Fraught, but cut through with humour, *The Price* is one of Arthur Miller's finest plays.

Penguin Modern Classics

THE SPY WHO CAME IN FROM THE COLD
JOHN LE CARRÉ

With a new Introduction by John le Carré

Alex Leamas is tired. It's the 1960s, he's been out in the cold for years, spying in Berlin for his British masters, and has seen too many good agents murdered for their troubles. Now Control wants to bring him in at last – but only after one final assignment.

He must travel deep into the heart of Communist Germany and betray his country, a job that he will do with his usual cynical professionalism. But when George Smiley tries to help a young woman Leamas has befriended, Leamas's mission may prove to be the worst thing he could ever have done.

In le Carré's breakthrough work of 1963, the spy story is reborn as a gritty and terrible tale of men who are caught up in politics beyond their imagining.

'He can communicate emotion, from sweating fear to despairing love, with terse and compassionate conviction. Above all, he can tell a tale' *Sunday Times*

'A portrait of a man who has lived by lies and subterfuge for so long, he's forgotten how to tell the truth' *Time Magazine*

PENGUIN MODERN CLASSICS

AMERICA
ANDY WARHOL

Andy Warhol carried a camera with him everywhere he went and, taken from ten years of extraordinary shots, his *America* aspires to the strange beauty and staggering contradictions of the country itself. Exploring his greatest obsessions – including image and celebrity – he photographs wrestlers and politicians, the beautiful wealthy and the disenfranchised poor, Capote with the fresh scars of a facelift and Madonna hiding beneath a brunette bob. He writes about the country he loves, wishing he had died when he was shot, commercialism, fame and beauty.

An America without Warhol is almost as inconceivable as Warhol without America, and this touching, witty tribute is the great artist of the superficial at his most deeply personal.

'He created his own universe and became its star' David Cronenberg, *Guardian*

'He understood our obsession with celebrity culture better – and sooner – than anyone else' Alison Jackson, *Sunday Telegraph*

PENGUIN MODERN CLASSICS

ESSAYS
GEORGE ORWELL

'Anyone who wants to understand the twentieth century will still have to read Orwell' Timothy Garton Ash, *New York Review of Books*

This outstanding collection brings together Orwell's longer, major essays and a fine selection of shorter pieces that includes 'My Country Right or Left', 'Decline of the English Murder', 'Shooting an Elephant' and 'A Hanging'. With great originality and wit Orwell unfolds his views on subjects ranging from a revaluation of Charles Dickens to the nature of Socialism, from a comic yet profound discussion of naughty seaside picture postcards to a spirited defence of English cooking. Displaying an almost unrivalled mastery of English plain prose style, Orwell's essays challenge, move and entertain.

With an Introduction by Bernard Crick

PENGUIN MODERN CLASSICS

HUMBOLDT'S GIFT
SAUL BELLOW

With an Introduction by Martin Amis

'Bellow at his best … funny, vibrant, ironic, self-mocking, and wise'
San Francisco Examiner

For many years, the great poet Von Humboldt Fleisher and Charlie Citrine, a
young man inflamed with a love for literature, were the best of friends. At the
time of his death, however, Humboldt is a failure, and Charlie's life has reached a
low point: his career is at a standstill, and he's enmeshed in an acrimonious
divorce, infatuated with a highly unsuitable young woman, and involved with a
neurotic Mafioso. But then Humboldt acts from beyond the grave, bestowing
upon Charlie an unexpected legacy that may just help him turn his life around.

WINNER OF THE NOBEL PRIZE FOR LITERATURE

Penguin Modern Classics

AFTER THE FALL
ARTHUR MILLER

'Endlessly fascinating, emotionally harrowing, and consumingly committed to telling the truth as Miller sees it' *Time*

Quentin is a successful lawyer in New York, but inside his head he is struggling with his own sense of guilt and the shadows of his past relationships. One of these is an ill-fated marriage to the charming and beautiful Maggie, who went from operating a switchboard to become a self-destructive star – a singer everyone wanted a piece of.

With tremendous psychological acuity and depth, and a brilliant, dreamlike structure, *After the Fall* is a literary masterpiece – the story of a man striving to comprehend his feelings for his friends, family and the women he has loved.

PENGUIN MODERN CLASSICS

FRANK SINATRA HAS A COLD: AND OTHER ESSAYS
GAY TALESE

Gay Talese is the father of American New Journalism, who transformed traditional
reportage with his vivid scene-setting, sharp observation and rich storytelling. His
1966 piece for *Esquire*, one of the most celebrated magazine articles ever published,
describes a morose Frank Sinatra silently nursing a glass of bourbon, struck down
with a cold and unable to sing, like 'Picasso without paint, Ferrari without fuel
– only worse'. The other writings in this selection include a description of a meeting
between two legends, Fidel Castro and Muhammad Ali; a brilliantly witty dissection
of the offices of *Vogue* magazine; an account of travelling to Ireland with hell-raiser
Peter O'Toole; and a profile of fading baseball star Joe DiMaggio, which turns into a
moving, immaculately-crafted meditation on celebrity.

'The best non-fiction writer in America' Mario Puzo

'The best American prose of the second half of the twentieth century' *Atlantic Monthly*

*Contemporary ... Provocative ... Outrageous ...
Prophetic ... Groundbreaking ... Funny ... Disturbing ...
Different ... Moving ... Revolutionary ... Inspiring ...
Subversive ... Life-changing ...*

What makes a modern classic?

At Penguin Classics our mission has always been to make the best
books ever written available to everyone. And that also means
constantly redefining and refreshing exactly what makes a 'classic'.
That's where Modern Classics come in. Since 1961 they have been an
organic, ever-growing and ever-evolving list of books from the last
hundred (or so) years that we believe will continue to be read over and
over again.

They could be books that have inspired political dissent, such as
Animal Farm. Some, like *Lolita* or *A Clockwork Orange*, may have
caused shock and outrage. Many have led to great films, from *In Cold
Blood* to *One Flew Over the Cuckoo's Nest*. They have broken down
barriers – whether social, sexual, or, in the case of *Ulysses*, the
boundaries of language itself. And they might – like *Goldfinger* or
Scoop – just be pure classic escapism. Whatever the reason, Penguin
Modern Classics continue to inspire, entertain and enlighten millions
of readers everywhere.

'No publisher has had more influence on reading habits than Penguin'
Independent

'Penguins provided a crash course in world literature'
Guardian

The best books ever written

P E N G U I N ⬤ C L A S S I C S

SINCE 1946

Find out more at www.penguinclassics.com